# GLOBAL CHALLENGES IN INTEGRATED COASTAL ZONE MANAGEMENT

Edited by

Erlend Moksness
Einar Dahl
and
Josianne Støttrup

WILEY-BLACKWELL
A John Wiley & Sons, Ltd., Publication

This edition first published 2013 © 2013 by John Wiley & Sons, Ltd

Wiley-Blackwell is an imprint of John Wiley & Sons, formed by the merger of Wiley's global Scientific, Technical and Medical business with Blackwell Publishing.

*Registered office:* John Wiley & Sons, Ltd, The Atrium, Southern Gate, Chichester, West Sussex, PO19 8SQ, UK

*Editorial offices:* 9600 Garsington Road, Oxford, OX4 2DQ, UK
The Atrium, Southern Gate, Chichester, West Sussex, PO19 8SQ, UK
2121 State Avenue, Ames, Iowa 50014-8300, USA

For details of our global editorial offices, for customer services and for information about how to apply for permission to reuse the copyright material in this book please see our website at www.wiley.com/wiley-blackwell.



*Library of Congress Cataloging-in-Publication Data*

Global challenges in integrated coastal zone management / edited by Erlend Moksness, Einar Dahl, and Josianne Støttrup.
pages cm
Includes bibliographical references and index.
ISBN 978-0-470-65756-0 (hardback: alk. paper) – ISBN 978-1-118-49648-0 (obook) (print) – ISBN 978-1-118-49649-7 (mobi) (print) – ISBN 978-1-118-49650-3 (epdf/ebook) (print) – ISBN 978-1-118-49651-0 (epub) (print) 1. Coastal zone management–International cooperation. I. Moksness, Erlend, editor of compilation. II. Dahl, Einar, editor of compilation. III. Støttrup, Josianne G., 1955– editor of compilation.
HT391.G544 2013
333.91′7–dc23

2012030609

A catalogue record for this book is available from the British Library.

Wiley also publishes its books in a variety of electronic formats. Some content that appears in print may not be available in electronic books.

Cover image: Øystein Paulsen, Institute of Marine Research, Norway
Cover design: www.hisandherdesign.co.uk

Set in 10/12pt Times by Aptara® Inc., New Delhi, India
Printed and bound in Malaysia by Vivar Printing Sdn Bhd

1 2013

# Contents


| | | |
|---|---|---|
| *About the Editors* | | vi |
| *Contributors* | | vii |
| *Referees* | | xi |
| *Preface* | | xiii |
| Chapter 1 | The Role of Science in the Transition to Sustainability: the Systems Approach Framework for Integrated Coastal Zone Management<br>*Tom Sawyer Hopkins and Denis Bailly* | 1 |
| **Section 1** | **Coastal Habitats and Ecosystem Services** | |
| Chapter 2 | The Norwegian Programme for Mapping of Marine Habitats – Providing Knowledge and Maps for ICZMP<br>*Trine Bekkby, Frithjof E. Moy, Heidi Olsen, Eli Rinde, Torjan Bodvin, Reidulv Bøe, Henning Steen, Ellen S. Grefsrud, Sigurd H. Espeland, Are Pedersen and Nina M. Jørgensen* | 21 |
| Chapter 3 | Monitoring Ecological Quality of Coastal Waters by the Nature Index (NI) – an Integrated Measure of Biodiversity<br>*Eivind Oug, Gro I. van der Meeren, Grégoire Certain and Signe Nybø* | 31 |
| Chapter 4 | The *Scomberomorus brasiliensis* Gill-net Production System in Northern Brazil; an "Invisible" and Mismanaged Small-scale Fishery<br>*Victoria J. Isaac, Roberto V.E. Santo, Bianca S. Bentes, Keila R.M. Mourão and Flávia Lucena-Frédou* | 49 |
| Chapter 5 | A Combination of Seagrass and Macroalgal Beds Enhances Survival of Young-of-the-Year Rockfish *Sebastes Cheni*<br>*Yasuhiro Kamimura and Jun Shoji* | 61 |
| **Section 2** | **Adaptation/Mitigation to Change in Coastal Systems** | |
| Chapter 6 | Natural Resource Potential of Macroalgae Harvesting in the Baltic Sea—Case Study Trelleborg, Sweden<br>*Emma Risén, Joseph Santhi Pechsiri, Maria E. Malmström, Nils Brandt and Fredrik Gröndahl* | 71 |

Chapter 7 Conserving Mangrove Ecosystem for Climate Change Adaptation in the Ganges Basin 85
*M. Shahadat Hossain*

Chapter 8 Monitoring Humpback Whale (*Megaptera novaeangliae*) Behaviour in a Highly Urbanised Coastline: Gold Coast, Australia 101
*Jan-Olaf Meynecke, Silje Vindenes and Daniella Teixeira*

**Section 3 Coastal Governance**

Chapter 9 An Overview of the Implementation of SAF Methodology Regarding the Stakeholder Response, in the Mussel-Farming Area of Chalastra, Thermaikos Gulf 117
*Zoi I. Konstantinou and Yannis N. Krestenitis*

Chapter 10 Knowledge Issues in ICZM and EBM Applied on Small Geographic Scales: Lessons from a Case Study in Risør, Norway 127
*Håkan T. Sandersen, Eirik Mikkelsen, Erlend Moksness and Jon Helge Vølstad*

Chapter 11 Governance Systems for Marine Protected Areas in Ecuador 145
*Vincent Gravez, Robert Bensted-Smith, Pippa Heylings and Taylor Gregoire Wright*

Chapter 12 The Battle for Space – the Position of Norwegian Aquaculture in Integrated Coastal Zone Planning 159
*Bjørn Hersoug*

**Section 4 Linking Science and Management**

Chapter 13 Integrated Coastal Zone Management – Preview and Evaluation of its Application on the Coast of Cyprus 171
*Nicholas Kathijotes and Athina Papatheodoulou*

Chapter 14 Framing a 'Post-Normal' Science–Policy Interface for Integrated Coastal Zone Management 179
*Scott Bremer*

Chapter 15 Co-Learning in Marine Protected Areas for Integrated Coastal Zone Management 192
*Mohammad A.L. Siddique, S. Zafrin, S. Myers, T. Smith, R. Babcock, and R.W. (Bill) Carter*

Chapter 16 Reflexively Mapping the Science–Policy Interface for Coastal Zones 206
*Anne Blanchard and Scott Bremer*

Chapter 17 Bridging the Science–Policy Divide in the Coastal Zone: Is There a Role for Learning Processes? 218
*Melissa Nursey-Bray and Nick Harvey*

Chapter 18 Perspectives of Social and Ecological Systems 229
*Paul Tett, Audun Sandberg, Anne Mette, Denis Bailly, Marta Estrada, Thomas Sawyer Hopkins, Maurizio Ribera d'Alcalà and Loraine McFadden*

*Index* 245

*Colour plates appear between pages 48 and 49*

# About the Editors

**Einar Dahl** is the head of the Coastal Zone Ecosystem Programme at the Institute of Marine Research (IMR) in Norway. His expertise is originally on phytoplankton taxonomy and ecology. His research has mainly been on harmful phytoplankton species, and possible causes to harmful blooms. More recently he has been working with coastal zone ecology in a broader perspective, including the implementation of the EU's Water frame directive. He has contributed to about 50 papers in peer review journals.

**Erlend Moksness** is a Research Director at the Institute of Marine Research (IMR) in Norway. He has been responsible for establishing a research and management advice programme on the coastal zone ecosystem at the same institute. His background is in recruitment in marine fish, fish ageing, stock enhancement of marine fishes and aquaculture of marine fishes. He has published more than 100 scientific articles and has been co-editor of 11 proceedings and scientific books.

**Josianne Støttrup PhD** is a senior scientist at National Institute of Aquatic Resources (DTU Aqua, previously Danish Institute for Fisheries Research) since 1984. Her research focus has been on marine larval nutrition initially, but evolved to coastal juvenile fish habitats, flatfish ecology and more recently to integrated coastal management. She has led the marine coastal fisheries management programme since 1994 and now heads the Section of Coastal Ecology.

# Contributors

**Maurizio Ribera d'Alcalà**, Stazione Zoologica Anton Dohrn, 80121 Napoli, Italy

**R. Babcock**, Marine and Atmospheric Research, Commonwealth Scientific and Industrial Research Organization (CSIRO), 233 Middle Street, Cleveland, QLD 4163, Australia

**Denis Bailly**, Université de Bretagne Occidentale, UMR AMURE, Centre IFREMER de Bretagne, 29280 Plouzané, France

**Trine Bekkby**, Norwegian institute for Water Research, Gaustadalléen 21, NO-0349 Oslo, Norway

**Robert Bensted-Smith**, Fauna & Flora International, Alemania N31-118 y Mariana de Jesús, Apt 3b, Quito, Ecuador

**Bianca S. Bentes,** Universidade Federal do Pará, Instituto de Estudos Costeiros, IECOS – UFPA. Al. Leandro Ribeiro, s/n, Aldeia, CEP: 68600-000 Bragança – PA, Brazil

**Anne Blanchard**, Centre for the Study of the Sciences and the Humanities, University of Bergen, P.O. Box 7805, NO-5020 Bergen, Norway

**Torjan Bodvin**, Institute of Marine Research, Flødevigen Marine Research Station, NO-4817 His, Norway

**Reidulv Bøe**, Geological Survey of Norway, P.O. Box 6315 Sluppen, NO-7491, Trondheim, Norway

**Nils Brandt**, Teknikringen 34, Department of Industrial Ecology, KTH, SE-100 44, Stockholm, Sweden

**Scott Bremer**, Centre for the Study of the Science and the Humanities, University of Bergen, P.O. Box 7805, NO-5020, Bergen, Norway

**R.W. (Bill) Carter**, Associate Professor, Heritage Resource Management, Associate Director, Sustainability Research Centre, Co-editor, Australasian Journal of Environmental Management, University of the Sunshine Coast, Maroochydore, DC, QLD 4558, Australia

**Grégoire Certain**, Norwegian Institute for Nature Research (NINA), NO-7485 Trondheim, Norway. Present address: Institute of Marine Research (IMR), Tromsø, Norway

**Sigurd H. Espeland**, Institute of Marine Research, Flødevigen Marine Research Station, NO-4817 His, Norway

**Marta Estrada**, Institut de Ciències del Mar (CSIC), 08003 Barcelona, Spain

**Vincent Gravez**, Fundación Futuro Latinoamericano, Guipuzcoa E16-02 y Av. Coruña, Casilla 17-17-558, Quito, Ecuador

**Ellen S. Grefsrud**, Institute of Marine Research, P.O. Box 1870, NO-5817 Bergen, Norway

**Taylor Gregoire-Wright**, Yale University, USA

**Fredrik Gröndahl**, Teknikringen 34, Department of Industrial Ecology, KTH, SE-100 44, Stockholm, Sweden

**Nick Harvey,** Faculty of Humanities and Social Sciences, University of Adelaide, South Australia, 5005, Australia

**Bjørn Hersoug**, The Norwegian College of Fishery Science, University of Tromsø, NO-9037 Tromsø, Norway

**Pippa Heylings**, Fundación Futuro Latinoamericano, Guipuzcoa E16-02 y Av. Coruña, Casilla 17-17-558, Quito, Ecuador

**Tom Sawyer Hopkins**, Instituto per l'Ambiente Marino Costiero (CNR), 80133 Napoli, Italy: Dept. MEAS, North Carolina State University, Raleigh, North Carolina, USA

**M. Shahadat Hossain**, Institute of Marine Sciences and Fisheries, University of Chittagong, Chittagong 4331, Bangladesh

**Victoria J. Isaac**, Laboratório de Biologia Pesqueira e Manejo de Recursos Aquáticos, Federal University of Pará, Belém, Avenida Perimetral 2651, CEP 66077 830 Belém, PA, Brazil

**Nina M. Jørgensen**, Akvaplan-niva, Fram Centre, NO-9296 Tromsø, Norway

**Yasuhiro Kamimura,** Takehara Fisheries Research Laboratory, Hiroshima University, 5-8-1 Minato-machi, Takehara, Hiroshima 725-0024, Japan

**Nicholas Kathijotes,** International Ocean Institute (Cyprus) FP, Professor, Department of Civil Engineering and Geodesy, Cyprus University of Technology, Limassol, Cyprus CY3036

**Zoi I. Konstantinou**, Division of Hydraulics & Environment, Department of Civil Engineering, Aristotle University of Thessaloniki, 54124, Greece

**Yannis N. Krestenitis**, Division of Hydraulics & Environment, Department of Civil Engineering, Aristotle University of Thessaloniki, 54124, Greece

**Flávia Lucena-Frédou**, Universidade Federal Rural de Pernambuco – UFRPE, Departamento de Pesca e Aqüicultura – DEPAQ. Rua Dom Manoel de Medeiros, s/n, Dois Irmãos - CEP: 52171-900 - Recife/PE, Brazil

**Maria E. Malmström**, Teknikringen 34, Department of Industrial Ecology, KTH, SE-100 44, Stockholm, Sweden

**Loraine McFadden**, Flood Hazard Research Centre, Middlesex University, London, NW4 4BT, U.K.

**Gro I. van der Meeren**, Institute of Marine Research, Austevoll Research Station, NO-5392 Storebø, Norway

**Anne Mette**, KMGNE, Reichenberger Strasse 150, 10999 Berlin, Germany

**Jan-Olaf Meynecke**, Humpbacks and High-Rises Inc. Gold Coast 4222 and Australian Rivers Institute, Gold Coast 4222 Queensland, Australia

**Eirik Mikkelsen**, NORUT, P.O. Box 6434 Forskningsparken, NO-9294 Tromsø, Norway

**Frithjof E. Moy**, Institute of Marine Research, Flødevigen Marine Research Station, NO-4817 His, Norway

**Erlend Moksness**, Institute of Marine Research, Flødevigen Marine Research Station, NO-4817 His, Norway

**Keila R.M. Mourão,** Laboratório de Biologia Pesqueira e Manejo de Recursos Aquáticos, Federal University of Pará, Belém, Avenida Perimetral 2651, 66077 830 Belém, PA, Brazil

**S. Myers**, Sustainability Research Centre, University of the Sunshine Coast, Maroochydore, DC, QLD 4558, Australia

**Melissa Nursey-Bray,** Discipline of Geography, Environment and Population, University of Adelaide, South Australia, 5005 Australia

**Signe Nybø**, Norwegian Directorate for Nature Management, P.O. Box 5672 Sluppen, NO-7485 Trondheim, Norway/Norwegian Institute for Nature Research, NO-7485 Trondheim, Norway

**Heidi Olsen**, Geological Survey of Norway, P.O. Box 6315 Sluppen, NO-7491, Trondheim, Norway

**Eivind Oug**, Norwegian Institute for Water Research, Regional Office Sørlandet, NO-4879 Grimstad, Norway

**Athina Papatheodoulou**, Environmental Consultant, 3 Nikonos Street, Limassol, Cyprus CY4044

**Joseph Santhi Pechsiri**, Teknikringen 34, Department of Industrial Ecology, KTH, SE-100 44, Stockholm, Sweden

**Are Pedersen**, Norwegian Institute for Water Research, Gaustadalléen 21, NO-0349 Oslo, Norway

**Eli Rinde**, Norwegian Institute for Water Research, Gaustadalléen 21, NO-0349 Oslo, Norway

**Emma Risén**, Teknikringen 34, Department of Industrial Ecology, KTH, SE-100 44, Stockholm, Sweden

**Audun Sandberg**, University of Nordland, Faculty of Social Science, NO-8049 Bodø, Norway

**Håkan T. Sandersen**, University of Nordland, Faculty of Social Science, NO-8049 Bodø, Norway

**Roberto V.E. Santo**, Instituto Federal de Educação, Ciência e Tecnologia do Pará- IFPA Campus Belém. Coordenação de Biologia, Av. Almirante Barroso, 1155. CEP: 66093-020. Belém/PA, Brazil.

**Jun Shoji,** Takehara Fisheries Research Laboratory, Hiroshima University, 5-8-1 Minato-machi, Takehara, Hiroshima 725-0024, Japan

**Mohammad A.L Siddique**, Sustainability Research Centre, University of the Sunshine Coast, Maroochydore, DC, QLD 4558, Australia and Upazila Fisheries Officer, Department of Fisheries, Bangladesh

**T. Smith**, Professor and Director, Sustainability Research Centre, University of the Sunshine Coast, Maroochydore, DC, QLD 4558, Australia

**Henning Steen**, Institute of Marine Research, Flødevigen Marine Research Station, NO-4817 His, Norway

**Daniella Teixeira**, Humpbacks and High-Rises Inc. Gold Coast 4222 Queensland, Australia

**Paul Tett**, Scottish Association for Marine Science, Scottish Marine Institute, Oban, PA37 1QA, U.K

**Silje Vindenes**, Griffith University, Griffith Centre for Coastal Management, Gold Coast 4222 Queensland and Sea World Whale Watch, Main Beach 4217, Australia

**Jon Helge Vølstad**, Institute of Marine Research, NO-5817 Bergen, Norway

**S. Zafrin**, Sustainability Research Centre, University of the Sunshine Coast, Maroochydore, DC, QLD 4558, Australia

# Referees

**Colin Attwood**, University of Cape Town, RSA

**Dagný Arnarsdóttir**, University Centre of the West Fjords, Iceland

**Gulsen Avaz**, TUBITAK MRC Environment Institute, Turkey

**Trine Bekkby**, Norwegian Institute for Water Research, Norway

**Anne Blanchard**, Centre for the Study of the Sciences and the Humanities, Norway

**Peter R. Burbridge**, The Orchard, House of Ross, Comrie, Perthshire PH6 2JS, Scotland

**Donata Melaku Canu**, The National Institute of Oceanography and Experimental, Italy

**Hartvig Christie**, Norwegian Institute for Water Research, Norway

**Franciscus Colijn**, Institut für Küstenforschung, Germany

**Ragnar Elmgren,** Stockholm University, Sweden

**Ola Flåten**, University of Tromsø, Norway

**Jan Helge Fosså**, Institute of Marine Research, Norway

**Nici Gibbs**, New Zealand Seafood Industry Council, New Zealand

**Harald Gjøsæter**, Institute of Marine Research, Norway

**Jakob Gjøsæter**, Institute of Marine Research, Norway

**Alf Håkon Hoel**, Institute of Marine Research, Norway

**Thomas Sawyer Hopkins**, Institute for the Coastal Marine Environment, Italy

**Tore Haug**, Institute of Marine Research, Norway

**Frank Jacobsen**, Directorate of Fisheries, Norway

**Svein Jentoft**, University of Tromsø, Norway

**Silvia de Juan**, ICM-CSIC, Spain

**Michel Lample**, UBO-AMURE Centre IFREMER de Bretagne, France

**Finn Larsen**, National Institute of Aquatic Resources, Denmark

**Cathrine Legrand**, Linnæus University, weden

**Rachel Long**, Saint Mary's University, Canada

**Merle Looring**, University of Tartu Pärnu College, Estonia

**Patrick McConney**, University of the West Indies (UWI), Barbados

**Anne Mette**, Kolleg fuer Management und Gestaltung nachhaltiger Entwicklung gGmbH, Germany

**Jan-Olaf Meynecke**, Australian Rivers Institute, Australia

**Rosita Mileriene**, Coastal Research and Planning Institute, Klaipeda University, Lithuania

**Even Moland**, Institute of Marine Research, Norway

**Frithjof Moy**, Institute of Marine Research, Norway

**Melissa Nursey-Bray**, University of Adelaide, Australia

**Tor Næsje**, Norwegian Institute for Nature Research – NINA, Norway

**Dimitri Pavlov**, Moscow State University, Russia

**Søren Anker Pedersen**, International Council for the Exploration of the Sea, Denmark

**Nicole Pelton**, University of Adelaide, Australia

**Alf Ring Kleiven**, Institute of Marine Research, Norway

**Russell G. Richards**, Griffith University, Australia

**Audun Sandberg**, University of Nordland, Norway

**Jan Henrik Sandberg**, Norges Fiskarlag, Norway

**Håkan T. Sandersen**, University of Nordland, Norway

**Rafael Sardá**, Institut de Ciències del Mar (CSIC), Spain

**Jun Shoji**, Takehara Fisheries Research Station, Japan

**Henning Steen**, Institute of Marine Research, Norway

**Knut Stokke**, Norwegian University of Life Sciences, Norway

**Paul Tett**, Scottish Association for Marine Science, Scotland

**Arild Vatn**, Norwegian University of Life Sciences, Norway

**Jon Helge Vølstad**, Institute of Marine Research, Norway

**Victor Øiestad**, Post-box 72, Es-36620 Vilanova de, Arousa, Spain

# Preface

**The Second International Symposium on Integrated Coastal Zone Management took place in Arendal, Norway between 3 and 7 July 2011. The main objective of the Symposium 'Integrated Coastal Zone Management' was to present current knowledge and to address issues on advice and management related to the coastal zone.**

This international multidisciplinary conference intended to promote science and integration of knowledge for the sustainable management of coastal resources. It provided a venue for scientists, engineers, managers and policy-makers to discuss recent advances and innovative ideas, share experiences and develop networks. A total of 131 persons (including 19 students) from 31 countries participated in the symposium (Australia, Bangladesh, Brazil, Canada, Cuba, Cyprus, Denmark, Estonia, France, Germany, Greece, Iceland, India, Indonesia, Iran, Italy, Japan, Lithuania, Malaysia, New Zealand, Norway, Portugal, Republic of Korea, Russia, Spain, Sweden, The Netherlands, Turkey, UK, USA). The symposium offered financial support to 17 persons from 11 countries: Australia (1), Bangladesh (2), Brazil (1), Canada (1), Cuba (2), India (2), Iran (2), Malaysia (1), Philippines (2), Russia (3), and USA (1). Many of the conference participants (and several of the presentations given) were from the EU integrated project 'SPICOSA' (www.spicosa.eu). The project developed a 'System Approach Framework' (SAF), which aims at incorporating the ecological, social and economic dimensions for the assessment of policy options for the sustainable management of complex coastal zone systems. The project involved 54 partners and encompassed 18 study site applications, whereby the SAF was tested in different geomorphologic systems, governance structures, cultural conditions and economic characteristics.

During the Symposium, a total of 133 presentations (8 key-note, 55 oral and 70 posters) addressed issues within the following four themes:

- **Coastal habitats and ecosystem services**
- **Adaptation/mitigation to change in coastal systems**
- **Coastal governance**
- **Linking science and management**

Coastal waters around the globe suffer from strain due to a wide range of human activities. The situation calls for a holistic approach, combining expertise from nature and socio-economic sciences, to reach a balanced and sustainable development of the coastal zone. The concept Integrated Coastal Zone Management (ICZM) was born in 1992 during the United Nations Conference on Environment and Development (UNCED). ICZM is defined as a decision-making process focused on sustainable use, development and protection of seaside terrestrial and coastal marine areas and their resources, in continuous realization. The coastal zone includes both coastal waters and the part of land separating the marine and terrestrial environment. The coastal zone thus offers a myriad of goods and services ranging from production to protection. Despite its relatively small coverage in percentage of global surface area, this zone is highly productive and sustains a high biodiversity.

Growing pressure from increasingly diverse human activities coupled with climate change impacts threaten the functional integrity of these coastal ecosystems. A multi-disciplinary approach towards understanding drivers, pressures and impacts in the coastal zone requires effective integration of data and information in policy and management. Competing claims to the coastal areas of a given nation are nothing new, but they are likely to intensify in the future. Currently, about 60% of the world population lives within 60 km of the coast, but at the current rate of growth, this percentage is predicted to increase to 75% within just two decades. In addition, tourism is growing in importance worldwide, adding its share of pressure on the coastal zone and increasing pressure on these coastal resources. Aquaculture, yet another competitor for coastal resources, is also on the increase as capture fisheries stagnate or decline globally. Competition for resources can result in either cooperative or conflicting responses. In the case of coastal resources, the expansion of the aquaculture industry represents the reallocation of common pool resources such as coastal and sea areas in to private hands – converting once common areas into private areas from which others are excluded. In addition, aquaculture carries the potential to impact those resources that remain held in common – wild fish stocks, the seascape and the coastline. The specific mechanisms for handling distribution of coastal goods are to be found within the institutions that exist for ICZM. One key idea behind the concept of ICZM is to create cooperative management solutions for the use of the areas, making them both politically as well as ecologically sustainable. A range of plans may be ecologically sustainable, but for any plan to work, stakeholders must be invested in it. This implies that stakeholders, whether they are local fishermen or local municipalities, have access to and understand the problems and issues with which they are confronted and the choices with which they are faced.

One of the prominent key-note speakers at the conference, Professor Tom Hopkins, science coordinator of SPICOSA, asked why we are moving too slowly towards Sustainability. The answer to the question, of who is responsible for Sustainable Development, is no one and everyone. This may give us a clue as to why we won't get there in time. We are starting to study complex human systems and we know that they have the capacity to reorganize to lower entropic states and that this capacity depends on the health and diversity of its components. The unfortunate reality is instead that these systems are degrading to higher entropic states. We know some scenarios of change that could make the degradation worse, but we don't seem to have any complete scenarios that would reverse this degradation trend. This is where everyone comes in and everyone needs to cooperate. The chain of interacting components that constitutes the circle of responsibility necessary for a sustainable society may start or end with appropriate science and public awareness, but in between there is a lot of inertia against change in the societal components that needs to be overcome in order to make the cycle complete. According to Professor Hopkins, the SPICOSA project has arguably demonstrated a method by which science might help jump-start this reorganization process. It would require accelerating two coupled trends that are already in place, an expansion to complex systems science, and a more interactive role between science and society. To provide specific material for this discussion, Professor Hopkins presented how the 'systems approach' was applied to policy issues in eighteen European coastal sites and what was learned about how scientific research, coupled with local societies, can better address coastal issues and guide coastal management for the transition to Sustainable Development.

The strength of the first and present symposium was that there were no parallel sessions, resulting in that all participants had to learn from each other independent of their own background and interest. This resulted in a wide thematic spread, from microalgae to big whales within natural science, governance issues, stakeholder interactions and socioeconomic issues.

Chapter 1

# The Role of Science in the Transition to Sustainability: the Systems Approach Framework for Integrated Coastal Zone Management

*Tom Sawyer Hopkins and Denis Bailly*

**Abstract**

Sustainability science is emerging as the transdisciplinary research field that provides the information and technical support needed for sustainable development (SD). It has complex and far reaching implications for research and for society. SD is a learning process that requires difficult social changes in paradigms and behaviors. This process involves a circle of responsibility that must start and continue with public awareness, appropriate science, and supporting leadership. In this chapter we present perspectives gained, through the development of the Systems Approach Framework (SAF) SPICOSA project, that emphasize the need to accelerate two coupled research trends already initiated: an expansion to complex systems science, and a strong participatory role between science and society. We present how the SAF was applied to policy issues in eighteen European coastal sites and what was learned about how scientific research, coupled with local societies, can better address coastal issues, and guide integrated coastal zone management in the transition to sustainable development.

## 1.1 Introduction

### *1.1.1 A sustainability experiment*

Sustainability, although an evolving and illusive concept, is well established within the fabric of priorities for both European science and governance. It is in this context that a recent EU Integrated Project (IP), Policy Integration for Coastal Zone Assessments (SPICOSA) was realized. As an Integrated Project, its objectives were to produce a product that would stimulate a research restructuring in the European Research Area, have a multidisciplinary

*Global Challenges in Integrated Coastal Zone Management*, First Edition. Edited by Erlend Moksness, Einar Dahl and Josianne Støttrup.

value to society, and be inclusive of new members of the EU. Its specific research objectives were: "Development of decision-making tools to identify options for Sustainable Development through ICZM (Integrated Coastal Zone Management) and to monitor the implementation of specific strategies based on forecasting scenarios, cost-effectiveness and cost–benefit analysis, consistency with policies, including precautionary principle, environmental regulation and economic instruments evaluation, technological assessment. This shall be based on the description of the human activity in the coastal zones and structuration of the various societal and environmental functions of these zones" (DG Environment, 2006). Rather than respond to this list of objectives separately, the IP-SPICOSA Project developed and tested a Systems Approach Framework (SAF) that would address systematically all of them together in 18 Study Site Applications (SSA) (Hopkins et al., 2011; www.spicosa.eu).

On the basis of these experiments, we present several perspectives on how the SAF as an operational methodology provides an example of sustainability science and of its utility and support for ICZM. The development of the SAF presupposes that there exists a methodology gap between the general concepts of societal sustainability and its practical implementation. Thus, the SAF is conceptually based on the hypothesis that an expanded role of science is emerging in response to an urgent societal need to transition to sustainable development (SD); and that this role will require a greater capability to simulate problems in complex systems involving natural and human subsystems. Companion hypotheses are that:

1. The expansion of science must include a functional integration of the ecological, social, and economic sciences and engage in collaborative partnerships with policy and the public;
2. Good examples of transitional efforts towards sustainable development are an essential element of the process;
3. The complexity of EU Coastal Zones together with its well-established ICZM community presents a timely opportunity to generate these examples.

The super-complex challenge that modern society must deal with is that of reversing the unsustainable trends, which threaten the stability of our global society, and strengthening the sustainable trends. These two trends represent a global reorganizational bifurcation for human civilization. The historical fact that human civilizations have confronted such bifurcations numerous times, with greater success at smaller scales than at larger ones (Diamond, 2005), does not ensure that it can confront the present one on a global scale for which there would likely be no safe refuge. Hence several strong questions arise:

1. Is the path bifurcation true and how immediate is it?
2. Are we really so interconnected as to include all global societies?
3. Is SD possible on local to global scales and how?
4. How can science and technology be better integrated and be more focused on benefiting society than on destabilizing it?
5. Does the systems approach, in a research framework, provide an appropriate basis for investigating complex systems?

## 1.1.2 Challenges of SD restructuring

Answering these questions requires a mix of objective and subjective arguments. The objective assessments regarding the seriousness of the bifurcation and its ultimate direction, depend strongly on the methods used, for example compare the cautious optimism of Goklany (2009) with the experimental concern of the MDG Report (2011) or the serious concern of Homer-Dixon (2006). The subjective argument concerns a learning process for the human behavioral and the institutional changes needed for sustainability (Romeiro, 2000; Kemp and Martens, 2007). In this chapter, we only glance at the first three questions and focus on question four, where we have acquired a clearer view on the content of Sustainability Science, and on question five, where we have new experimental knowledge about the connectivity of science and policy, albeit at the scale of coastal zone systems (CZS).

The global bifurcation is a result of the inherent nature of complex systems to reorganize (Kauffman, 1995) when exposed to stresses as expressed in trends and/or extraordinary events that can induce non-linear shifts in the systems' dynamics. The reorganizing process has two potential outcomes: to a higher negentropic or a higher entropic state. The corresponding question is whether we choose a path of proactively reorganizing our societies towards greater sustainability, or a laissez-faire path of ignoring the indicators of increasing unsustainability. Currently, reorganizational efforts in both directions are evident in all three sectors of natural environment, social wellbeing, and economic stability. This stark issue is the impetus of the Millennium Assessment of Human Behavior (MAHB), which "aims to promote rapid change in human behavior to avoid the collapse of global civilization" (Ehrlich, 2010).

Scale and connectivity are important dimensions. The social and ecosystem spheres have a spectral peak at the human-to-biome scale, of say ~1 m to ~100 km, which will be determinant in the SD process. In contrast, the current trends in economy, resources, communication, and climate change will require sustainability at regional to global scales. This suggests that the larger-scale requirement might be satisfied through the integration of complex, nodal networks consisting of smaller-scaled sustainable systems (Hopkins et al., 2012). Also important is the coincidence of the time scales that characterize change in our society, for example within a range of a several decades are those that are generational, those associated with the decline of global resources per capita, those required for institutional and cultural changes, and those resulting from adjustments to large disturbances (war) or environmental shifts (climate change) (Ehrlich, 2010; Levin, 1999). Certainly, a critical question regarding SD, in the context of its complexity in scale and connectivity, is whether the progress and knowledge gained through incipient and different experiments can contribute to a synergistic acceleration of SD regardless of scale (e.g. Sustainable Cities International Network (sustainablecities.net)).

## 1.1.3 Sustainability science

Sustainable development is the process of improving the state of sustainability. The definition has broadened since the Brundtland Commission Report (1987) by supplementing its strong emphasis on environmental degradation with an equal attention regarding the responses in the social and economic sectors (Adams, 2006). Here we define sustainability as a social state that maintains equitable balances between the three types of capital (natural, social, and economic), which are not always replaceable, not always reversible, and each differing in their optimal scale of sustainability (Daly, 1991). Each of these spheres differs in the manner by which it achieves sustainable states, and each cycles between extremes; for example, ecological, through growth and collapse cycles on genetic to adaptive time scales (Gunderson and Holling, 2002); social, through dominance and cooperation (Werner, 1999); and economic, through command, free-market, and mixed, economies (Friedman 1962; Mandel 1986; Costanza 1991; Costanza et al., 1997a). Historically, the time scales of the latter two range from generational to centurial. Understanding the temporal incongruences between these spheres, evaluating their mutual interactions, providing objective assessments of their trends, and transferring this knowledge through participatory actions to decision-makers and the public are all the subject matter of sustainability science.

Historically, scientific research evolved with a bias toward the descriptive investigations of objects, species, and human behavior, and gave less attention to the dynamics of interactions, processes, and social reorganization. This tended to establish disciplinary boundaries in research that disfavor the multidisciplinary investigations of the sustainable configurations of these disciplinary spheres and their optimal symbiotic overlap. Klein et al. (2002) argue that sustainability science is transdisciplinary in the sense that its investigations must engage directly the society being studied. Similarly, Clark and Dickson (2003), defined the goal of sustainability science as: "creating and applying knowledge in support of decision making for sustainable development" and with "a focus on the dynamic interactions between nature and society". Our contention here is that the SAF represents a good practical example of this emerging area of science by incorporating into its methodology the systems approach (Fiksel, 2006) to facilitate and evaluate problems in complex socio-environmental coastal systems and to improve the interactive capacity with policy and with the public.

## 1.2 SAF methodology

### *1.2.1 Systems Approach Framework (SAF)*

The crucial tenet of the SAF is the necessity for an objective mechanism to test policy effectiveness towards the goals of sustainable development, including its public acceptance. To provide this, research needs the capacity to monitor and simulate dynamic changes in complex systems in a manner that we can answer management questions such as: How would this action/policy promote sustainability? What are the short-term costs and the long-term benefits? How would a policy change or affect public wellbeing? What conflicts might arise? How might existing social structures be modified to further the public acceptance of SD?

The Systems Theory (Bertalanffy, 1968) contends that the interactions between objects are more important than the objects themselves, that is, the objects are formed and controlled by their interactions with other objects. This implies a continuum of connections throughout the universe, a concept a little difficult to assimilate when contrasted with our tendencies to separate ourselves in order to seek our own individual identities. When identifying systems, whether it is our society, our environment, or ourselves, we must acknowledge that in every case there exist interconnections with other systems. This collaborates with

**Table 1.1** The principal features of the Systems Approach Framework for coastal zones.

1. **It was developed to evaluate complex coastal systems**
   *but could be applied to other systems.*
2. **It simulates and assesses dynamic changes in a complex coastal system**
   *in response to research, policy, or other questions.*
3. **It is holistic in perspective and reductionist in resolution**,
   *is hierarchical in structure, and seeks an iterative compromise between effort and resolution.*
4. **It integrates environmental–economic–social dimensions**
   *and focuses on improving non-market assessments of relevant social and environmental values.*
5. **It simulates policy options to correct unsustainable practices**
   *through optimizing benefit and minimizing damage to essential components.*
6. **It requires the participation of policy, stakeholders, and public**
   *to form a collaborative partnership for sustainable development.*
7. **It is an open methodology, constructed to evolve through each application**,
   *with support tools, web-handbook, model library, data portal, user community, etc.*

Godel's Theorem: that is, that a system cannot be understood from within itself – its external connections must also be understood. The SAF is also based on recent findings concerning the principles of complex systems (Kauffman, 1995; Capra, 1996).

The SAF is a problem-driven methodology, based on the systems approach, that simulates and appraises policy issues in coastal zone systems in order to provide higher order analyses (than description) of policy scenarios concerning the issue resolution (Hopkins et al., 2011; Tett et al., 2011; www.coastal-saf.eu). As illustrated in Table 1.1, the SAF initiates with a Policy Issue and a set of Scenarios specific to an observable dysfunction (Impact) in the coastal zone, which then provides the focus for the simulation analysis (Figure 1.1). During the SAF development, the Policy Issue is generated through negotiation between the Study Site Team (Researchers) and volunteers (Stakeholders) from the environmental-management sector and representatives from organizations having direct interests in the local coastal zone. An iterative simulation-analysis then investigates the causes of and responses to the Impact in order to quantify and interpret policy options for the selected Issue.

## *1.2.2 Decision making*

The trajectory of sustainable development depends critically on the psychology of decision-making, both individual and collective (policy). Because of the urgency posed by global unsustainability, we cannot afford wrong decisions or a trial-and-error approach to get ourselves on a sustainable track. An important aspect of decision-making has to do with timing, which requires an understanding of the time-dependency of change in a system. Often policy decisions are made after a degradation event, which then inspires a surge of policy efforts to "not let this happen again" and that are later abandoned in favor of a cheaper, insufficient reconstruction due to lesser up-front costs and a lack of sustainable planning options. On the other hand, because the recovery costs (removing pollution or recovering production) can increase exponentially with time, our approach should be more precautionary to avoid even larger costs (Figure 1.2), particularly where irreversibility is a strong factor (e.g. biodiversity or sea-level rise).

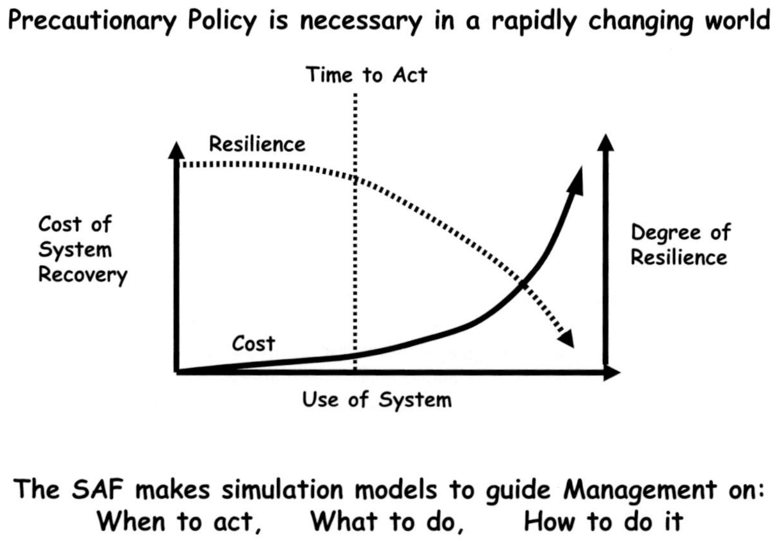


**Figure 1.1** Schematic of the SAF implementation for an ICZM Policy Issue. The SAF in application for ICZM is an information loop, which begins and ends with policy. The loop initiates with a policy issue and ends with information in the form of scenarios related to that policy issue. The process is participatory by a group of stakeholders and the information output is formatted for them and the public. The quantification is iterative in order to meet practical constraints and optimize the information quality. The first and last steps are more holistic and the second and third steps more mechanistic.

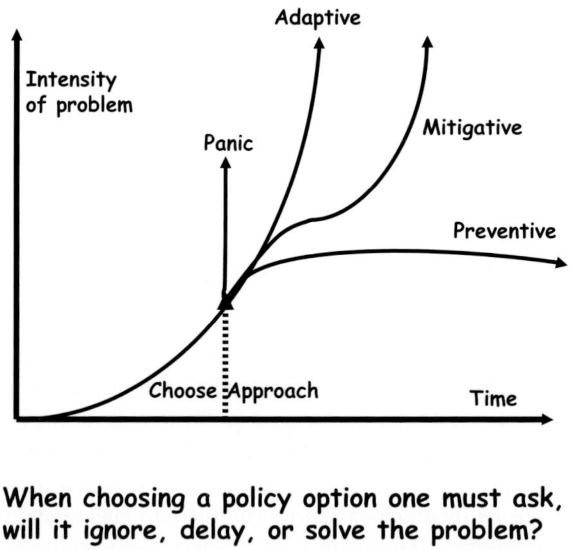


**Figure 1.2** A schematic showing the case of a degrading system where the resilience is decreasing and the cost of recovery increasing non-linearly. Precautionary management anticipates a problem before it becomes too expensive to remedy or irreversible.

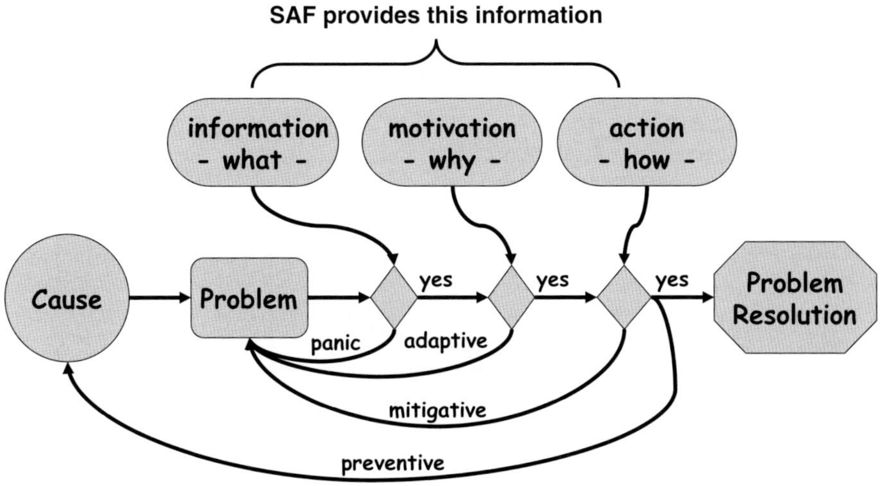


**Figure 1.3** Schematic trajectories of different decision strategies. Wrong decisions made without information can make problems worse or create other problems, adaptive strategies allow a problem to continue to grow while the system adapts to it, mitigative strategies provide a temporary abatement of a problem, and preventive strategies cause a problem to diminish.

All major controversial decisions should be evaluated to avoid Type I or Type II errors, that is rejecting a true hypothesis or accepting a false hypothesis, respectively (Shermer, 2002). To calculate the cost of a Type I error, one compares the difference between the products formed by the (probability of being true) $\times$ (the cost of being true) with the (probability of being false) $\times$ (the cost of being false). For example, rejecting climate change, even at low probabilities and at inestimable costs of being true, would be a far more grievous error than accepting climate change and paying for the up-front costs of converting to sustainable energy sources. This example also illustrates how the propensity of short-term economics can favor business-as-usual decisions.

Problems in complex systems are due to imbalances that the system cannot correct – at least on a time scale suitable to some of its inhabitants. Policy has several options in terms of the type of decision it makes to resolve such problems (Figure 1.3). By way of illustrating this point, we can breakdown the decision-making process into three phases: information, motivation, and action (Figure 1.4). A specific decision can be summarized as moving sequentially through these three phases, but to be efficient, information on each phase must be available in an iterative manner to the deciding mind. The fact that the process must begin with the information phase underlines the importance of the source and quality of information that initiates the sequence. The motivation phase requires the decider to make an unbiased assessment of both objective and subjective information sources. The former can be provided by social and natural science research assessments. However, the decider might override the former due to his subjective conviction that derives from his personal information base and/or the social milieu of those he represents. Thus, the decider has veto power over information from rational-based research, particularly when the rationale is unfamiliar to him. This makes a case for better education and translation

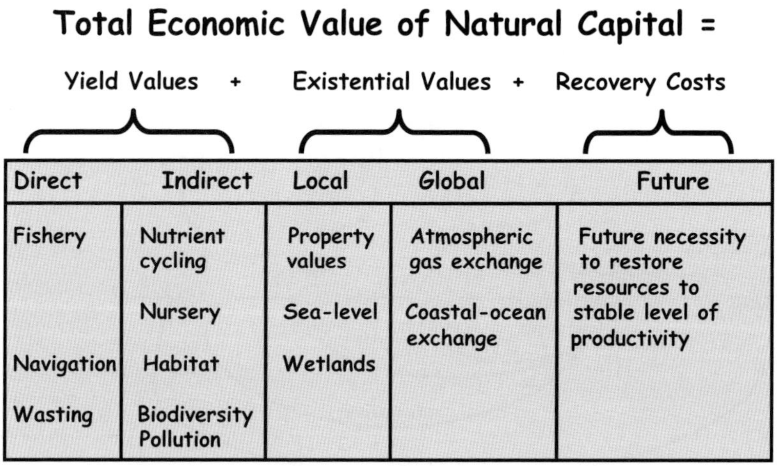


**Figure 1.4** Schematic of relative effectiveness of different decision strategies and the associated feedback loop response for each strategy: adaptive, mitigative, and preventive. The diamond shapes indicate decision points, i.e. indicating whether the decision process stops or continues. In practice, these three steps become iterative. From SCGP (2001).

of the research results about a problem and its consequences, including better, objective cost/risk evaluations. It is also argues for increasing public awareness such that it can exert a stronger influence on the motivation of the decision-maker.

The selection in the Action phase is interdependent with the Motivation phase and with the level of Action selected. If motivation is weak the decider might opt for adaptive measures, medium for mitigative, and strong for preventive action. On the other hand, if the Action phase is clear, simple, or inexpensive, the motivation to act will be greater than if not. For an Adaptive strategy, much less further information is needed, i.e. only that needed for avoidance of the consequences. For a Mitigative strategy, more additional information is needed on how to temporarily reduce consequences of non-action. For a Preventive strategy, information on the Cause, Problem, and Action must be understood and subjected to a long-term cost-benefit analysis. Greater public awareness creates a pressure for as many preventive policies as possible.

## *1.2.3 Social and economic values*

The most important areas for improved assessments are in the social and economic dimensions. There are many benefits recognized in ecosystems: direct and indirect, effective and potential, present and future, Some are grounded in uses (use-value), some concern a potential use (option-value), and others concern moral claims (aesthetic, existential, and other non-use values), all of which support the importance of maintaining high levels of ecosystem goods and services. Only the monetary values of the use-values, commercial or non-commercial, enter directly into market pricing. They are easy to process in numerical modeling and very informative about preferences in allocation of scarce resources and in the efficiency of their use.

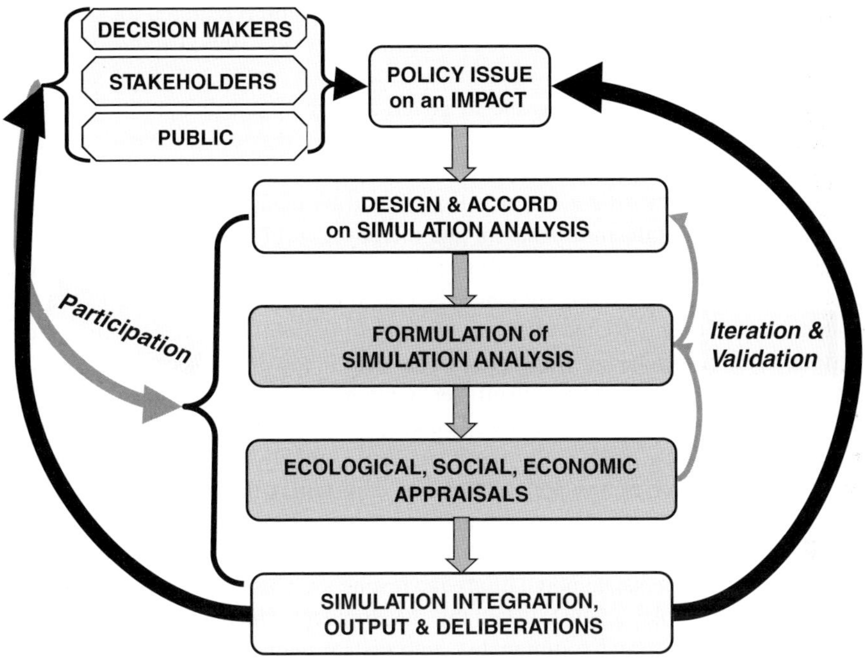


**Figure 1.5** Examples of the total economic value of natural capital of a coastal estuary. The specific values may be positive (benefits) or negative (costs). From SCGP (2001).

These monetary values do not reflect the total costs of all the use-components and the processes that maintain ecosystem functionalities (Figure 1.5). For example, runoff from intensive agriculture degrades the viability of rivers and drives eutrophication of coastal waters without affecting the financial budgets of agriculture, unless a specific mechanism is designed to reintegrate the costs of these externalities into the accounting of the farming. Expanded to global scales, the recovery costs of these externalities have created an enormous environmental debt (Srinivasan, 2008). Furthermore, the global economy promotes unsustainable practices by skirting the market feedback loop designed for human-scaled interactions between consumer and producer, such that the price does not convey the distant damage generated in the form of resource depletion, biodiversity losses, health risks, and morally questionable labor practices. In addition, the market emphasis on use-values favors short-term assessments, and disfavors long-term benefits, because use-values have a preference for the present over the future (discounting).

The efforts of the society to counterbalance the cumulative external effects and short-term decision making calls for accurate cost–benefit and cost-efficiency analysis to design practical targets and better assessments for public policies. It is important that the cost of externalities be accounted for as well as the private and public expenditures needed to adapt, mitigate, or prevent the consequences of environmental degradation. Also important is the willingness of people and decision-makers to invest more in the conservation and productivity of natural capital, as well as the study of the relative benefit from different institutional and technical options. Generally speaking, better and more diversified valuation methodology employing the larger suite of evaluations required for sustainability assessments is badly needed (Costanza, 1997b; Hawken et al., 1999; Brown, 2001; TEEB, 2010; ten Brink et al., 2012).

Many social values also play a role in governance, although this is often considered controversial or discounted in policy evaluations because they are not easily reducible to Newtonian dimensions or that they are inherently non-definable. Recent thinking (Eisenstein, 2007) suggests that these social values should be regarded as emergent properties of the system that can influence behavior and thereby control fluxes of matter and energy, for example a vegetarian that values animals doesn't eat meat, one who values sustainable solutions advocates the recycling of materials, etc. Many of these social values are culture dependent and are deeply rooted in the social devices of family, school, religion, etc., and are historically slow to change. They play a role both in the public acceptance of a policy and in the decision-making process itself (see Section 1.2.2).

The social values in modern societies are changing more rapidly through education, the media, and the Internet. Exposure to the Internet may prove to be a critical factor in the success of SD. However, as a vector for social reorganization, the globalization of communication can generate contradictory influences relative to SD. For example, higher living standards and increased supply of goods and services through markets tends to weaken the sense of dependency on distant populations and resources. This tends to reinforce individualism and egoism that play against sustainable development. The Internet can also reinforce the sense of dependency and communality with distant human societies and ecosystems. The case for sustainability requires a concern for other human beings (social responsibility) and other living things (niche responsibility) and it requires a level of altruism sufficient to ensure common survival.

In practice, social-value representations can be observed through behavior or discourse observations (e.g. dynamic surveys), which then provide helpful insights into how they interact with public policy. The relationship between policy and public awareness (and values) is one of mutualism that can be abetted by research and education through refining the common needs of both. Without this effort, policy efforts suffer and compromise is made difficult. For example, a human group that highly values the freedom of access to nature will strongly oppose any attempt to restrict access. In a similar way, commercialization of nature through markets of tradable rights (tradable emission permits, fishing quotas, etc.) may be rejected simply on moral grounds. For a SAF application, such observed information on social values can be analyzed qualitatively in narrative format and some portions can be rendered into a numerical format as input into the simulation models to improve the information for effective policy-making.

## 1.3 Results

### *1.3.1 Policy concerns*

Because of the voluntary nature of the Stakeholder participation during the SAF development, the composition was selective in favor of those with the time and interest to participate. Consequently, the issues chosen were not necessarily representative of the most urgent or serious issues. As might be expected, volunteers representing the industrial sectors and national environmental departments were very low; however, in many cases interest was generated at higher levels through presentations at the project's end.

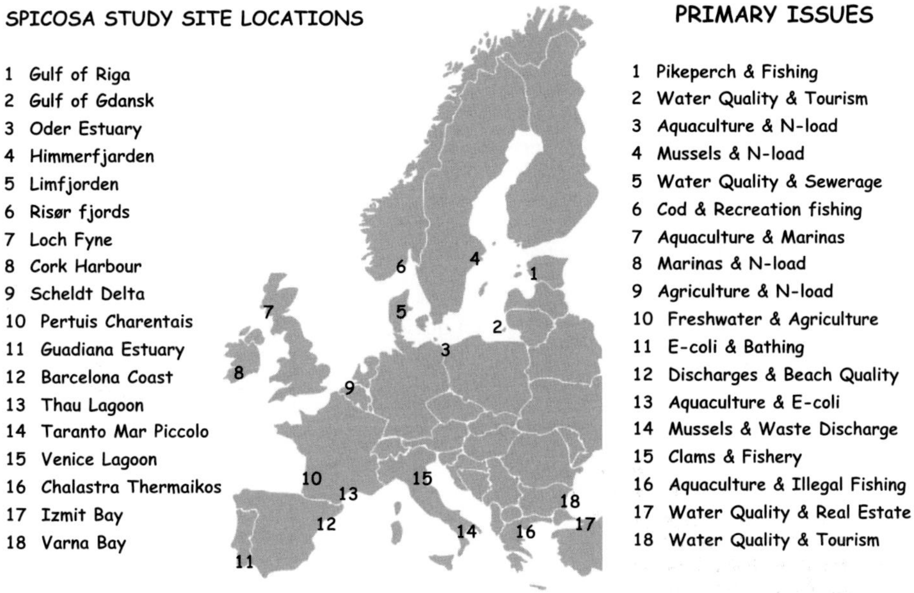


**Figure 1.6** The locations of the Study Sites Applications (SSAs) of the SPICOSA Project indicated by a counter clockwise sequence. The names of SSAs are on the left and their corresponding primary policy issues are on the right.

However, the goal of examining a wide range of multi-issue scenarios was ensured by the very diversity of the coastal zones themselves (Figure 1.6). For each SSA policy issue, a set of corollary issues arose, in the selection of scenarios, which related to each of the ecological, social and economic dimensions. For example, the Policy Issue of nitrogen loading, which reflected a general concern for compliance to the EU Water Framework Directive, provoked dissimilar scenarios and corollary issues in different coastal zones. On an average, each SSA had three major environmental issues, two economic issues, and two social issues. Many of these had conflicting aspects to them, for example conflicts over use, with governance, or through competition. None of the SSAs had a priority for resource planning, although at the end of the SAF application this became an obvious priority.

## *1.3.2 How do the issues interact within the coastal system?*

Since the requested focus (DG Environment, 2006) involved better management of coastal resources, the priority questions concerned what human activities (HAs) cause CZS degradation and what corrective policy changes are needed. By necessity, the SAF also asks what HAs are affected by such degradations. We therefore categorized the causal HAs for each Issue into four groups according to how they interact with the ecosystem, as in Table 1.2. Even though all the SSAs had a primary causal environmental issue, they also had corolllary issues in other environmental categories and in their associated responses in the social and economic dimensions, as summarized in Table 1.3. No two SSAs having the same environmental issue had the same collary issues and none had conservation as a primaty issue, although some considered this as secondary issue, e.g. SSAs 6 and 15.

**Table 1.2** A grouping of the primary issues of the SSAs by their interactions with the resident ecosystem.

| Category | Description | SSAs |
|---|---|---|
| **Wasting 8** | By inputting mass, energy, or information to a system in a manner different relative to some previous reference state, e.g. nutrient and detritus loading, toxins, heavy metals, hormones, synthetic chemical discharges, atmospheric deposition, etc. | Gdansk, Oder, Himmerfjarden Fyne, Guadania Thau, Izmit, Varna |
| **Harvesting 6** | By extracting mass or energy in an amount exceeding the production capacity of its reference state, e.g. overfishing, bycatch trawling, mining, etc. | Riga, Limfjord, Risør, Taranto, Venice, Chalastra |
| **Modifying 4** | By intervening with some internal process or component, e.g. habitat destruction, shoreline development, invasive species, etc. | Cork, Scheldt, Charentais, Barcelona |
| **Conserving 0** | By eliminating damaging interactions, building resilience, protecting areas, regulating use, etc. | |

The left column gives the type and the number of the SSAs in that category, the middle column gives the corresponding definition of the category, and in right column the associated geographic names of the SSAs (see Figure 1.5).

### *1.3.3 The risks of ignoring connections between human activities (HAs)*

A further look at how these HAs interact with an ecosystems helps to explain the complexity facing any resource management attempting to solve environmental problems. Each causal category has a different type of interaction with other components of the ecosystem system, such that if the primary issue were treated independently, the result could negatively influence any attempts to implement sustainable solutions.

Importantly, the HAs causing a disturbance are not always those experiencing the response, which renders impossible any direct feedback loop that would reduce its ability to create disturbances. Generalized examples from the SSAs follow:

1. Those HAs wasting substances in a river may have no direct connection, or cognizance, with those HAs harvesting the estuarine biological production. An industrial HA discharging toxins may not voluntarily regulate the discharge or issue a health warning, which then is left to a researcher or policymaker. Often the policymaker is forced to officially recognize a pollution problem due to public pressure or media exposure, but then may seek delaying actions when faced with the uncertainties of proof of source, established risk to humans, or costs for cleanup.
2. Those HAs harvesting an estuary may be incognizant of the risk of exceeding sustainable yield, or may ignore it, because components of the socio-economic system act to buffer, disperse, or absorb the response to loss of a specific fishery. For example, industrial fishery has access to technology that can make the catch more efficient, cheap energy (petroleum) and often large subsidies (fishsubsidy.org) to supplement the effort with the result that overfishing is facilitated (Pauly, 2006).

**Table 1.3** The frequency of occurrence and diversity of dominant policy issues addressed by the 18 SSAs.

| Wasting | | | | | |
|---|---|---|---|---|---|
| Ecological | 24 | Economic | 16 | Social | 16 |
| Pollution | 6 | Public costs of wwt | 6 | Trans-boundary conflicts | 3 |
| Nitrogen loading | 5 | Tourist income | 4 | Ecosystem health | 2 |
| Aquaculture | 3 | Costs of N-loading | 2 | Public costs of wwt | 2 |
| Eutrophication | 3 | Employment potential | 2 | Recreational benefits | 2 |
| Transparency | 3 | Fishery income | 1 | Seafood contamination | 2 |
| Urban/storm runoff | 3 | Habitat conservation | 1 | Tourist employment | 2 |
| Harmful algae | 1 | | | Directives | 1 |
| | | | | Public costs of N-loading | 1 |
| | | | | User conflicts | 1 |
| Harvesting | | | | | |
| Ecological | 18 | Economic | 12 | Social | 12 |
| Fish population | 4 | Fishery income | 5 | Ecosystem health | 3 |
| Aquaculture shellfish | 3 | Habitat conservation | 2 | Habitat conservation | 2 |
| Fishing practices | 3 | Public costs of wwt | 2 | Public costs of N-loading | 2 |
| Benthic habitat | 2 | Costs of N-loading | 1 | Public costs of wwt | 2 |
| Nutrient loading | 2 | Public costs of wwt | 1 | Shore property values | 1 |
| Harmful algae | 1 | Tourist income | 1 | User conflicts | 1 |
| Pollution | 1 | | | Seafood contamination | 1 |
| Transparency | 1 | | | | |
| Storm runoff | 1 | | | | |
| Modifying | | | | | |
| Ecological | 12 | Economic | 8 | Social | 8 |
| Ecosystem health | 3 | Agricultural income | 2 | Recreational benefits | 3 |
| Employment | 2 | Costs of N-loading | 2 | User conflicts | 3 |
| User conflicts | 1 | Employment potential | 2 | Directives | 1 |
| Habitat conservation | 1 | Freshwater scarcity | 1 | Trans-boundary conflicts | 1 |
| Seafood contamination | 1 | Costs of wwt | 1 | | |
| Cultural values | 1 | | | | |
| Property values | 1 | | | | |
| Recreation potential | 1 | | | | |

They are listed in vertical panels according to the three areas of human influence. Each panel lists the issues in descending order of occurrence for the three ESE dimensions. Each SSA had at least three issues pertaining to the ecological component and two issues pertaining to the economical and social components. For example, an SSA dealing with wasting (N-loading) might have two other ecological concerns (eutrophication, transparency), two economic concerns (public costs of WWT, tourist income), and two social concerns (tourist employment, ecosystem health).

3. Marina development modifying the shoreline may have a high-end economic advantage but its construction in a nursery habitat may weaken fish recruitment or pollute the pelagic habitat. Conventional environmental impact studies may not include important non-market consequences of such tradeoffs. For example, long-term loss of habitat (e.g. loss of biodiversity) may be omitted in cost-benefit assessments where the short-term benefits (e.g. increased economic activity) are preferred.

When more than one HA is involved, the lack of constructive connections between them becomes a main source of conflicts. In each of the above generic examples, conflicts arise in the sense of one activity benefiting at the expense of harm to another CZS activity or component. In addition, when any given primaty issue has multiple connections with other issues, the resolution of that primary issue in one CZS is not necessarily valid for the same issue in another CZS. Furthermore, multi-issue assessments of long-term benefits/losses, which often are made even more difficult when diffused among interconnected issues. These caveats impugn the effectiveness of single policy directives and even management efforts to enforce the same policy over a set of coastal systems. They also require that ICZM develop statuatory links with larger-scaled governance in order that HAs disturbing the system become autonomously self-regulating respect to their environmental damage.

## 1.4 Discussion

### *1.4.1 Sustainability science*

We have presented aspects concerning the need for further develop of sustainability science. We have used the SAF as an example of how researchers can more effectively interact with decision-makers, stakeholers, and the public to the benefit of pursuing sustainable development. The product that they bring to planning discussions constitutes a higher level of explanation on how a local system functions and how it reacts to policy changes. For ICZM planning, the SAF places an emphasis on the simulation and monitoring of dynamic changes in a CZS in a manner that management questions can be answered. A SAF application offers an example of how management could have a continuous, objective mechanism to test policy effectiveness towards a common goal.

However, from a systems' point of view, our current manner of monitoring the environment or social conditions needs upgrading. Present data acquisition has a skewed bias toward recording environmental and human conditions and their trends, which are essential but are not adequately coordinated in time and space that would better enable systems simulations to track changes at the time scale of dominant processes. This will require more intelligently designed observations to support systems models for simulating environmental processes and responses to disturbances, the behavior of societies exposed to resource stress,and for a precautionary defense against emerging problems. It will require innovative changes to transform the market economy to be more self-regulating and favor more equitable distribution of wealth based on social values.

A milestone for the maturity of sustainability science should be that of gaining acceptance within the research establishment and among the public as an instrument for SD. We recognize that this milestone is an iterative loop because a full intergration of scientific knowledge into societies and their govenance cannot be achieved without sufficient funding support in research and academia and without corporate concurrence and societal acceptance of sustainability as a guiding vision for humanity. The assoicated goal of greater social responsibility will require change in both individual and societal behaviors, with the latter being dependent on the former. The challenge of changing the social milieu (e.g. awareness about unsustainability and the need for collective action to reach human sustainability) is an interwoven, complex dynamic that requires attention to several factors, which are well discussed by Werner (1999) and outlined in Table 1.4.

**Table 1.4** Factors fundamental to instituting social behavioral change (adapted from Werner, 1999).

1. **Social milieu – Social awareness and concern about a problem,** *requires conviction based on the necessity and willingness to abandon competing social milieus; its dependence on peer pressure accelerates acceptance.*
2. **Motivation – Knowledge motivation to engage in behavioral change,** *requires the internalization of knowledge about what behavior is needed for promoting sustainability up to the degree that it becomes automated.*
3. **Memory – Recalling situational prompts that make the motivation salient,** *the motivation for sustainability is reinforced by repeated incidents of unsustainability, repeated conversations with others, feedback from situations or behaviors that bring to mind the value of sustainability.*
4. **Opportunities – occasions to follow through with the behavior change,** *supportive measures that facilitate sustainable habits by making them easier and more convenient.*
5. **Skills and ability or perceived confidence to make a behavior change correctly,** *social organizations can provide individual help in attaining certain new skills needed for sustainability change.*

## *1.4.2 Utility of SAF for ICZM*

GESAMP (1996) reported that "There is a great need for an accepted ICZM evaluation methodology ... to document trends, identify their likely causes, and objectively estimate the relative contributions ... to the observed social and environmental change." The fact that the SAF has provided a wide experiential exposure and that it is an open, self-evolving methodology makes it a good candidate to start fulfilling this need.

The SAF very deliberately is *not about* policy management, that is recommending procedures on how management is conducted or how it makes decisions. Instead, it *is for* policy management, that is about providing a higher-level of information, quantifying scenarios, providing a guide for planning. This is an important boundary, one that must be observed to preseve the scientific objectivity of the SAF. For the case of the SSA experiments, the information produced was for policy scenarios directed towards SD as a common ICZM goal.

The SSA experiments have demonstrated a strong utility as an ICZM evaluation methodology in three areas:

1. As a dynamic indicator. Operational simulation models offer a quasi-real-time monitoring of the ecosystems health (SSA14), related economic input/output flows, and social values concerning the coastal resources (SSAs17, 18).
2. As an operational tool. Many of the activities influencing the CZs require flexible operational management, for which the SAF is excellently suited, for example allocating freshwater supplies to various users (SSA10), regulating point and non-point nitrogen loading (SSA9), optimizing fishery/aquaculture practices (SSAs 5, 6, 14, 16), and controlling bacterial exposure on beaches (SSAs 11, 12, 13).
3. As an incentive for collaborative planning. The holistic characteristic of the SAF, its specific evaluation capacity, and the stakeholder participation all served to demonstrate a workable mechanism for bringing together multiple interests and to explore the potential for objective testing of policy options in the broad areas of land-use, urban waste, pollution abatement, and user conflicts so necessary for SD. This advantage was better demonstrated in some of the northern SSAs where there already exist stakeholder–researcher relationships (SSAs 4, 5, 6, 7).

Based on this SAF experience, we would like to mention several recommendations as focal points for further ICZM consideration. General reports on the progress of ICZM focus more on acceptance of ICZM than on actual progress toward sustainable development (ETC-TE, 2002). That is to say, ICZM initiatives are an essential step, and its non-statuatory status provides more flexibility, but they generally do not address sufficiently the methods needed for policy effectiveness toward implementing sustainable practices (EC Rec. 413, 2002).

1. A focus on monitoring 'state' variables is helpful, but is insufficient for adequate dynamic representations of the system, which require better observations of processes, feedback linkages, and dynamic indicators.
2. An ICZM focus on the natural systems cannot provide balanced information regarding coastal sustainability unless it includes social and economic dimensions both in the simulation analysis and in the monitoring efforts.
3. Progress on SD in CZS is seriously hampered by unsustainable HAs. To resolve this, ICZM will need to collaborate with the HAs and with a broader range of governance.

## 1.5 Conclusions

Globally the reorganizational effort towards SD is increasingly apparent through scientific research, green technology, and public awareness, but is still far from the 'tipping point' where its progress has the momentum to survive likely disturbances and set-backs. Until science can effectively raise the public and policy awareness of the 'what-and-how' involved in SD, environmental management cannot be very effective in the transition. Science (natural and social) itself should proactively foster the growth of systems science, elimination of confining disciplinary boundaries, and its ability to outreach to the public. The intention of this article is to stress the need for greater integration of Sustainable Science into ICZM through precautionary, systematic assessments that can provide for guidance on policy effectiveness towards SD, and the value of experiential examples in promoting SD regardless of scale.

The generalized needs for ICZM related to the goal of SD as demonstrated by the SAF experiments are:

1. Developing the capacity to simulate and assess complex social-ecological systems;
2. Establishing working partnerships with policy-makers, stakeholders, and the public within a CZ community;
3. Promoting good examples of transitional efforts towards sustainable development;
4. Instituting monitoring programs dedicated to data and information needed for SAF applications and formatted for public participation;
5. Encouraging human activities to be self-regulating with respect to resource degradation and to tolerable limits of social-economic equality;
6. Understanding that policy cannot manage these CZ systems without a SD plan that integrates the previous five requirements into a continuous assessment process.

## Acknowledgements

The authors wish to acknowledge the 6th FP of the European Commission (RTD Climate change and Ecosystems call) for its foresight and funding in the form of farsighted objectives to stimulate research towards the needs of the evolving EU community. Most of all we thank the several hundred researchers who have contributed, persevered, and brought an optimistic enthusiasm to the completion of the Integrated Project SPICOSA. We wish to recognize the cooperation and willingness to support that all the partners brought to the project, and we especially thank Maurice Heral and Daniel Roy of IFREMER for their expert and conscientious attention to the administration, and Bruno D'Argenio of IAMC/CNR for his continuous support, of the Project.

## References

Adams, W.M. (2006) The future of sustainability: Rethinking environment and development in the twenty-first century. Report of the IUCN Renowned Thinkers Meeting, 29–31 January 2006.

Bertalanffy, L. von (1968) *General Systems Theory*. George Braziller, New York, 295 pp.

Brown, L. (2001) *Eco-Economy*. W.W. Norton and Company, New York, 333 pp.

Brundtland Commission Report (1987) World Commission on Environment and Development. *Our common Future*. Oxford University Press, p. 8.

Capra, F. (1996) *The Web of Life*. HarperCollins, London, 320 pp.

Clark, W.D. and Dickson, N.M. (2003) *Sustainability Science: The Emerging Research Program*. National Academic Press, Washington, DC.

Costanza, R., ed. (1991) *Ecological Economics: The Science and Management of Sustainability*. New York: Columbia Univ. Press, 525 pp.

Costanza, R., Cumberland, J., Daly, H., Goodland, R. and Norgaard, R. (1997a) *An Introduction to Ecological Economics*. CRC Press, 288 pp.

Costanza, R., D'Arge, R., de Groot, R., Farber, S., Grasso, M., Hannon, B., Limburg, K., Naeem, S., O'Neill, R.V., Paruelo, J., Raskin, R.G., Sutton, P. and Belt, M. van den (1997b) The value of the world's ecosystem services and natural capital. *Nature* **38115**, 253–260.

Daly, H.E. (1991) *Steady-State Economics*. Island Press, Washington, DC, 297 pp.

DG Environment (2006) DG Environment FP6, "Global Change and Ecosystems" Sub-Priority 1.1.6.3.

Diamond, J. (2005) *Collapse*. Viking Books.

EC Rec. 413 (2002) Recommendation of the European Parliament and of the council of 30 May 2002 concerning the implementation of Integrated Coastal Zone Management in Europe. *Official Journal of the European Communities*, 6.6.2002

Ehrlich, P.R. (2010) The MAHB, the culture gap, and some really inconvenient truths. *PLoS Biology* **8**(4): e1000330. doi:10.1371/journal.pbio.1000330.

Eisenstein, C. (2007) *The Assent of Humanity*. Panenthea Press, Harrisburg, PA, 589 pp.

ETC-TE (2002) Measuring Sustainable Development on the Coast. In Report to the EU ICZM Expert Group by the Working Group on Indicators and Data led by the European Topic Center on Terrestrial Environment, version 312.doc.

Fiksel, J. (2006) Sustainability and resilience: toward a systems approach. *Sustainability: Science, Practice, & Policy*, **2**(2). http://sspp.proquest.com/static_content/vol2iss2/SSPP-v2.2.pdf (accessed 9 Spetember 2012).

Friedman, M. (1962) *Capitalism and Freedom*. University of Chicago Press, 208 pp.

GESAMP (1996) The Contributions of Science to Coastal Zone Management. Rep. Stud. GESAMP, IMO/FAO/UNESCO-IOC/WMO/WHO/IAEA/UN/UNEP *Joint Group of Experts on the Scientific Aspects of Marine Environmental Protection.* **61**, 66.

Goklany, I.M. (2009) Have increases in population, affluence and technology worsened human and environmental wellbeing? *The Electronic Journal of Sustainable Development* **1**(3).

Gunderson, L.H. and Holling, C.S. (Eds.) (2002) *Panarchy: Understanding Transformations in Human and Natural Systems.* Island Press, Washington, DC.

Hawken, P., Lovins, A. and Lovins, H. (1999) *Natural Capitalism: Creating the Next Industrial Revolution.* Little, Brown & Co., New York, 396 pp.

Hopkins, T.S., Bailly, D. and Støttrup, J.G. (2011) A systems approach framework for coastal zones. *Ecology and Society* **16**(4): 25. http://dx.doi.org/10.5751/ES-04553-160425

Hopkins, T.S., Bailly, D., Støttrup, J.G., Standberg, A., Elgren, R. and Glegg, G. (2012) Experimental Basis and Potential of the SAF as a Transition Methodology for Sustainable Development. In Systems Approach for Sustainable Development in Coastal Zones. *Ecology and Society* **17**(3).

Homer-Dixon, T. (2006) *The Upside of Down.* Island Press, 448 pp.

Kauffman, S. A. (1995) *At Home in the Universe: The Search for Laws of Self-Organization and Complexity.* Oxford University Press.

Kemp, R. and Martens, P. (2007) Sustainable development: how to manage something that is subjective and never can be achieved? *Sustainability: Science, Practice, & Policy*, **3**(2).

Klein, J.T., Grossenbacher-Mansuy, W. and Häberli, R. (2002) Transdisciplinarity: Joint Problem Solving among Science, Technology, and Society: An Effective Way for Managing Complexity.Birkhäuser, Basel 2001, and Basarab Nicolescu, Manifesto of Transdisciplinarity.State University of New York Press, Albany.

Levin, S. (1999) *Fragile Dominion: Complexity and the Commons.* Perseus Books, 256 pp.

Mandel, E. (1986) In Defence of Socialist Planning. *New Left Review* I/159, September–October 1986.

MDG Report (2011) The Millennium Development Goals Report, United Nations. http://www.un.org/millenniumgoals/11_MDG%20Report_EN.pdf (accessed 10 September 2012).

Pauly, D. (2006) Major trends in small-scale marine fisheries, with emphasis on developing countries, and some implications for the social sciences. *Maritime Studies (MAST)* **4**(2), 7–22.

Romeiro, A.R. (2000) Sustainable development and institutional change: the role of altruistic behavior, *IE/UNICAMP, Campinas*, **97**.

SCGP (2001) Scientific Concepts and Global Problems, MEA 430, CoursePak, North Carolina State University, 177 pp.

Shermer, M. (2002) *The Skeptic Encyclopedia of Pseudoscience.* APC-CLIO, Oxford, 455 pp.

Srinivasan, U.T., Carey, S.P., Hallstein, E., Higgins, P.A.T., Kerr, A.C., Koteen, L.E., Smith, A.B., Watson, R., Harte, J. and Norgaard, R.B. (2008) The debt of nations and the distribution of ecological impacts from human activities. *PNAS* **105**, 1768–1773.

TEEB (2010) The Economics of Ecosystems and Biodiversity: Mainstreaming the Economics of Nature: A synthesis of the approach, conclusions and recommendations of TEEB, 39 p.

ten Brink P., Mazza L., Badura T., Kettunen M. and Withana S. (2012) Nature and its Role in the Transition to a Green Economy. http://www.teebtest.org/wp-content/uploads/2012/10/Green-Economy-Report.pdf.

Tett, P., Sandberg, A. and Mette, A., Eds. (2011). *Sustaining Coastal Zone Systems.* Edinburgh, Dunedin Academic Press, 173 pp.

Werner, C.M. (1999) Psychological perspectives on sustainability. In Becker, E. and Jahn, T. (Eds) *Sustainability and Social Sciences.* Zed Books, London.

Section 1

# Coastal Habitats and Ecosystem Services

Chapter 2

# The Norwegian Programme for Mapping of Marine Habitats – Providing Knowledge and Maps for ICZMP

*Trine Bekkby, Frithjof E. Moy, Heidi Olsen, Eli Rinde, Torjan Bodvin, Reidulv Bøe, Henning Steen, Ellen S. Grefsrud, Sigurd H. Espeland, Are Pedersen and Nina M. Jørgensen*

**Abstract**

Globally, there is a growing pressure on coastal ecosystems from human activities. To make well-founded management decisions, managers and policy makers need information (i.e. maps) on the location of different habitats and key areas. This information has generally been lacking in the marine environment. In 2003, a Norwegian programme for mapping and monitoring of marine biodiversity began integrating data on habitat and species distribution in coastal areas. The field mapping commenced in 2007, and at present, selected habitats and key areas have been mapped for approximately 50% of the municipalities and 11 of the 17 coastal counties in Norway. The Norwegian coast is long and complex and mapping all habitats in all areas is not feasible. Consequently, different methodological approaches (including spatial predictive modelling) have been applied. This chapter presents some of the methods developed, the main findings of the programme and lessons learned from linking science and management.

## 2.1 Introduction

Globally, there is a growing pressure on coastal ecosystems from human activities. According to the UN Rio convention (1992), all countries are obliged to understand and preserve their biological diversity. In Norway, this was addressed with a White Paper (Report to the Storting, 1996–1997), describing the policy for achieving a sustainable development of the environment and instructing all municipalities to collect knowledge on biodiversity.

---

*Global Challenges in Integrated Coastal Zone Management*, First Edition. Edited by Erlend Moksness, Einar Dahl and Josianne Støttrup.

This work was followed by the White Paper on biodiversity (Report to the Storting, 1995) and the establishment of the Norwegian national programme for mapping and monitoring of biodiversity (initiated in 2003). The marine component of the programme is financed by the Ministry of the Environment, the Ministry of Fisheries and Coastal Affairs and the Ministry of Defence. The Steering Group consists of one representative from each of the following management authorities: the Norwegian Directorate for Nature Management (programme leader), the Directorate of Fisheries, the Norwegian Defence Estates Agency, the Climate and Pollution Agency and the municipalities. The scientific work is carried out by Norwegian Institute for Water Research (NIVA, scientific coordinator), Institute of Marine Research (IMR), Geological Survey of Norway (NGU) and Akvaplan-niva (ApN).

To make well-founded management decisions for the coastal zone, managers and policy makers need information (i.e. maps) on where important habitats and key areas are found and how valuable they are. When the mapping programme was initiated, little information was available for the marine environment. Consequently, a national guide describing key habitats and areas in the coastal zone was published (DN Håndbok, 2001) and a national mapping programme was initiated. During 2003–2006, the marine component of the programme focused on developing methods for mapping, modelling and integrating existing data. The information was integrated in guidelines for municipalities (Rinde et al., 2007) and resulted in a revision of the handbook for mapping (DN Håndbok, 2007). The field mapping began in 2007 and is still on-going.

This chapter is based on the Final report for the period 2007–2010 of the national marine mapping programme (Bekkby et al., 2011a) and presents the areas and habitats mapped, the methods used, the main findings and the lessons learned from linking science and management.

## 2.2 Mapped habitats and key areas

The mapping is limited to coastal areas (i.e. within 1 nautical mile outside the base line), as these areas are under much pressure, are covered by the EU Water Framework Directive and are getting most of the attention from managers and planning authorities. A selection of habitats and key areas for selected species and populations were mapped in the period 2007–2011 (Table 2.1). These habitats and key areas were selected because they (DN Håndbok, 2007):

- have high species diversity
- have unique physical or chemical conditions and function as a habitat for characteristic biological communities
- are habitats for species that require special attention
- are habitats for special populations
- are highly exposed to human activities and impacts.

The mapping activity has from 2007 to 2011 covered the Oslofjord region (five counties), two Agder counties, Hordaland, the two Trøndelag counties and Troms (i.e. 11 of the 17 coastal counties in Norway; Figure 2.1). These regions were selected because of the urgent need for knowledge in areas of high human pressure, because of the eagerness of managers

**Table 2.1** A list of the habitats and key areas mapped as a part of the National programme for mapping and monitoring of marine biodiversity in the period 2007–2011.

| Habitats/key areas | Mapping institute[a] | Method used |
|---|---|---|
| **Habitats** | | |
| Large kelp forests | NIVA/IMR/ApN | Spatial predictive modelling |
| Ice marginal deposits | NGU | Demarcation from bathymetric data |
| Soft sediments in the littoral zone | NIVA/ApN | Orthophoto[b] verification of digital models |
| Eelgrass meadows | NIVA/IMR/ApN | Model-assisted field mapping |
| Carbonate sand deposits | NGU/NIVA | Spatial predictive modelling |
| **Key areas** | | |
| Oyster areas | IMR | Field mapping |
| Scallop populations | IMR | Underwater video survey and bathymetric data |
| Spawning areas for fish | IMR | Density estimation and egg drift modelling |

All habitats and key areas covered by the programme are presented in DN Håndbok (2007). Methods used are described in more detail in the main text.

[a]NIVA, Norwegian Institute for Water Research; IMR, Institute of Marine Research; NGU, Geological Survey of Norway; ApN, Akvaplan-niva.

[b]Orthophoto is a geometrically corrected aerial photograph.

and planners in these regions to initiate mapping and because of the synergy that existed with ongoing projects. According to the programme plan, the remaining counties will be mapped in the period 2011–2015.

## 2.3 Methods used and main findings

The coast of Norway is long (approximately 100 000 km including mainland and islands) and complex, with large environmental and tidal gradients. Consequently, field mapping is time consuming and costly. Alternative methods have therefore been developed as a tool for mapping. Table 2.1 presents the methods used for the different habitats and key areas.

### *2.3.1 Large kelp forests*

Kelp forests are highly productive systems (Kain, 1971; Sjøtun et al., 1995; Abdullah and Fredriksen, 2004), with a diverse associated flora and fauna (Moore, 1972; Norton et al., 1977; Schultze et al., 1990; Christie et al., 1998, 2003). *Laminaria hyperborea* is the dominant kelp species along the Norwegian coast, providing habitat for several fish species of commercial interest (Norderhaug et al., 2005). This kelp species was mapped using spatial predictive modelling based on field data. Semi-quantitative data on kelp forest density were sampled along geophysical gradients using underwater camera and GPS positioning. Based on the statistical relationships and the GIS layers of the environmental factors, probability maps at a 25 m spatial resolution were developed, see Bekkby et al. (2009) for a more detailed description of the methods and references. Areas with more than

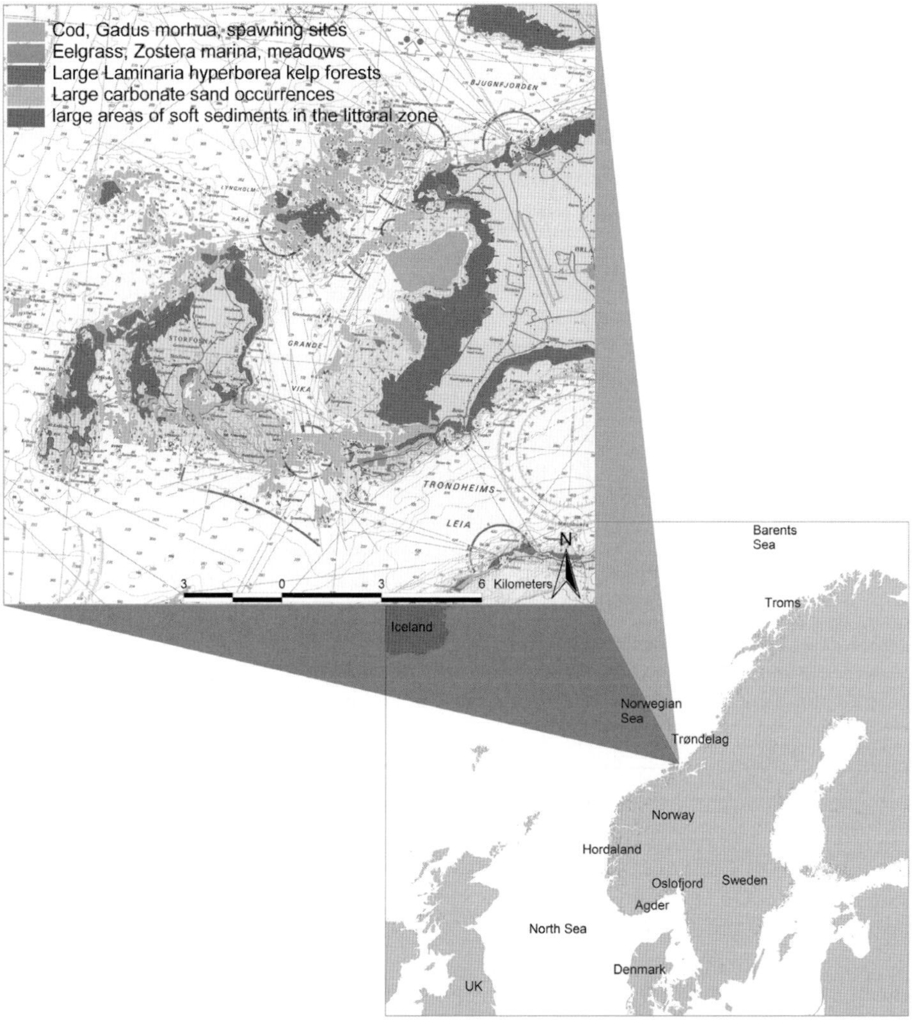


**Figure 2.1** From 2007 to 2011, the mapping activity covered the Oslofjord area (five counties), the two Agder counties, Hordaland, the two Trøndelag counties and Troms (left panel shows area names). The zoomed in area to the left shows an example of a map including a selection of habitats and key areas for parts of Trøndelag (Ørland-Storfosna close to the outer part of the Trondheimsfjord, 63°N 9°E). (see Plate 2.1 for colour version)

60% probability of finding kelp forest were selected for more in-depth description and for estimation of value (see section 2.4).

The kelp forests were found on rocky substrate in moderately exposed and exposed areas at depths down to 20–25 m (Skagerrak) or 30 m (from Hordaland and northward). Current speed was found to be less important than wave exposure. In Troms, the general observed pattern was of healthy and dense kelp forests in the outer coastal and high current areas, whereas the kelp vegetation in more sheltered areas was commonly reduced by sea urchin grazing. In Trøndelag the kelp forests seem to have recovered after being heavily grazed for decades (Sivertsen, 2006; Norderhaug and Christie, 2009), although sea urchins were still dominant in some sheltered parts. In the Oslofjord region and the Agder counties, large kelp forests were less numerous and the individual kelp plants smaller than in Hordaland, Trøndelag and Troms.

### 2.3.2 Ice marginal deposits

Ice marginal deposits (terminal moraines) are found in many Norwegian fjords, along the coast and on the continental shelf, with the most characteristic features being 13 500–11 000 years old. This habitat is characterised by a mixture of sediment grain sizes (clay, silt, sand, gravel, boulders) and provides, together with local increase in current velocity, a special fauna composition (including coral reefs). This habitat is mapped with varying detail along the coast based on available bathymetric data and literature studies.

### 2.3.3 Soft sediments in the littoral zone

Soft sediments in the littoral zone have high production and a high number of species. These areas are of high importance to sea and wading birds. Consequently, many areas are covered by the Convention on Wetlands of International Importance (the Ramsar Convention). This habitat has been mapped and delineated using orthophoto (aerial or satellite photograph geometrically corrected, provided by www.norgeibilder.no) to verify modelled areas. The modelled areas identify flat terrain in shallow areas (i.e. depth values between $+1$ and $-2$ m according to lower astronomical tide (LAT) level). Areas of soft sediments in the littoral zone are relatively small in southern Norway (the Oslofjord region, Agder and Hordaland), whereas they are large in the Trøndelag counties and Troms, due to the large tidal fluctuations in the latter two regions. The habitat often overlaps with eelgrass meadows.

### 2.3.4 Eelgrass meadows

Eelgrass, *Zostera marina,* is the most abundant seagrass in Norway, even though other species, such as *Ruppia* spp. and *Potamogeton* spp., have occasionally been recorded. *Z. marina* meadows are highly productive, have several associated faunal groups and are regarded as of great ecological importance, especially as nursery areas for fish (den Hartog, 1970; Gotceitas et al., 1997; Duarte and Chiscano, 1999; Fredriksen et al., 2004). *Z. marina* meadows are most often found in shallow, sheltered and relatively flat areas (Bekkby et al., 2008), and the depth distribution generally reflects the water transparency. Areas prioritised for field mapping of this habitat were selected using GIS models, historical data, literature surveys and interviews of fishermen with local knowledge. The mapping was carried out using underwater cameras, aqua scopes and by direct visual observations at low tide (the latter in the northernmost areas). In Troms (the northernmost county mapped in 2011), eelgrass meadows were found down to about 4.5 m (standardised to mean sea level), with short plants in the intertidal parts, increasing in size with increasing depth. The separation between the two species described from these areas, *Z. marina* and *Z. angustifolia* (Benum, 1958), is unclear. In Trøndelag and Hordaland, the meadows were commonly scattered and moderately dense, although dense meadows did occur. Depth distribution was down to 5–7 m. In Agder and the Oslofjord region, a few large and a high number of small meadows were observed in bays and straits both in the outer archipelago and in sheltered fjords. Eelgrass was observed down to 8–10 m in Agder and down to 4–5 m in the Oslofjord area.

### 2.3.5 Carbonate sand deposits

Carbonate sand is bioclastic deposits composed of skeletal fragments from marine organisms, the most important being shells, snails, barnacles, sea urchins and calcareous algae. Carbonate sand deposits in Norway have been produced and accumulated during the past 10 000 years (Ottesen et al., 1995) and are common in Trøndelag, Hordaland, western part of Agder and in parts of Nordland and Troms. Carbonate sand areas with onging production and accumulation were mapped using spatial predictive modelling (the method is presented in Bekkby et al., 2011b) using GIS layers of environmental conditions and presence–absence data of carbonate sand deposits from grab samples. Depth, wave exposure and current speed were the most important predictors (carbonate sand is usually found in wave-exposed and high-current areas).

### 2.3.6 Oyster areas

The flat oyster (*Ostrea edulis* L.) populations in Norway are at its northern Nordic border. The species is considered strongly endangered (EN) in the Norwegian red list (Artsdatabanken, 2010), and OSPAR (The Convention for the Protection of the Marine Environment of the North-East Atlantic) considers it to be in regression throughout Europe, both as a species and as a habitat (biogenetic reef, OSPAR Commission, 2008). We located possible oyster areas based on interviews, local information and historical data (Bodvin et al., 2011). Field mapping was carried out using an underwater video camera (for locations deeper than 1.5 m) and visual observations (in more shallow locations with low densities). At locations with high oyster densities, the number of oysters was counted and averaged within a frame. The oyster bed area and number of oysters per $m^2$ were estimated. We also recorded maximum density, area of maximum density, mean density for the total location, area of the total location, number of generations and how rare the oyster was in the area. The mapping has shown large and dense populations in Aust-Agder (Bodvin et al., 2011), with the main stocks found in Arendal ($> 50$ oysters $m^{-2}$). Along the coast from Sweden to Rogaland (west coast of Norway), we found locations with $> 5$ oysters $m^{-2}$, i.e. biogenetic reefs according to OSPAR (OSPAR Commission, 2008).

### 2.3.7 Scallop populations

The mapping programme has focused on two scallop species; the great scallop (*Pecten maximus*) and Iceland scallop (*Clamys islandica*). In Norway, the largest populations of great scallop are found in Hitra, Frøya and Froan, on the coast of Sør-Trøndelag. The presence of great scallop is strongly associated with carbonate sand slopes, and V-shaped troughs most likely provide optimal current velocity and food abundance. The largest populations of Iceland scallop are found around Jan Mayen and in the Svalbard region (Wiborg, 1963, 1973). In mainland Norway the highest densities of the species occurs in Finnmark and Troms. From Nordland and south to Rogaland only a few relict populations remain after the last ice age. Consequently, Troms and Finnmark were prioritised for mapping of the Iceland scallop. Iceland scallops are found on gavel-sand to hard bottom in high-current areas.

Scallop beds were mapped using a vessel-towed camera platform. In addition to video records, Iceland scallops were collected using a small triangular dredge to estimate size, age and sex distribution. The great scallop was collected by SCUBA diving for selected areas. The scallop bed area was defined and the average number (or coverage) of scallops was calculated based on video interpretation.

The distribution of scallops is patchy within a bed, making the mapping very time-consuming. In order to develop a more efficient method of mapping scallop populations, detailed bathymetry and sonar data (Geoswath interferometric sonar) was collected in an area of about 6 km$^2$ in Froan (Sør-Trøndelag County) (Olsen et al., 2009). The aim was to assess whether interferometric sonar could be used to map stocks of the great scallop. Analyses of the relationship between the number of scallops and the sediment types (i.e. shell-sand grain size) are still ongoing.

### 2.3.8 *Spawning areas for fish*

Spawning areas for cod (*Gadus morhua*) were mapped based on interviews of fishermen (carried out by municipalities and the Directorate of Fisheries). Field verification was conducted through annual net hauls, mapping egg densities in the water masses. Spawning areas have been defined and delineated using measured egg densities and oceanographic models. Results show that the Agder counties, with their short fjords and small bays, have many small spawning areas. In the Oslofjord area, the inner part is a good spawning area, whereas outer areas are almost without eggs. Hordaland at the west coast has few eggs, but several high egg densities were found in Trøndelag. Troms had high occurrences of North East Arctic cod (i.e. cod feeding in the Barents Sea) eggs in outer areas.

## 2.4 Linking science and management and making data available for spatial planning

The maps provided by this programme are highly sought after by managers and planners, as making the right choice for coastal zone development is difficult, maybe even impossible, without them. Habitats such as eelgrass meadows and soft sediment in the littoral zone are often located in high conflict areas, as they occur in shallow and sheltered areas that are also attractive for coastal exploitation (e.g. urban development and small craft harbours). Eelgrass meadows are of special importance to coastal zone managers and planners, as angiosperms are one of the obligate biological quality elements of the EU Water Framework Directive.

The habitats and key areas mapped in this programme are visualised and made available through www.naturbase.no, a portal developed by the Norwegian Directorate for Nature Management. An example map from Trøndelag is shown in Figure 2.1. Through this web site, information can be downloaded as vector (shape) or SOSI files. These maps provide managers and planners with knowledge of valuable marine areas with respect to biodiversity, and have to be considered in order to achieve a sustainable management in the coastal zone. Considerable effort has been dedicated to describing the areas and habitats, and the programme has implemented a method for estimating value to each of the occurrences of the selected habitats and key areas based on specified criteria (Rinde et al., 2007). Providing the occurrences with a value is important for giving legitimacy to the planning process, and the result is a more active management. The criteria are, as far as possible, objective and transparent. If subjective criteria are used, these are thoroughly described. All occurrences are classified as either A (nationally very important), B (regionally important) or C (locally important). Habitat size, density, diversity, age, production rate and intact ecological function are regarded as the most important criteria. Size, density and overlap with other species/habitats are the most commonly used criteria, as information on the others is often lacking.

Our experience is that C areas get very low protection by local planners compared to A and B areas. This raises a question about the need for additional classes in order to visualise different levels of local importance. Revised criteria for setting ecological value are under development based on new knowledge obtained through the mapping programme.

Much of the field work has been planned based on previous interviews with local users, fishermen, etc. This is a challenge, as different people have different knowledge of an area, different reasons for providing the information and a different understanding of what is meaningful information. Including models as a planning tool for field surveys helps us to choose areas and collect data suitable for statistical analyses.

Educating managers and planners in how to apply the habitat maps and associated information to practical management and planning is challenging, but nevertheless important. The fact that the maps represent a static view (a 'snap shot' picture) of a dynamic reality must be kept in mind. Abundance and distribution may change with time and the maps must never replace actual surveys prior to making decisions on coastal development. However, the maps do represent areas in which the habitats and key species are most likely to be found at any given time, even though they may vary temporally.

## Acknowledgements

The Norwegian Ministry of the Environment, the Ministry of Fisheries and Coastal Affairs and the Ministry of Defence have financed the programme for the period 2007–2011. All research institutes involved have contributed with additional financing, for instance through synergy with other projects. We thank researchers, research assistants and technical staff at the Norwegian Institute for Water Research, the Institute of Marine Research, the Geological Survey of Norway and Akvaplan-niva for all their effort.

## References

Abdullah, M.I. and Fredriksen, S. (2004) Production, respiration and exudation of dissolved organic matter by the kelp *Laminaria hyperborea* along the west coast of Norway. *Journal of the Marine Biological Association of the UK* **84**, 887–894.

Artsdatabanken. (2010) *The Norwegian Redlist for Species*. 490 pp.

Bekkby, T., Rinde, E., Erikstad, L., Bakkestuen, V., Longva, O., Christensen, O., Isæus, M. and Isachsen, P.E. (2008) Spatial probability modelling of eelgrass *Zostera marina* L. distribution on the West coast of Norway. *ICES Journal of Marine Science* **65**, 1093–1101.

Bekkby, T., Rinde, E., Erikstad, L. and Bakkestuen, V. (2009) Spatial predictive distribution modelling of the kelp species *Laminaria hyperborea*. *ICES Journal of Marine Science* **66**(10), 2106–2115.

Bekkby, T., Bodvin, T., Bøe, R., Moy, F.E., Olsen, H. and Rinde, E. (2011a) National program for mapping and monitoring of marine biodiversity in Norway. Final report for the period 2007–2010. NIVA Report LNR 6105, 31 pp. ISBN 978-82-577-5840-0. (In Norwegian with English abstract.)

Bekkby, T., Rinde, E., Olsen, H., Bøe, R., Henning, H. and Jørgensen, N.M. (2011b) Habitat suitability modelling as a tool in the Norwegian programme for mapping of marine habitats. *ICES CM 2011*\ G:07.

Benum, P. (1958) *The Flora of Troms Fylke*. Tromsø museums skrifter Vol. VI. 402 pp., 546 maps.

Bodvin, T., Mortensen, S., Jelmert, A., Strand, Ø and Grefsrud, E.S. (2011). Utredning av europeisk flatøsters *Ostrea edulis L.* Kunnskapsoversikt med forslag til handlingsplan. DN-utredning 10. 19 s.

Christie, H., Fredriksen, S. and Rinde, E. (1998) Regrowth of kelp and colonization of epiphyte and fauna community after kelp trawling at the coast of Norway. *Hydrobiologia* **375/376**, 49–58.

Christie, H., Jørgensen, N.M., Norderhaug, K.M. and Waage-Nielsen, E. (2003) Species distribution and habitat exploitation of fauna associated with kelp (*Laminaria hyperborea*) along the Norwegian coast. *Journal of the Marine Biological Association of the UK* **83**, 687–699.

den Hartog, C. (1970) *Seagrasses of the World*. Amsterdam, London: North-Holland Publishing Company, 275 pp.

DN Håndbok (2001) Direktoratet for naturforvaltning Håndbok 19–2001.

DN Håndbok (2007) Direktoratet for naturforvaltning Håndbok 19–2001 (revidert 2007). 51 pp.

Duarte, C.M. and Chiscano, C.L. (1999) Seagrass biomass and production: area assessment. *Aquatic Botany* **65**, 15–174.

Fredriksen, M., Krause-Jensen, D., Homer, M. and Laursen, J. S. (2004) Long-term changes in area distribution of eelgrass (*Zostera marina*) in Danish coastal waters. *Aquatic Botany* **78**, 167–181.

Gotceitas, V., Fraser, S. and Brown, J.A. (1997) Use of eelgrass beds (*Zostera marina*) by juvenile Atlantic cod (*Gadus morhua*). *Canadian Journal of Fisheries and Aquatic Science* **54**(6), 1306–1319.

Kain, J.M. (1971) The biology of *Laminaria hyperborea*. 6 Some Norwegian populations. *Journal of the Marine Biological Association of the UK* **51**, 387–408.

Moore, P.G. (1972) The kelp fauna of North East Britain. 1 Introduction and the physical environment. *Journal of Experimental Marine Biology and Ecology* **13**, 97–125.

Norderhaug, K.M., Christie, H., Fosså, J.H. and Fredriksen, S. (2005) Fish–macrofauna interactions in a kelp (*Laminaria hyperborea*) forest. *Journal of the Marine Biological Association of the UK* **85**, 1279–1286.

Norderhaug, K.M. and Christie, H. (2009) Sea urchin grazing and kelp re-vegetation in the NE Atlantic. *Marine Biology Research* **5**, 515–528.

Norton, T.A., Hiscock, K. and Kitching, J.A. (1977) The ecology of Lough Ine. 20. The Laminaria forest at Carrigathorna. *Journal of Ecology* **65**, 919–941.

Olsen, H.A., Bøe, R., Totland, O. and Grefsrud, E.S. (2009) Bunnforhold og kamskjellforekomster – kartlegging av to områder i Froan. *NGU Report* 2009.044. 43 pp.

OSPAR Commission (2008). Background Document for *Ostrea edulis* and *Ostrea edulis* beds. 17 pp.

Ottesen, D., Bøe, R. and Grøsfjeld, K. (1995) Carbonate sand deposition along the coast of southern Norway. *NGU Bulletin* **427**, 56–59.

Report to the Storting (1996–1997) White paper 58. Miljøvernpolitikk for en bærekraftig utvikling [Environmental politics for a sustainable future].

Report to the Storting (2000–2001) White paper 42. Biologisk mangfold [Biodiversity].

Rinde, E., Rygg, B., Norderhaug, K.M., Nygaard, K., Longva, O., Olsen, H., Bodvin, T. and Steen, H. (2007) Instruction manual for mapping of marine biological diversity. NIVA reports for all counties included in the project. (In Norwegian.)

Schultze, K., Janke, K., Krüß, A. and Weidemann, W. (1990) The macrofauna and macroflora associated with *Laminaria digitata* and *L. hyperborea* at the island of Helgoland (German Bight, North Sea). *Helgoländer Wissenschaftliche Meeresuntersuchungen* **44**, 39–51.

Sivertsen, K. (2006) Overgrazing of kelp beds along the coast of Norway. *Journal of Applied Phycology* **18**, 599–610.

Sjøtun, K., Fredriksen, S., Rueness, J. and Lein, T.E. (1995) Ecological studies of the kelp *Laminaria hyperborea* (Gunnerus) Foslie in Norway. In: Skjoldal, H.R., Hopkins, C., Erikstad, K.E. and Leinaas, H.P. (eds) *Ecology of Fjords and Coastal Waters*. Elsevier Science, Amsterdam, pp. 525–536.

UN Rio Convention (1992) www.un.org/documents/ga/conf151/aconf15126-1annex1.htm (accessed 10 September 2012).

Wiborg, K.F. (1963) Some observations on the Iceland scallop *Chlamys islandica* (O.F. Muller) in Norwegian waters. *Fiskeridirektoratets Skrifter, Serie Havundersøkelser* **13**(6), 38–53.

Wiborg, K.F. (1973) Distribution of *Chlamys islandica* on Bear Island Banks. *Translation of the Fisheries Resources Board Canada* **2314**, 7 pp.

Chapter 3

# Monitoring Ecological Quality of Coastal Waters by the Nature Index (NI) – an Integrated Measure of Biodiversity

*Eivind Oug, Gro I. van der Meeren, Grégoire Certain and Signe Nybø*

**Abstract**

At the request of the Norwegian Government, a comprehensive index aimed at monitoring the state of and trends in biodiversity across and within all aquatic and terrestrial ecosystems has been developed and implemented in Norway. The Nature Index (NI) aggregates information from a number of standardised indicators representing species, diversity measures or biodiversity surrogates. For coastal waters, a total of 65 indicators were entered, covering algae, invertebrates, fish, seabirds and sea mammals. The index indicated that the overall status of coastal waters was good, with a slight worsening from 1990 and 2000 to 2010. The pelagic status improved in south-eastern Norway following reduced nutrient inputs, but deteriorated in mid Norway due to a strong decrease of seabirds. The benthic status had deteriorated due to reductions in kelp, invertebrates and fish. Compared to terrestrial systems, the coastal waters showed a better status than forests, mires, wetlands and open lowland. The NI is intended to become a permanent tool for use in Norwegian nature management and political planning, with yearly updates.

## 3.1 Introduction

Present environmental management is deeply concerned with the sustainable use of natural resources to preserve ecosystem goods and services. It is widely accepted that ecosystem services depend on the diversity of biological species communities in terms of the functions of the organisms living there and their abundance and distribution (Hooper et al., 2005). There is currently much concern about the relationships between biodiversity, loss of

*Global Challenges in Integrated Coastal Zone Management*, First Edition. Edited by Erlend Moksness, Einar Dahl and Josianne Støttrup.

species and ecosystem services. It is generally agreed, however, that maintaining a diversity of organisms is essential for stabilising ecosystem processes in response to disturbances and changing abiotic conditions (Loreau et al., 2001; Hooper at al., 2005; Thrush and Dayton, 2010). Protecting biodiversity is therefore considered essential for maintaining ecosystem functions and ensuring the capacity of ecosystems to adapt to changing conditions. Measuring and monitoring ecosystem state and services is an overwhelming task. Traditionally, monitoring and assessments have focused on the composition and diversity of species assemblages, or on changes in species populations, as indicators of the ecological state. Even though a limited number of measurement variables are used, the data may be voluminous and complex and is often not compatible between either different ecosystem levels or different ecosystems. With regard to management, it is a major challenge to translate scientifically based information into forms which can be easily understood by environmental managers and policy makers, and which can therefore form the basis for actions to safeguard environmental quality and ecosystems.

At the request of the Norwegian Government, a comprehensive index aimed at measuring development patterns in ecological quality within and across all major ecosystems in Norway has recently been developed and implemented. The Nature Index (NI; Certain et al., 2011), synthesises information on biodiversity from a variety of sources into simple measures of status and trends. The main aim is to provide a measure that easily illustrates the overall development of biodiversity in Norway. The political background is a need for tools to assess progress in halting the loss of biodiversity, an essential objective in the implementation of the 'Strategic Plan for the Convention on Biological Diversity' (CBD, 2002). The EU, and Norway, has reinforced this goal in separate agreements (see Nybø et al., 2011). The NI should therefore become an instrument for nature monitoring and management by making ecosystem changes visible, identifying areas and ecosystems where management is required, and pointing toward gaps in knowledge for particular areas, ecosystems or trophic levels (Certain et al., 2011).

The NI is composed of a series of indicators, representing a range of elements from species to species groups and diversity measurements (Nybø, 2010a, b; Certain et al., 2011). Each indicator is scaled relative to a reference state which corresponds to an ecological situation where the indicator is not threatened and is not threatening any part of the ecosystem (see Certain et al., 2011 for a complete definition). NI results can be presented at several aggregated levels according to underlying questions and objectives. A global main index is calculated by combining data for all indicators. Alternatively, indicators related to a given theme can be aggregated to produce thematic indices, for specific geographical regions, effects of pollution, habitat degradation, exploitation of resources, and climate changes. Separate calculations are made for different time periods in order to illustrate trends. The NI has been developed for all major aquatic and terrestrial ecosystems in Norway (Nybø, 2010a, Nybø et al., 2011; Certain et al., 2011). The data were assembled from past and present studies covering all fields of biomonitoring and ecological research through a network of scientists. Practical work was largely carried out by a group of experts that met at regular intervals to discuss principles and methodological issues. It has been a prerequisite that the index should build on ongoing activities and use existing data and information.

This paper presents the NI for coastal waters. The data for the indicators were gathered from plankton and benthos trend monitoring programmes, environmental impact assessments, coastal habitat mapping, commercial fish stock assessments and seabird and sea mammal monitoring.

## 3.2 Concepts and calculations of the Nature Index (NI)

### *3.2.1 The NI framework*

All details concerning the NI methodology are provided in Certain et al. (2011), including definitions, assumptions, and mathematical and statistical formulations. Therefore, the following section will only briefly review the main principle behind the method and the specific choices made for coastal waters. The NI framework builds on previous nature indices such as the Dutch Natural Capital Index, the Living Planet Index and the Biological Intactness index (RIVM, 2002; Loh and Wackernagel, 2004; Scholes and Biggs, 2005). The main asset of NI is that each indicator, irrespective of type of data or data structure, is standardised to a common scale before being aggregated to depict overall patterns. In practice, it is important that as many independent features as possible, representing all ecosystem levels from primary producers to top predators, are characterised.

The NI was calculated separately for nine broad types of natural system, termed 'major ecosystems', respectively ocean pelagic, ocean bottom, coastal pelagic, coastal bottom, freshwater, mires and wetland, open lowland, forests, and mountains. The coastal zone was defined to comprise all coastal and fjord systems, including brackish water, reaching seawards one nautical mile from the baseline.

### *3.2.2 Indicator scores and reference states*

Most indicators represent species or species groups measured by abundance or biomass, species communities measured by diversity indices, or biodiversity 'surrogates' which, for instance, express the occurrence or area of particularly important habitats. The standardisation is performed using a ratio between estimated indicator values and defined values for reference states. All indicators, and subsequently the calculated NIs, will therefore take values between 0 (completely degraded situation) to 1 (optimal situation for biodiversity). It is unrealistic to expect that all indicators can equal 1 at the same time, but a Nature Index >0.8 is considered to reflect a good ecological state. Statistical uncertainty is estimated for each standardised indicator, and this is further used to calculate confidence intervals around aggregated NI indices (see Certain et al., 2011 for details).

The reference state for the indicators was established according to different principles, all relating to a sustainable ecological state (see Certain et al., 2011 for an exhaustive list of definitions). For several non-harvested systems and organisms, pristine or near-pristine natural conditions were used, whereas for managed systems or commercial species, concepts such as past knowledge, precautionary levels or carrying capacity were preferred. Indicator values and reference states were generally estimated from data by expert judgement or model predictions, or were based on monitoring data.

Data were recorded at four points in time – 1950, 1990, 2000 and 2010. However, few indicators had values for 1950, and many also lacked values for either or both 1990 or 2000. The NI was therefore best documented in 2010.

### 3.2.3 *Use of weights*

The set of indicators is in many ways unbalanced for the ecological information NI is intended to communicate. A weighting system was therefore designed, in agreement with the group of experts, to balance the influence of indicators (Certain et al., 2011). Indicators that were considered to be particularly important for biodiversity were treated separately and it was agreed they would contribute 50% of the NI value. These 'extra-representative indicators' included, for instance, species community indices and ecological 'keystone' species. For the other 50% of the NI value, it was decided that all trophic levels should contribute in equal parts (primary producers should not count either more or less than top predators). Therefore, no single indicator might overrule information from other indicators.

### 3.2.4 *Area representation*

Data collection and NI calculations were performed using 430 Norwegian municipalities as spatial units. For several indicators, especially fish, birds and mammals, the data represented populations within larger areas, in which case the same values were entered for groups of adjacent municipalities or larger geographical regions. Reference states, which generally do not vary on a small scale, were established for regions or the whole country.

## 3.3 Indicators for coastal waters

In total, 29 indicators were entered for the free water masses (coastal pelagic) and 36 for bottom areas (coastal bottom). Indicators were selected by experts on the respective organism groups. It was a requirement that the indicators were measured at several relevant time points, were geographically representative at least on a regional scale, and that their reference states could be defined. The set of indicators essentially represented all possible indicators that complied with the requirements and that could also be processed within the time period allocated for the development of the NI.

The indicators for the free water masses encompassed planktonic algae, zooplankton, jellyfish, fish, seabirds and sea mammals (Table 3.1). The seabirds constituted the most comprehensive group with 20 indicators, representing species or sub-species, whereas there were few indicators for algae and zooplankton. Lack of coherent and geographically well-distributed data for algae and invertebrates made it impossible to provide a broader range of indicators for these groups. Most data for fish, birds and mammals originated from population monitoring programmes where information on previous population sizes or estimates of critical population parameters could be used for setting reference levels.

Several of the species of birds, fish and sea mammals spend more or less of their lives in other major ecosystems, such as open ocean or land. In order to avoid duplication in the NI, their contribution was split in proportion to their relationship to each of the major ecosystems (Table 3.1: percentage specificity). Planktonic algae and zooplankton were treated as extra-representative and hence up-weighted in the calculations as they represent groups of species, and also to compensate for the low number of indicators at low trophic levels. Herring (*Clupea harengus*), sand eel (*Ammodytes* sp.) and the lion's mane jellyfish

**Table 3.1** Indicators of the coastal pelagic major ecosystem.

| Indicator | English name | Measurement | Reference | Occurrence | Specificity % | Extra—repr. |
|---|---|---|---|---|---|---|
| **Plankton** | | | | | | |
| Chlorophyll *a* | — | Concentrations | Past knowledge | All | 100 | C |
| Zooplankton | — | Biomass | Pristine | All | 100 | C |
| *Cyanea capillata* | Lion's mane | Relative density | Pristine | All | 100 | K |
| **Fish** | | | | | | |
| *Clupea harengus* | Herring | Percent of reference | Precautionary level | All | 100 | K |
| *Ammodytes* spp. | Sand eel | Percent of reference | Precautionary level | ESWC | 30 | K |
| *Cyclopterus lumpus* | Lumpfish | CPUE | Management | All | 35 | — |
| **Birds** | | | | | | |
| *Fulmarus glacialis* | Northern fulmar | Demographic trend | Precautionary level | SWCN | 30 | — |
| *Morus bassanus* | Northern gannet | Demographic trend | Precautionary level | WCN | 30 | — |
| *Phalacrocorax carbo carbo* | Great cormorant | Demographic trend | Precautionary level | WCN | 100 | — |
| *Phalacrocorax carbo sinensis* | Great cormorant | Demographic trend | Precautionary level | ES | 100 | — |
| *Phalocrocorax aristotelis* | European shag | Demographic trend | Precautionary level | WCN | 100 | — |
| *Stercorarius skua* | Great skua | Demographic trend | Precautionary level | CN | 45 | — |
| *Larus canus* | Common gull | Demographic trend | Precautionary level | All | 50 | — |
| *Larus fuscus fuscus* | Lesser black-backed gull | Demographic trend | Precautionary level | WCN | 90 | — |
| *Larus fuscus intermedius* | Lesser black-backed gull | Demographic trend | Precautionary level | ES | 90 | — |
| *Larus argentatus* | Herring gull | Demographic trend | Precautionary level | All | 50 | — |
| *Larus marinus* | Great black-backed gull | Demographic trend | Precautionary level | All | 50 | — |
| *Rissa tridactyla* | Black-legged kittiwake | Demographic trend | Precautionary level | WCN | 80 | — |
| *Sterna hirundo* | Common tern | Demographic trend | Precautionary level | All | 90 | — |
| *Sterna paradisaea* | Arctic tern | Demographic trend | Precautionary level | SWCN | 90 | — |
| *Alca torda* | Razorbill | Demographic trend | Precautionary level | WCN | 30 | — |
| *Uria aalge* | Common murre | Demographic trend | Precautionary level | WCN | 30 | — |
| *Uria lomvia* | Thick-billed murre | Demographic trend | Precautionary level | N | 30 | — |
| *Fratercula arctica* | Atlantic puffin | Demographic trend | Precautionary level | WCN | 30 | — |
| *Gavia adamsii* | Yellow-billed loon | Percent of reference | Pristine | All | 100 | — |
| *Mergus serrator* | Red-breasted merganser | Percent of reference | Pristine | All | 100 | — |
| **Mammals** | | | | | | |
| *Halichoerus grypus* | Grey seal | Abundance | Pristine | WCN | 70 | — |
| *Orcinus orca* | Killer whale | Percent of reference | Past knowledge | All | 60 | — |
| *Phoca vitulina* | Harbour seal | Abundance | Past knowledge | All | 70 | — |

Table entries show type of measurement, basis for establishing reference state, occurrence in Norwegian coastal regions, specificity to the coastal pelagic, and indicators weighted as extra-representative (Extra-repr.). Occurrence: E, south-eastern Norway; S, south coast (Skagerrak); W, western coast (North Sea); C, central Norway; N, northern Norway. Extra-representative: C, species community; K, keystone species.

(*Cyanea capillata*) were considered to be keystone species in fjord ecosystems and were also up-weighted.

The coastal bottom indicators comprised benthic algae, vascular plants, corals, bivalves, crustaceans, sea urchins, fish, seabirds and mammals (Table 3.2 ). The invertebrates constituted the most comprehensive group with 15 indicators, followed by fish with 10 indicators and seabirds with five indicators. The set of indicators was better balanced between organism groups than for the free water masses, but lower trophic levels were still underrepresented, and among the invertebrates there were rather a large number of commercially exploited species. For most indicators, past knowledge or precautionary levels were used for setting reference states, reflecting that the coastal zone has been used by man for thousands of years. The majority of the indicators were confined to coastal waters (100% specificity), either because of presence only in coastal areas (many invertebrates), or because the data were recorded for coastal stocks (several fish). Several indicators were up-weighted (extra-representative) in the calculations because they represented species communities (macroalgae and invertebrate diversity indices), groups of species (Gobiidae), or are of key ecological importance (kelp, corals and coastal cod). Kelp, sponges and corals are all large sessile upright-growing species that create habitats for a number of associated species. The sea urchin *Strongylocentrotus droebachiensis* is a negative indicator, as high abundances are only found on barren grounds where the kelp is grazed down (Norderhaug and Christie, 2009).

Several of the indicators, especially among the seabirds, are not relevant along the whole coast (Table 3.1 and Table 3.2). For natural reasons, regional NIs will therefore be based on different sets of indicators. In addition, for several widely distributed indicators, data were limited to parts of the coast only. This was, for instance, the case for the macroalgae diversity indices, several invertebrates, non-commercial fish and many birds. The number of indicators used for calculation of NIs in small geographical areas was therefore always lower than the total number for coastal waters. Inner fjord areas were, with a few exceptions, omitted. This was mainly because very few indicators contained data from inner fjord areas. In addition, it may be assumed that reference values for many indicators will differ from open coastal waters due to different hydrophysical conditions and water exchange dynamics, which, however, could not be ascertained due to the sparse data.

## 3.4 Status and trends in coastal waters

### *3.4.1 The pelagic system*

The overall status of the pelagic system was generally good with mostly small or moderate deviations from reference conditions (Figure 3.1). The strongest deviations were recorded for south-east Norway, which is the most densely populated area of Norway, and parts of central and northern Norway. In south-east Norway, this reflected eutrophication of coastal waters due to inputs of nutrients from surrounding land areas and long-transported current-borne nutrients from the Baltic and the southern North Sea (OSPAR, 2010a). The eutrophication was particularly expressed by increased values of chlorophyll *a*, following enhanced algal production. In northern Norway, the somewhat less good status was due to declining populations of species of seabirds and fish. The strongest decline occurred

**Table 3.2** Indicators of the coastal benthic major ecosystem.

| Indicator | English name | Measurement | Reference | Occurrence | Specificity % | Extra—repr. |
|---|---|---|---|---|---|---|
| **Macroalgae** | | | | | | |
| *Laminaria hyperborea* | Cuvie | Per cent of reference | Carrying capacity | All | 100 | K |
| *Saccharina latissima* | Sugar kelp | Concentration | Past knowledge | All | 100 | K |
| Intertidal macroalgae index[a] | — | Diversity | Pristine | All | 100 | C |
| Macroalgae depth limit index[a] | — | Species richness | Pristine | All | 100 | C |
| **Vascular plants** | | | | | | |
| *Eleocharis parvula* | Wetland sedges | Relative rank | Pristine | ESWC | 100 | — |
| **Invertebrates** | | | | | | |
| Porifera | Sponges | Occurrence | Pristine | All | 100 | C |
| *Lophelia pertusa* | Cold water coral | Relative rank | Pristine | EWCN | 100 | K |
| *Ostera edulis* | Common oyster | Relative rank | Past knowledge | ESW | 100 | — |
| *Mytilus edulis* | Blue mussel | Biomass | Pristine | All | 100 | — |
| *Chlamys islandica* | Iceland scallop | Per cent of reference | Past knowledge | N | 100 | — |
| *Pecten maximus* | King scallop | Relative rank | Past knowledge | ESWC | 100 | — |
| *Mya arenaria* | Sand gaper | Per cent of reference | Past knowledge | All | 100 | — |
| *Palaemonetes varians* | Atlantic ditch shrimp | Occurrence | Pristine | ES | 100 | — |
| *Homarus gammarus* | European lobster | Landings | Management | ESWC | 100 | — |
| *Carcinus maenas* | European shore crab | Landings | Pristine | All | 100 | — |
| *Cancer pagurus* | Edible crab | Biomass | Past knowledge | ESWC | 100 | — |
| *Strongylocentrotus droebachiensis* | Green sea urchin | Density | Pristine | All | 100 | K |
| Soft bottom fauna index[a] | — | Diversity | Pristine | All | 100 | C |
| Soft bottom fauna function[a] | — | Diversity | Pristine | All | 100 | C |
| Fjord crustaceans index | — | Species richness | Past knowledge | All | 100 | C |

(*Continued*)

**Table 3.2** (*Continued*)

| Indicator | English name | Measurement | Reference | Occurrence | Specificity % | Extra—repr. |
|---|---|---|---|---|---|---|
| **Fish** | | | | | | |
| *Anguilla anguilla* | European eel | Landings | Pristine | ESWC | 75 | — |
| *Labrus bergylta* | Ballan wrasse | Landings | Pristine | ESWC | 100 | — |
| *Ctenolabrus rupestris* | Goldsinny | Landings | Pristine | ESW | 100 | — |
| *Symphodus melops* | Corkwing wrasse | Landings | Pristine | ESW | 100 | — |
| *Hippoglossus hippoglossus* | Atlantic halibut | Per cent of reference | Past knowledge | All | 75 | — |
| *Pollachius pollachius* | Pollack | Landings | Pristine | All | 40 | — |
| *Pleuronectes platessa* | Plaice | Per cent of reference | Precautionary level | All | 100 | — |
| *Gadus morhua* | Atlantic cod | Landings | Pristine | All | 80 | K |
| Gobiidae | Gobies | Landings | Pristine | SWC | 100 | C |
| *Squalus acanthias* | Dogfish | Per cent of reference | Past knowledge | All | 30 | — |
| **Birds** | | | | | | |
| *Cepphus grylle* | Black guillemot | Demographic trend | Precautionary level | SWCN | 70 | — |
| *Somateria mollissima* | Common eider | Demographic trend | Precautionary level | All | 70 | — |
| *Melanitta fusca* | Velvet scoter | Per cent of reference | Pristine | All | 50 | — |
| *Polysticta stelleri* | Steller's eider | Per cent of reference | Pristine | N | 100 | — |
| *Melanitta nigra* | Common scoter | Per cent of reference | Pristine | All | 50 | — |
| **Mammals** | | | | | | |
| *Lutra lutra* (coastal stock) | Otter | Occurrence | Pristine | All | 100 | — |

Table entries show type of measurement, basis for establishing reference state, occurrence in Norwegian coastal regions, specificity to the coastal bottom, and indicators weighted as extra-representative (Extra-repr.). Occurrence: E, south-eastern Norway; S, south coast (Skagerrak); W, western coast (North Sea); C, central Norway; N, northern Norway. Extra-representative: C, species community; K, keystone species.

[a]Under development for EU Water Framework Directive.

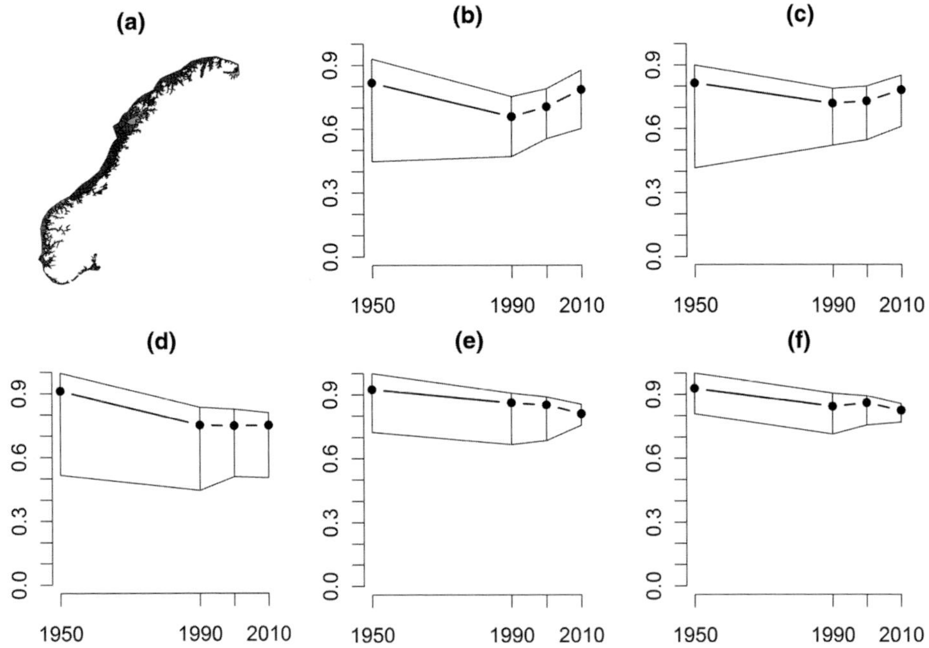


**Figure 3.1** (a) NI status for coastal pelagic 2010 at municipality level. (b–f) Trend 1950–2010 within geographical regions: (b) south-east; (c) south; (d) west; (e) central; (f) north. Vertical lines in trend diagrams indicate 95% confidence intervals. (see Plate 3.1a for colour version)

among seabirds feeding on small pelagic fish, for example common murre *Uria aalge*, puffins *Fratercula arctica* and kittiwake *Rissa tridactyla* (Barrett, 2002; Barrett et al., 2006). These species mainly feed in open ocean areas. In the NI calculations they were given a reduced specificity to coastal waters (Table 3.1), which essentially implied a down-weighting. Nevertheless, the coastal NI illustrated a less good state and thereby reflected that coastal waters are also affected by major changes in open ocean systems.

The trend diagrams calculated for coastal regions illustrate that the status improved from 1990 to 2010 in south-eastern Norway, whereas the status was stable or had declined weakly in western and northern Norway (Figure 3.1b–f). The positive trend in south-east Norway can be explained by the implementation of restrictions on nutrient discharges from municipal effluent water and agriculture brought into force during the 1980s and 1990s (Carstensen et al., 2006; OSPAR, 2010a), which have led to reduced algal production in nearshore waters. In northern Norway, the status reflects that seabird and fish stocks have not regained their previous population levels, but persist at lower or still decreasing levels. In all regions, the calculated NI values for 1950 were high. The values were, however, based on very few indicators and should be interpreted with caution.

The changes for seabirds are illustrated in more detail by thematic part-indices for birds aggregated according to feeding categories (Figure 3.2). The low and decreasing state for birds feeding on small pelagic fish is clearly shown (Figure 3.2b), whereas coastal fish-eating birds (e.g. cormorants *Phalacrocorax carbo* spp.) are in a good state (Figure 3.2a).

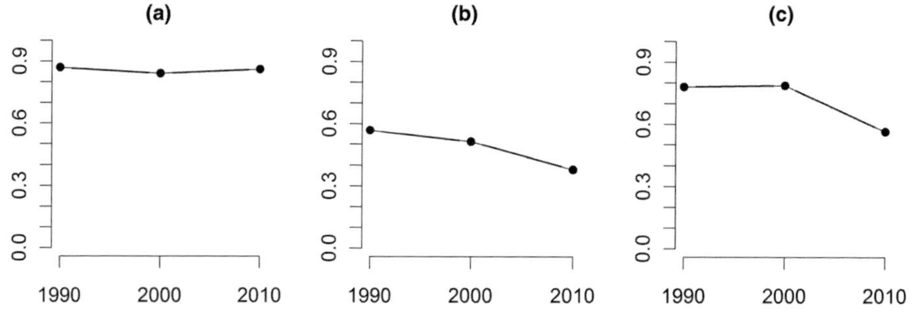


**Figure 3.2** Trends for seabirds according to feeding categories: (a) coastal fish-eating birds; (b) pelagic fish-eating birds; (c) surface feeding birds. Based on exhaustive bird colony counts without variance reported.

Surface-feeding birds, e.g. seagulls, have been declining in recent years (Figure 3.2c). The decline of common murre and puffins has been related to reductions in their food resource of small pelagic fish and to inadvertent drowning in fishing gear, but climatic changes may also play a part (Barrett et al., 2006). A particular problem for pelagic feeding birds is that several species have low fecundity rates and will only slowly resume previous stock levels, even with improved living conditions.

### *3.4.2 The benthic system*

The overall status of the benthic system was good in western and central Norway, whereas the status was less good in south-east and northern Norway (Figure 3.3). The less good status in south-eastern Norway was due to reduced populations for several species, for instance sugar kelp (*Saccharina latissima*), sand shell (*Mya arenaria*), lobster (*Homarus gammarus*) and some coastal fish stocks. It may seem that eutrophication from nutrient inputs and over-exploitation of commercial fish may be the most important factors, but it is increasingly acknowledged that loss of habitat caused by human activities in the coastal zone, stress from increased summer seawater temperatures and competition from alien species may also play important roles. In central and northern Norway, the deteriorated status is mainly a consequence of extensive reductions of kelp forests (*Laminaria hyperborea*) due to grazing by the sea urchin *Strongylocentrotus droebachiensis.* At the same time, reductions in cold water corals (*Lophelia pertusa*) and fish (eel *Anguilla anguilla* and coastal stocks of cod *Gadus morhua*) contribute to the overall status.

The trend diagrams for coastal regions indicate that the status decreased or remained stable from 1990 to 2010 in all regions except for central Norway where the status improved (Figure 3.3b-f). The improvement mainly reflected re-growth of kelp in areas previously grazed by sea urchins. Figure 3.4 shows kelp indicator values for central Norway from 1950 to the present, to illustrate the development patterns. Kelp grazing was first observed in central Norway in the 1970s and spread northwards during the 1980s. In the southern areas, a recession of sea urchins and regrowth of kelp started in the late 1980s, which has gradually moved northwards to the present northern limit close to the Arctic Circle. The

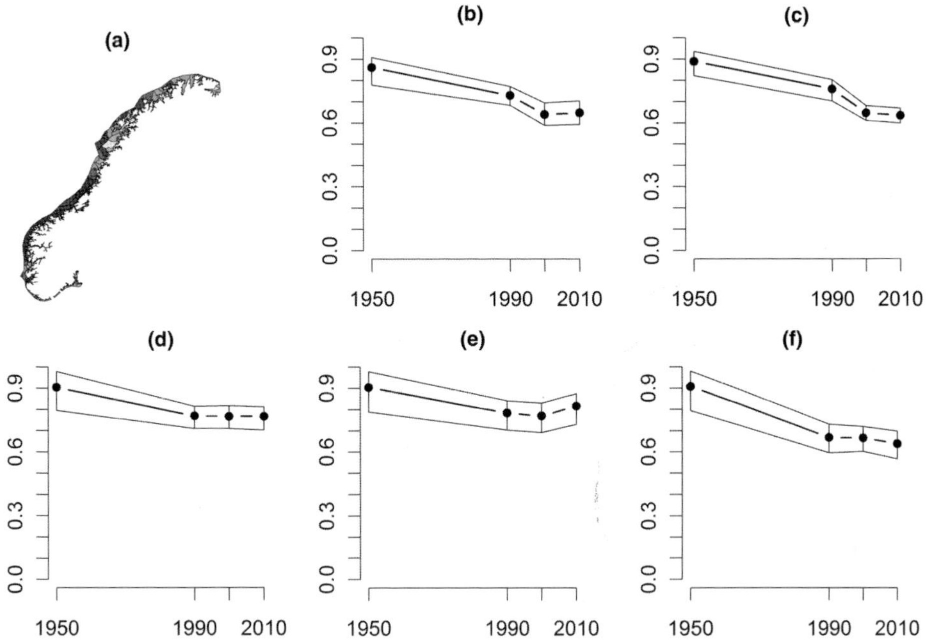


**Figure 3.3** (a) NI status for coastal bottom 2010 at municipality level. (b-f) trend 1950–2010 within geographical regions: (b) south-east; (c) south; (d) west; (e) central; (f) north. Vertical lines in trend diagrams indicate 95% confidence intervals. (see Plate 3.3a for colour version)

causes for the regrowth and the ecological factors behind the sea urchin grazing are poorly understood (Norderhaug and Christie, 2009).

It could be assumed that the extensive kelp grazing may have consequences for species at higher trophic levels which are dependent on resources in the kelp forests. Figure 3.5 illustrates changes in indicator values for three predator species – cod (*Gadus morhua*), eider duck (*Somateria mollissima*) and coastal seals (harbour seal *Phoca vitulina* and grey seal *Halichoerus grypus*). Both cod and eider duck showed declining populations in central and northern Norway since 1990 in the period when the kelp forests have been grazed down. In particular, the coastal stock of cod is presently far below the reference level (Figure 3.5a).

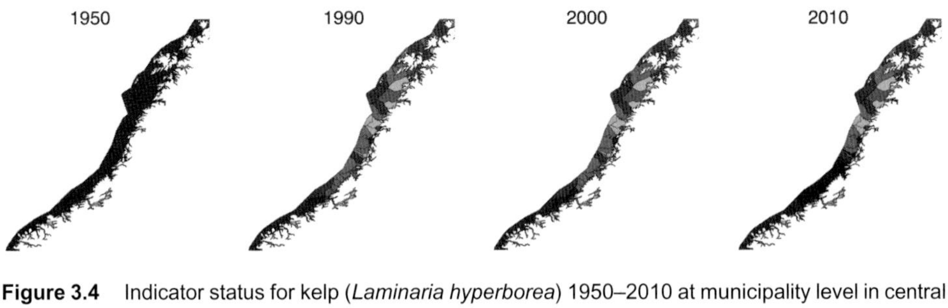


**Figure 3.4** Indicator status for kelp (*Laminaria hyperborea*) 1950–2010 at municipality level in central Norway. (see Plate 3.4 for colour version)

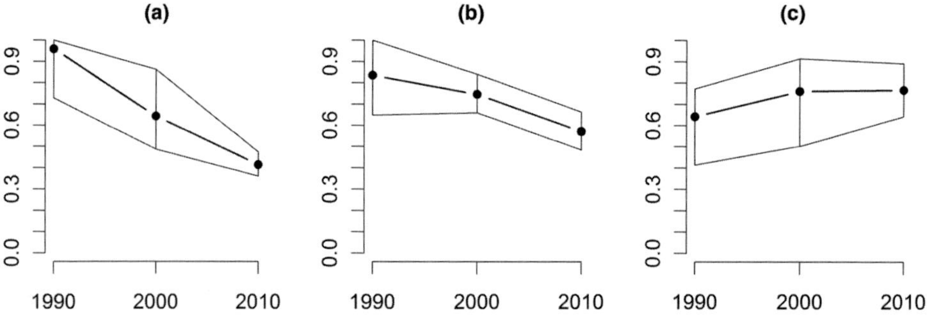


**Figure 3.5** Trends in indicator values 1990–2010 for species at higher trophic levels in central and northern Norway: (a) coastal stocks of Atlantic cod; (b) common eider; (c) harbour seal and grey seal.

These trends may suggest relationships, but the data presently used for calculation of indicator values were assembled from separate sources and cannot be further used to examine cause–effect relationships. Seals increased slightly. Seals, however, are hunted and the trend most probably reflects changes in hunting pressure rather than food availability.

## *3.4.3 Coastal waters compared to other major ecosystems*

Coastal waters showed approximately the same status as the ocean pelagic and bottom ecosystems (Figure 3.6). The overall trend for coastal pelagic and coastal bottom was

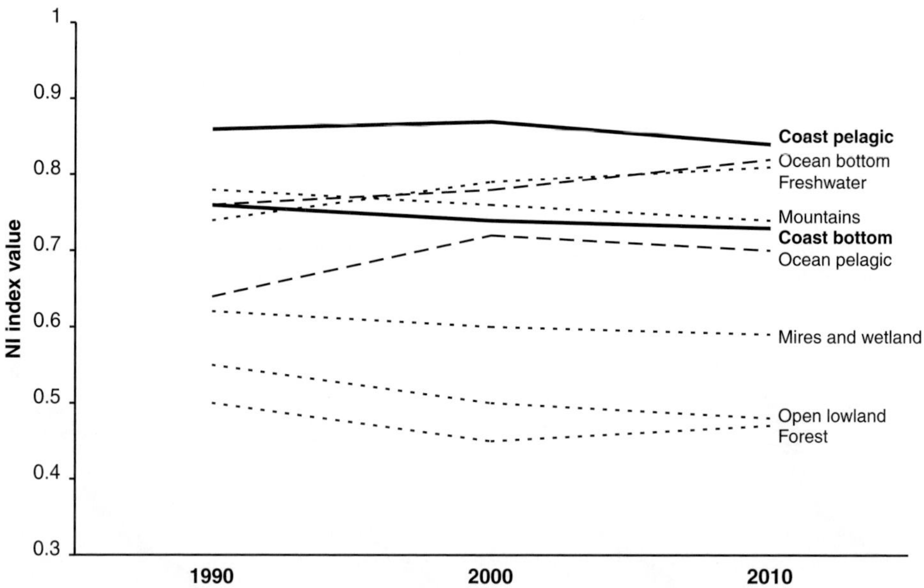


**Figure 3.6** NI trend lines 1990–2010 for coastal pelagic and coastal bottom compared to other major ecosystems. Confidence intervals omitted for clarity. Modified from Certain et al. (2011).

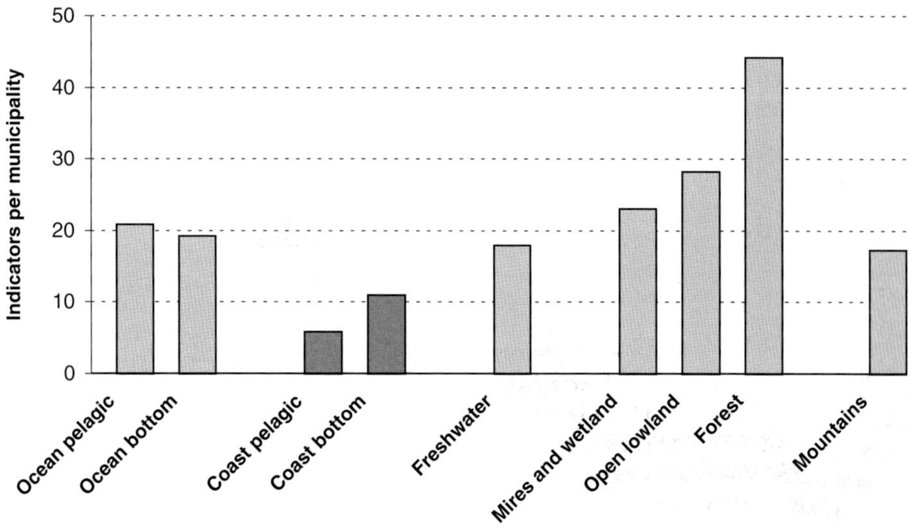


**Figure 3.7** Average number of indicators per municipality for the nine major ecosystems. Modified from Certain et al. (2011).

slightly decreasing during the period 1990–2010, whereas for ocean ecosystems the trend was generally increasing. The increase can be explained by a better status for many commercial fish species, for which extensive regulations and better management practice have ensured that more species are kept at sustainable population levels (OSPAR, 2010b). Freshwater and mountain major ecosystems also had a generally good status, with freshwater showing an improving trend due to reduced impacts of acidification.

The terrestrial lowland major ecosystems of mires and wetland, open lowland and forests all showed lower status than the coastal, ocean and freshwater ecosystems. The low status can be explained by the extensive anthropogenic use of these areas in the past or present, leading to considerable departures from a reference state of pristine conditions. Open lowland, in particular, includes cultured areas such as semi-natural grasslands and coastal heaths, which are dependent on continuous management to maintain their status. The low status and decreasing trend reflect the termination of traditional forms of agriculture, leading to continuous landscape changes and loss of many especially species-rich habitats (Nybø et al., 2011).

The number of indicators was low for the coastal zone compared with all the other major ecosystems. At the lowest spatial level, the municipality, there was an average of 5.8 indicators for the pelagic ecosystem and 10.9 indicators for the bottom ecosystem (Figure 3.7). Compared with the open ocean, the average number of pelagic indicators was not more than about one-third, and for the bottom about one-half. Both ocean pelagic and ocean bottom incorporates a number of commercial fish species which are monitored and reported in routine stock assessment programmes. In addition, sea mammals and some commercial invertebrates are included. This difference essentially illustrates the strong importance of fisheries in continental shelf and open ocean areas, and the resources put into monitoring of harvested stocks.

The number of indicators in terrestrial and freshwater habitats was approximately the same or higher than in the open ocean (Figure 3.7). With regard to organism groups, however, the terrestrial and freshwater major ecosystems had a more equivalent representation than was the case for the marine major ecosystems with more indicators for plants and invertebrates. In open lowland and forests, in particular, fungi and lower plants (mosses) constituted significant groups (Nybø, 2010a, b). It appears therefore that the ecological representation was better balanced in terrestrial and freshwater major ecosystems than in the marine systems.

## 3.5 Discussion

The development of NI represents the largest and most consistent assembling of knowledge on biodiversity and ecological status in Norway. A wide range of monitoring data from quite different programmes has been taken into account as well as data from separate environmental assessment studies and field inventories. The index in itself is a means of communicating the scientifically based ecological assessments to managers, policy makers and the general public. The index framework and routines are of general applicability that can be used in any context, at any scale and by all countries (Certain et al., 2011). Detailed documentation covering the selection of indicators, status and trends for all major ecosystems has been described by Nybø (2010a, b).

The mere process of developing the index, synthesising the data and carrying out the evaluations across all major ecosystems, has given new insight into which types of data are needed and how monitoring programmes should be structured to meet the needs for ecologically based nature management. Hence, an important aspect of the index, aside from the results, is as a means of structuring environmental assessments and monitoring activities for management purposes. Furthermore, it has been a major and crucial challenge to establish reference states for the indicators. Essentially, this task implies a formalising of the basic knowledge on natural systems by giving numerical expressions for the sustainable state of ecosystem components. Environmental scientists are repeatedly confronted with questions about the state of natural systems, whether they are good or bad, which often have to be answered with reference to quite general perceptions of what constitutes the 'normal' state. During the process of developing the index, there has been a series of discussions and considerations within the group of experts about the basic principles used for setting reference states, the estimated values themselves and the geographical relevance of the values. This has given a far more solid basis for answering questions about state and trends which, in addition, can be communicated in simple ways.

The principle of scoring indicator values relative to a reference value on a 0 to 1 scale is basically similar to the principles used for assessing water quality in the EU Water Framework Directive (WFD). In the WFD this relationship, termed the ecological quality ratio (EQR), is calculated to compare different water quality parameters (de Bund and Solimini, 2007). The similarity between the two systems implies that the water quality elements in WFD representing biodiversity can be entered more or less as they are as indicators in NI. A main difference between WFD and NI, however, is that WFD is designed as a highly structured classification system for environmental quality assessment using a particular selection of quality elements, whereas NI seeks to represent an overall status for biodiversity, with no class gradations, based on the most different types of indicators as possible.

The presently calculated NI trendlines and maps should be viewed with some caution, however. For several major ecosystems it was problematic to provide a sufficiently balanced and representative selection of indicators. This could not be fully remedied by the weighting system. In addition, the data quality varied between different geographical regions or over time for a number of indicators (Nybø et al., 2011). However, the statistical uncertainty associated with the estimated indicator values is fully taken into account and quantified within the NI framework, and is communicated as part of the NI results. It is crucial to stress that it is equally important both to identify areas where management is needed to reduce environmental impact and to identify areas where research is needed to better detect potential environmental impacts. The NI framework allows both (Certain et al., 2011). Therefore, the index seems to meet the general objective of being a concise and easily communicated measure of the status of biodiversity in Norway.

Some of the shortcomings in the data supply were quite apparent in the NI for coastal waters. The number of indicators was rather low, several important ecological groups were poorly represented, particularly in the pelagic, and the geographical coverage was unsatisfactory. In general terms, the poor data supply mirrors the status for trend monitoring in coastal waters in Norway. Compared with the open ocean, there are few commercially important species being routinely monitored for stock management; and compared with terrestrial and freshwater habitats, there are few routine programmes recording occurrence and distribution of species at lower trophic levels. This reflects that monitoring activities have to a large extent been aimed at resource management or directed at species and habitat systems of particular public concern. For terrestrial systems, it is also of concern that monitoring is far less resource-demanding than in the sea, where most investigations are labour intensive and often extremely expensive. To a certain extent, the coastal zone has become 'squeezed' and overlooked between the major systems in open oceans and on land with regard to monitoring species and the state of the environment.

In contrast to the low supply of trend monitoring data in the coastal zone, there is an abundance of data from environmental assessment studies, especially for soft-bottom fauna and rocky-shore algal vegetation. Most of the assessment study data, however, refer to strongly polluted harbours and fjords which receive effluents of municipal and industrial discharge water. In Norway, the environmental authorities have requested assessment studies to be carried out since the 1980s as a basis for administering measures to clean up polluted areas. Strongly polluted areas are usually of very limited geographical extension, however. Assessment data are therefore not representative for the municipality as a whole. As a consequence, strongly polluted areas were omitted from the NI calculations to avoid undue influence from local problems. It nevertheless remains an open question as to where to set the limit for exclusion of such small perturbed environments. In addition, it became clear during data compilation that few assessment studies encompassed structurally sampled data depicting trends for the NI period 1990–2010.

All the aquatic NIs showed a better status than the major terrestrial ecosystems, with the exception of mountains. This can be explained by the extensive use of land areas, but the differences are still larger than would be expected from general knowledge about the systems. A number of the indicators for coastal waters used precautionary level or past knowledge as the basis for the reference state, whereas in the forest a large number of the indicators used pristine nature as the reference. When making comparisons between different major ecosystems, the use of different principles for setting indicator reference states should be considered (Certain et al., 2011). It should also be taken into account that inner fjord areas, which include some of the most polluted sites in Norway, were omitted due to inadequate numbers of indicators and very local problems. It is also a matter for speculation as to whether a broader selection of indicators for coastal waters could have resulted in lower index values, especially if more disturbance- and pollution-sensitive plants and invertebrates had been taken into account. Addressing these issues remains a challenge for further development of the NI. The trends and geographical variation within each major ecosystem – including the coastal zone –however, seem to agree with common views.

In parallel with the development of the NI, a fairly similar attempt to assess the environmental state from indicators of biodiversity has been made for the Baltic Sea (HELCOM, 2009). In this approach, termed BEAT (the HELCOM Biodiversity Assessment Tool), selected biodiversity indicators have been scored against reference states in much the same manner as in NI, but the selection of indicators, the principles for weighing and the setting of reference levels were different. The approach was considered successful, but it was realised that there was a need to develop more indicators, determine reference conditions and harmonise monitoring programmes on biodiversity for future assessments (HELCOM, 2009). It is worth noting that the experiences for BEAT correspond to a large extent with those for NI, for Norwegian coastal waters.

## 3.6 Conclusions and perspectives

The NI comprises the most comprehensive summary of status and trends for biodiversity in Norway. The 2010 index should, however, be considered a 'first-generation' index that suffers from several shortcomings, particularly in the selection and quality of indicators. The NI for coastal waters is based on too few and insufficient indicators to be able to provide a thorough and balanced picture of the status and development of diversity at all ecosystem levels. Nevertheless, the index illustrates geographical patterns and trends which agree with present views. The general conclusion from the group of experts involved in the work is that NI conveys a reasonably good picture of the status and development of biodiversity in all major ecosystems of Norway (Nybø et al., 2011). Coastal water was highlighted as one of the major ecosystems where monitoring is crucially needed (Certain et al., 2011).

The Nature Index is intended to become a permanent tool for use in Norwegian nature management and political planning. The Norwegian Ministry of Finance has suggested that NI should become an official indicator for sustainable development, which will then be regularly updated. Presently, steps are being taken to develop the set of indicators further and to improve the basis for determining reference states. For the coastal zone, this implies that more indicators for the lower trophic levels will be added. In addition, several indicators will be developed as part of the monitoring system for water quality under the EU WFD, which will easily be integrated into NI. Norway has adopted the WFD as part of a trade agreement with the EU. In the context of targeting future research on the state of the environment, key elements are to identify knowledge gaps and poorly documented ecosystem components. The NI seems to have the capability to become an instrument for integrating science and management in the coastal zone, aimed at preserving biodiversity and ensuring sustainable use of the environment.

## Acknowledgements

We are indebted to about 40 colleague experts at our institutes for cooperation and willingness to share their expertise in defining and documenting the indicators for coastal waters. We thank two anonymous reviewers for helpful criticisms of the manuscript. The development of the Nature Index has been funded by the Norwegian Ministry of Environment and administered by the Norwegian Directorate of Nature Management.

## References

Barrett, R.T. (2002) Atlantic puffin *Fratercula arctica* and commom guillemot *Uria aalge* chick diet and growth as indicators of fish stocks in the Barents Sea. *Marine Ecology Progress Series* **230**, 275–287.

Barrett, R.T., Lorentsen, S.H. and Anker-Nilssen, T. (2006) The status of breeding seabirds in mainland Norway. *Atlantic Seabirds* **8**(3), 97–126.

Carstensen, J., Conley, D.J., Andersen, J.H. and Aertebjerg, G. (2006) Coastal eutrophication and trend reversal: a Danish case study. *Limnology and Oceanography* **51**, 398–408.

CBD (2002) Strategic Plan for the Convention on Biological Diversity. COP 6 Decision VI/26. Sixth Ordinary Meeting of the Conference of the Parties to the Convention on Biological Diversity. The Hague, Netherlands. Available at www.cbd.int/decisions (accessed 10 September 2012).

Certain, G., Skarpaas, O., Bjerke, J.W., Framstad, E., Lindholm, M., Nilsen, J.E., Norderhaug, A., Oug, E., Pedersen, H.C., Schartau, A.K., Meeren, G.I., Aslaksen, I., Engen, S., Garnåsjordet, P.A., Kvaløy, P., Lillegård, M., Yoccos, N.G. and Nybø, S. (2011) The Nature Index: a general framework for synthesizing knowledge on the state of biodiversity. *PlosOne* **6**(4), 1–14.

de Bund, W. and Solimini, A.G. (2007) Ecological Quality Ratios for ecological quality assessment in inland and marine waters. European Commission, DG Joint Research Centre Report 22722 EN. 24 pp.

HELCOM (Baltic Marine Environment Protection Commission) (2009) Biodiversity in the Baltic Sea: an integrated assessment on biodiversity and nature conservation in the Baltic Sea. Baltic Sea Environment Proceedings No. 116B. Helsinki Commission, 188 pp.

Hooper, D.U., Chapin, F.S., Ewel, J.J., Hector, A., Inchausti, P., Lavorel, S., Lawton, J.H., Lodge, D.M., Loreau, M., Naeem, S., Schmid, B., Setälä, H., Symstad, A.J., Vandermeer, J. and Wardle, D.A. (2005) Effects of biodiversity on ecosystem functioning: a consensus of current knowledge. *Ecological Monographs* **75**, 3–35.

Loh, J. and Wackernagel, M. (Eds.) (2004) Living Planet Report 2004. Gland, Switserland: WWF.

Loreau, M., Naeem, S., Inchausti, P., Bengtsson, J., Grime, J.P., Hector, A., Hooper, D.U., Huston, M.A., Raffaelli, D., Schmid, B., Tilman, D. and Wardle, D.A. (2001) Biodiversity and ecosystem functioning: current knowledge and future challenges. *Science* **294**, 804–808.

Norderhaug, K.M. and Christie, H. (2009) Sea urchin grazing and kelp re-vegetation in the NE Atlantic. *Marine Biology Research* **5**, 515–528.

Nybø, S. (ed.) (2010a) Naturindeks for Norge 2010. DN-utredning 3-2010. Norwegian Directorate for Nature Management. Available at www.dirnat.no (accessed 10 September 2012). [In Norwegian, English summary.]

Nybø, S. (ed.) (2010b) Datagrunnlag for Naturindeks 2010. DN-utredning 4-2010. Norwegian Directorate for Nature Management. Available at www.dirnat.no (accessed 10 September 2012). [In Norwegian.]

Nybø, S., Certain, G. and Skarpaas, O. (2011) The Norwegian Nature Index 2010. Norwegian Directorate for Nature Management, Report no. 1–2011. Available at www.dirnat.no (accessed 10 September 2012).

OSPAR (The Convention for the Protection of the Marine Environment of the North-East Atlantic) (2010a) Quality Status Report. Chapter 4: Eutrophication. OSPAR Commission. Available at www.ospar.org (accessed 10 September 2012).

OSPAR (The Convention for the Protection of the marine Environment of the North-East Atlantic) (2010b) Quality Status Report. Chapter 8: Use of living marine resources. OSPAR Commission. Available at www.ospar.org (accessed 10 September 2012).

RIVM (2002). Technical design Natural Capital Index framework and implementation for the Nature Outlook 2. In Ten Brink, B (ed.) RIVM report 408657007. National Institute of Public Health and the Environment (RIVM). Bilthoven, The Netherlands. [In Dutch].

Scholes, R.J. and Biggs, R. (2005) A biodiversity intactness index. *Nature* **434**, 45–49.

Thrush, S.F. and Dayton, P.K. (2010) What can ecology contribute to ecosystem-based management? *Annual Revue of Marine Science* **2**, 419–441.

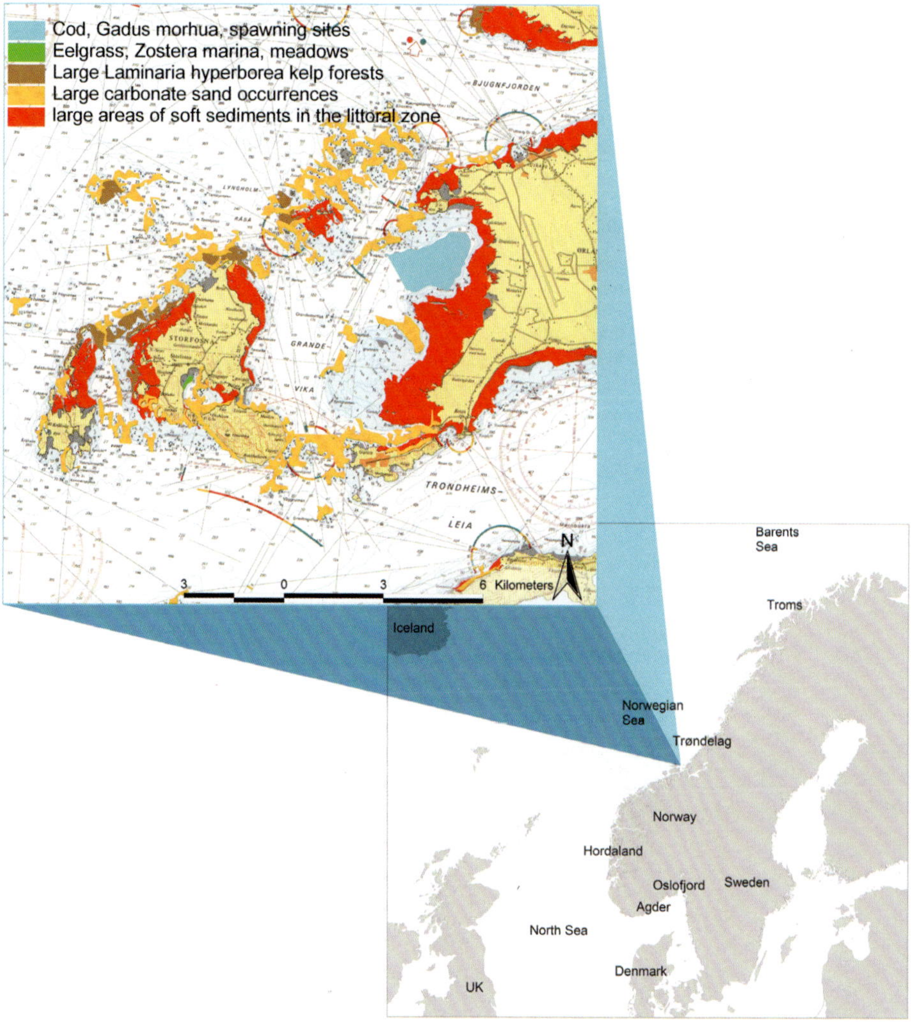


**Plate 2.1** From 2007 to 2011, the mapping activity covered the Oslofjord area (five counties), the two Agder counties, Hordaland, the two Trøndelag counties and Troms (left panel shows area names). The zoomed in area to the left shows an example of a map including a selection of habitats and key areas for parts of Trøndelag (Ørland-Storfosna close to the outer part of the Trondheimsfjord, 63°N 9°E).

*Global Challenges in Integrated Coastal Zone Management*, First Edition. Edited by Erlend Moksness, Einar Dahl and Josianne Støttrup.

**Plate 3.1** (a) NI status for coastal pelagic 2010 at municipality level.

**Plate 3.3** (a) NI status for coastal bottom 2010 at municipality level.

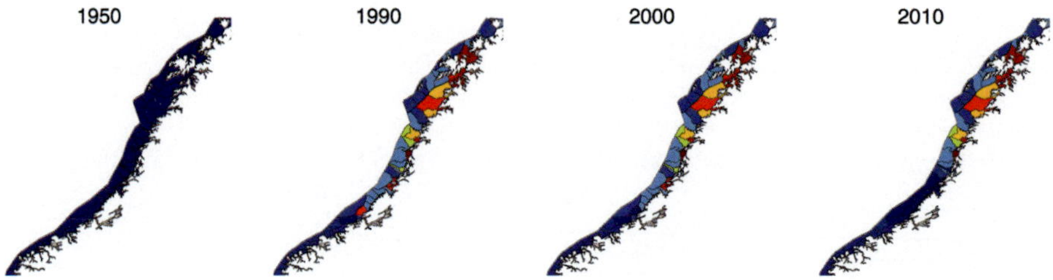


**Plate 3.4** Indicator status for kelp (*Laminaria hyperborea*) 1950–2010 at municipality level in central Norway.

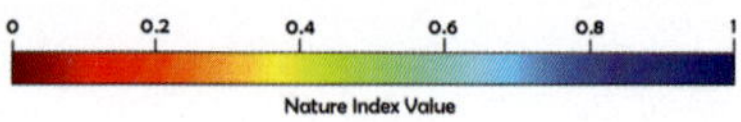

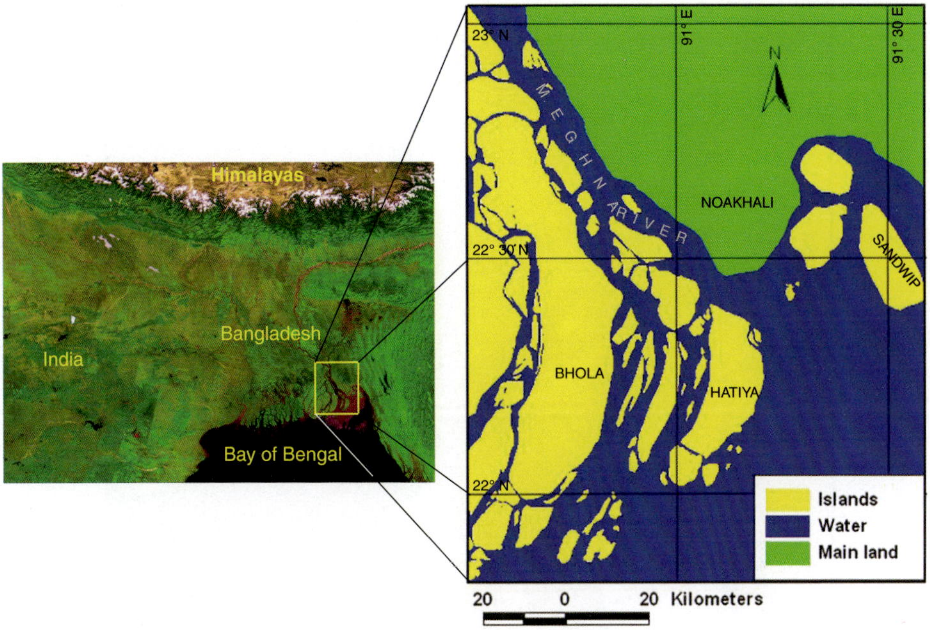


**Plate 7.1** MODIS image of December 2010 shows Ganges-Brahmaputra-Meghna river systems in Bangladesh. Classified Landsat TM image (right enlarged) shows geographical location of the islands in the Ganges basin of Bangladesh.

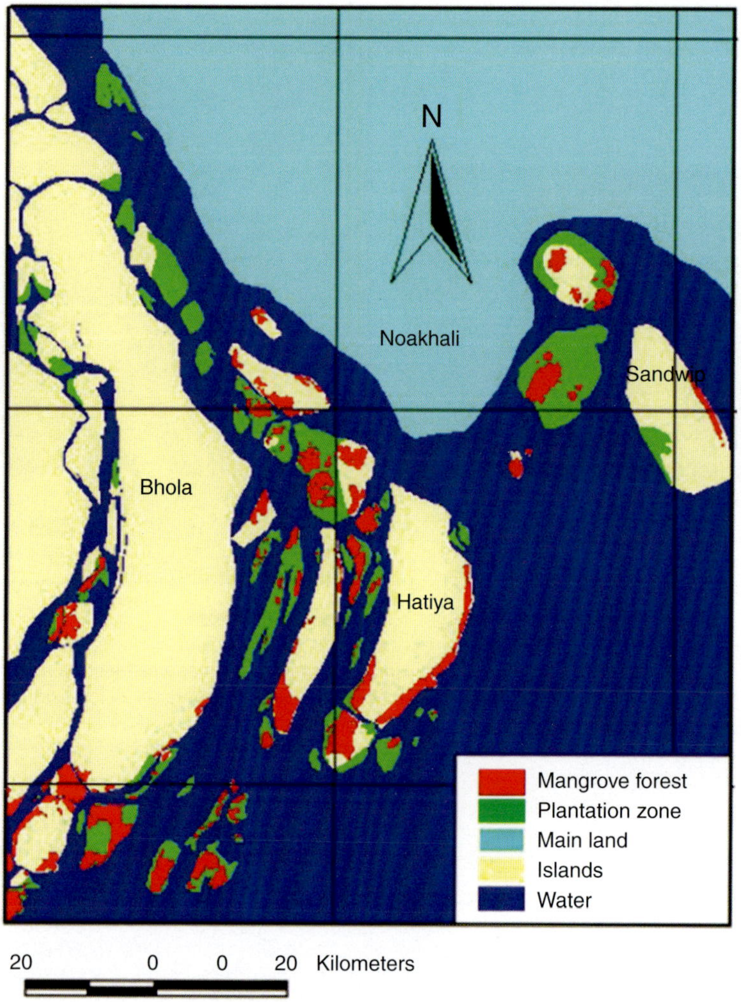


**Plate 7.3** Spatial distribution of mangrove forest and accreted stable zone in the Ganges basin from the supervised classification of the 2007 Landsat TM image.

Chapter 4

# The *Scomberomorus brasiliensis* Gill-net Production System in Northern Brazil; an "Invisible" and Mismanaged Small-scale Fishery

*Victoria J. Isaac, Roberto V.E. Santo, Bianca S. Bentes, Keila R.M. Mourão and Flávia Lucena-Frédou*

**Abstract**

The small-scale gill-net fishery is one of the most important fisheries along the Brazilian coast, and on the Amazonian coast this fishery accounts for over 60% of the boats and 70% of the marine production. *Scomberomorus brasiliensis* represents 6% of catches. Statistical data on effort and catches by fishing grounds were not available prior to this study. Data obtained by port-sampling in the municipality of Bragança, Para State, from 2000 to 2010, demonstrate the increase of capture per unit effort over time but also that the boat size and number of fishermen per trip and, hence, fishing power has increased. Also, the average length of gillnets has increased, and is now 6 km. To sustain high catches and ensure higher incomes, fishing effort has moved to grounds that are further away from the landing ports. This indicates possible depletion of fish stocks. *Scomberomorus brasiliensis* was already harvested at maximum sustainable yield in 2000, and yet appropriate management measures were not implemented. A program to reduce harvesting combined with measures for better use of the products could be sustainable alternatives. However, the government continues to finance renewal of gear and vessels in an attempt to increase the global catch, without considering the carrying capacity of the stocks.

## 4.1 Introduction

Gill-netting is a common practice in Brazilian coastal waters. On the country's northern coast, a diversity of techniques are used, with vessels varying considerably in terms of their autonomy (days of fishery) and cargo capacity (Brito and Furtado Junior, 2010). The region's

*Global Challenges in Integrated Coastal Zone Management*, First Edition. Edited by Erlend Moksness, Einar Dahl and Josianne Støttrup.

gill-net fleet typically harvests some 40 fish species, and is the most important artisan fishery in coastal areas and the neighboring continental shelf. The Serra Spanish mackerel, *Scomberomorus brasiliensis* (Collette et al., 1978) is one of the principal species harvested using gill-nets, accounting for just over 6% of the total catch of the marine/estuarine zone of the northern region of Brazil (MPA, 2010). This species is caught together with a number of other species, in particular rays and sharks, *Carcharhinus leucas, Ginglymostoma cirratum, Sphyrna lewini, Dasyatis guttata,* and *Aetobatus narinari* (Almeida, 2008), some of which are included in the Brazilian Environment Ministry's list of endangered species.

*Scomberomorus brasiliensis* is a neritic pelagic species found in coastal areas of the western Atlantic, between the Caribbean and Central America, southwards along the South American coast as far as southern Brazil. This slow-growing fish ($K = 0.189$ cm/year; Nóbrega and Lessa, 2009) may reach a body length of 130 cm, with first sexual maturity occurring between 41 and 46 cm (3.0–3.4 years) (Lima et al., 2009). The species typically migrates along the coast, but may also migrate offshore to feed or reproduce (Batista and Fabre, 2001). In the coastal waters off the Brazilian state of Maranhão, reproduction is continuous, with peaks between March and June (Lima, 2000, 2004; Silva et al., 2005). This species has high mortality rates, and harvesting rates were already close to the sustainable limit in 2000 (Lucena-Frédou and Asano-Filho, 2006). Studies in the northern and northeastern Brazilian Exclusive Economic Zone have indicated the presence of a single stock of this species (Cunha et al., 2008; Nóbrega et al., 2009), although no assessment of this stock is available.

Despite the economic and social importance of artisan gill-net fisheries and the intense exploitation of *S. brasiliensis* in Brazil's northern region (Lucena-Frédou and Asano-Filho, 2006), this activity, like other small-scale fisheries (Mourão et al., 2009), is not included in the strategic planning of measures designed to protect the target species or the associated fauna, or the habitats in which they occur. The authorities justify the lack of any systematic official management of this fishery system by the scarcity of reliable data on either the species or catches. However, the lack of data should not be a limiting factor for the development of management strategies if the FAO's (1995) criterion of precaution is adopted. In Brazil, this sector of the fishery industry is clearly "invisible" to, and thus mismanaged by, the relevant authorities and political institutions.

The present study aims to provide a critical assessment of the data available on the coastal gill-net production of the Serra Spanish mackerel (*S. brasiliensis*) in northern Brazil. It is hoped that this analysis will provoke further discussion on management issues and contribute to the implementation of sustainable practices in this sector of fishery production.

## 4.2 Methods

For this study two data sources were used. Data on the overall fishery production of Brazil's northern region were obtained from the Federal Environment Institute, IBAMA (www.ibama.gov.br). For the state of Pará, the data covered the period between 2000 and 2007, whereas for Amapá and Maranhão, the recorded period was between 2001 and 2007. Because the official data are not reliable enough and do not include effort estimates, gear type or fishing ground, additional data were collected directly in nine fishing ports in

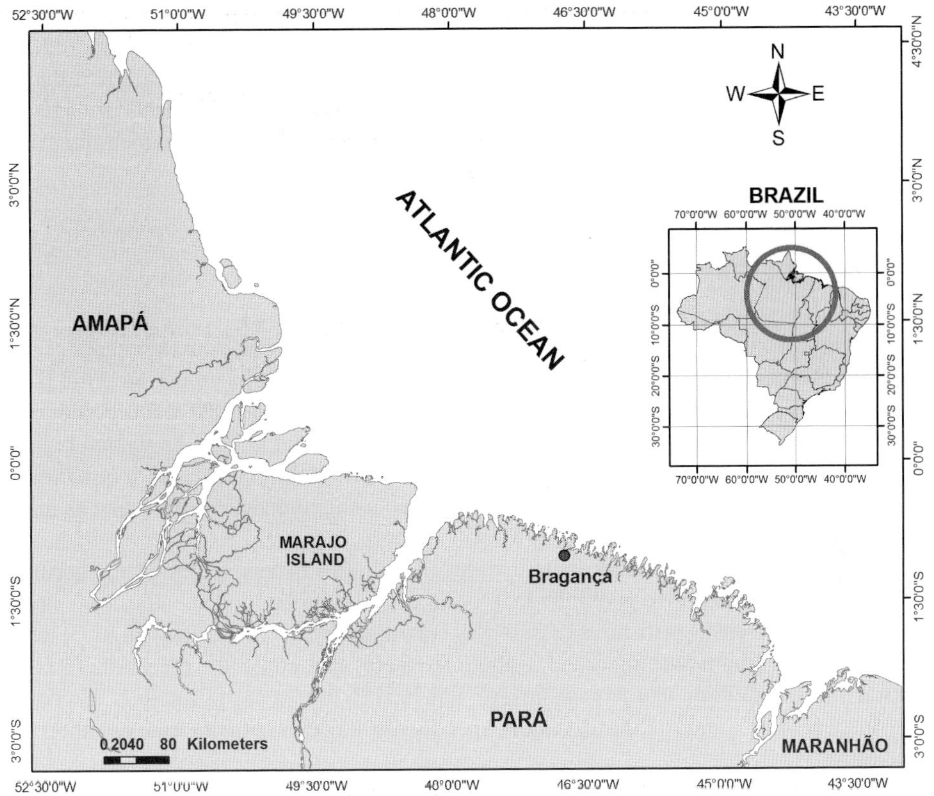


**Figure 4.1** Northern coast of Brazil, including the states of Maranhão, Pará, and Amapá, where the Serra Spanish mackerel is harvested.

the municipality of Bragança, Pará (Figure 4.1) between May 2000 to June 2001, March 2006 to April, 2007, and April 2008 to March 2010, giving 1738 records. The data were recorded daily, by interviewing all fishers or boat owners during the landings, focusing on the motorized vessels of the Serra Spanish mackerel gill-netting fleet. The information recorded for each trip was: landed catch in volume of each species, the area fished, total fishing effort, gear size. Bragança is one of the region's most important fishing ports and can be considered representative of this fishing sector for the purposes of the present study.

Based on this last data, fishery productivity was estimated using the capture per unit effort (CPUE), which is given by:

$$\mathrm{CPUE} = \frac{\text{Total volume (kg) captured per trip}}{\text{Number of days at sea}^{*}\text{Number of meters of gill} - \text{net}^{*}1000}$$

For 2009 additional data were collected to estimate the cost of each trip, which was calculated by summing the expenditure on ice for the conservation of the catch, provisions for the crew, and the fuel used to reach the fishing grounds based only on data from 2009. The evolution of the productivity of the fleet over the study period was analyzed using descriptive statistics (mean ± 95% confidence interval) and one-way analysis of variance (ANOVA).

Two field trips were also conducted in September 2004 and December 2005, during which 15 qualified informants (experienced fishermen and longtime residents of the communities) were interviewed in order to define the opinion of local fishermen about the current perspective and demands of the region's fisheries.

## 4.3 Results

According to the official numbers, the total marine/estuarine catches from the coastal region and proximal continental shelf of the northern coast of Brazil oscillated between 145 000 t in 2000 and 114 000 t in 2007. Small-scale fisheries perform almost 90% of this production. The majority of this catch (±70%) was landed in Pará, which is Brazil's second biggest fishery state, and included between 11 000 t (2000) and 6000 t (2007) of *S. brasiliensis* (Figure 4.2). The production of this species followed a slight tendency of decline as in the last years.

The total number of vessels or fishermen involved in this fishery sector off the northern coast of Brazil is unknown, given the lack of an official registry and specifications in the official data. An census of fishing vessels was conducted by the authorities in 2005, when a total of 16 678 units was recorded in the three states that constitute the northern coast of Brazil (Amapá, Pará, and Maranhão), of which 5136 (31%) are gill-netters. However, it is unclear how many of these vessels specifically target *S. brasiliensis* or other fish species.

According to the Bragançaç's data, the Serra Spanish mackerel is fished mainly by small boats (mean length = 11.32 ± 0.12 m), with the larger vessels using barrels of ice to

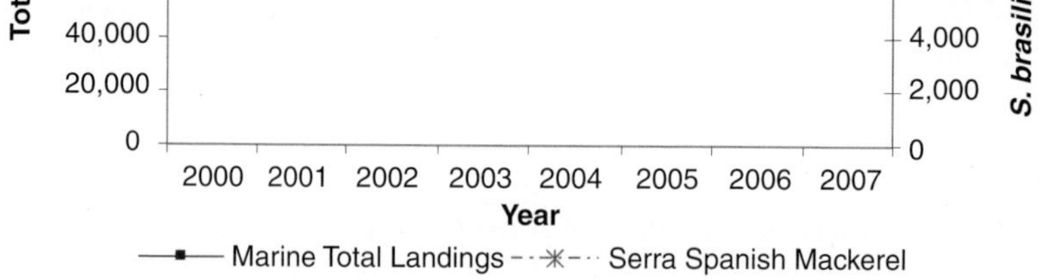


**Figure 4.2** Total marine and estuary fishery catch (solid line) landed in the Brazilian state of Pará between 1997 and 2007, and the catch of *Scomberomorus brasiliensis* (dotted line) during the same period.
Source: www.ibama.gov.br.

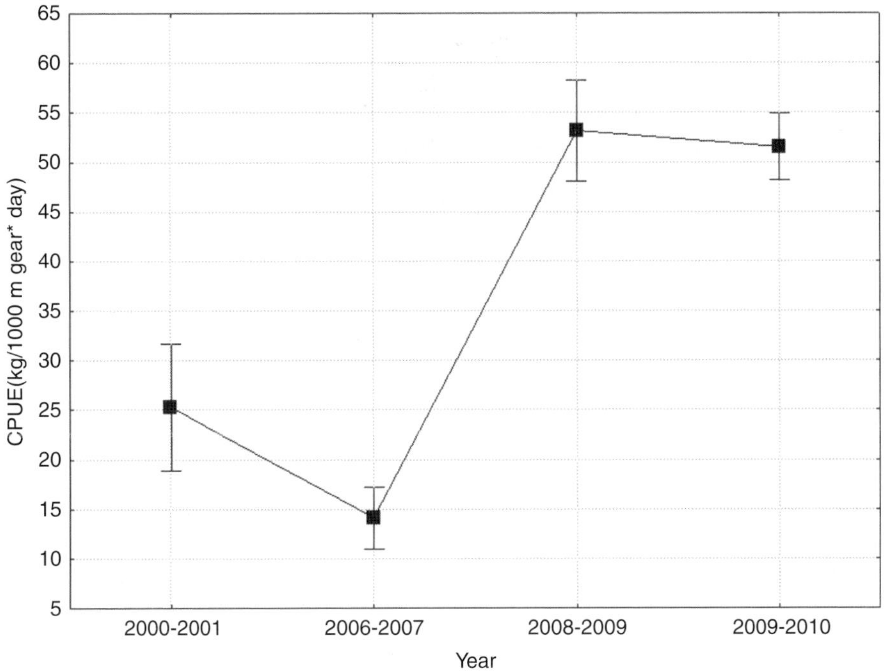


**Figure 4.3** Variation in the capture per unit of effort (CPUE) recorded for the *S. brasiliensis* fishing fleet from Bragança, Pará (Brazil). ■ = mean; Ī = 95% confidence interval.

conserve the catch. The use of technology such as mechanical fishing gear, GPS, radio, and sonar is restricted to some of the larger boats. The gill-nets are made of monofilament nylon with a mesh of 50–60 mm (measured between opposite knots when fully stretched), with a very variable total length from 300 to 10 000 m. Mean trip duration is 15.29 ± 0.28 days, and crews average 5.9 ± 0.08 fishermen.

The results of the analysis of these records showed a worrying picture. A number of changes in the productive system were observed over the course of the study period, affecting fishing power. For example, the mean length of gill-nets increased significantly from 2.6 km in 2000 to 6.2 km in 2010 ($F_{(3,1128)} = 174.11$, $p = 0.00$), as did the mean CPUE, from 25 kg × 1000 $m^{-1}.day^{-1}$ in 2000, to 51 kg × 1000 $m^{-1}.day^{-1}$ in 2010 ($F_{(3,1128)} = 106.2$, $p = 0.00$) (Figure 4.3). The increase in the productive potential of the local fleet can also be seen in the significant variation in other parameters, such as the mean cargo capacity of the vessels ($F_{(3,1734)} = 70.95$, $p = 0.00$), mean vessel size ($F_{(3,1734)} = 172.66$, $p = 0.00$), mean number of fishermen per trip ($F_{(3,1734)} = 383.28$, $p = 0.00$), and the duration of trips, $F_{(3,1734)} = 91.01$, $p = 0.00$ (Figure 4.4).

The location of the fishing grounds exploited by the Bragança fleet also shifted over time. While the vessels initially concentrated on areas off the coast of Pará (Figure 4.5), in more recent years, attention has shifted to Maranhão, to the south-east, and principally to Amapá, to the north-west (Figure 4.1), fishing areas that were poorly exploited in the past. This shift in fishing grounds is supported by many of the changes observed in the characteristics

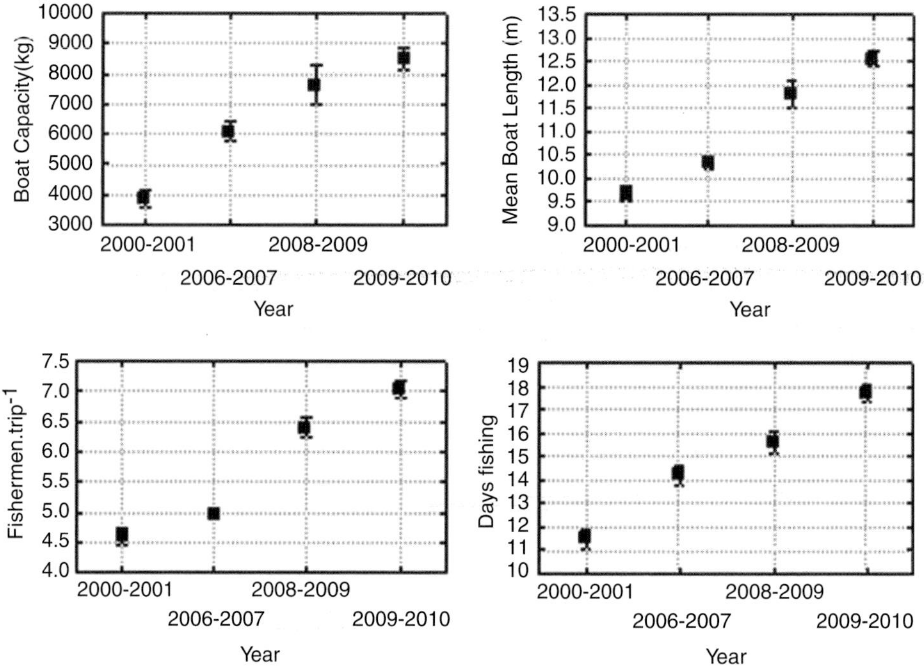


**Figure 4.4** Variation in mean cargo capacity (kg) and boat length (m), number of fishermen per trip, and trip duration (days) recorded for the *S. brasiliensis* fishing fleet from Bragança, Pará (Brazil) between 2000 and 2010. ■ = mean; Ī = 95% confidence interval.

of the fleet and the productive system, such as boat size, net length, and the number of crew members, which reflect the longer distances traveled to the more productive fishing grounds in neighboring states.

Unfortunately, the expansion do not generated economic advantage. While the increase in the fishing power of the vessels permitted the exploitation of more productive fishing grounds at greater distances from the home port, it also resulted in increasing costs (Figure 4.6).

The local fishermen interviewed during the study referred to the recent reduction in the catches near the coast of Pará and the size of the fish caught, which they relate to the unregulated increase in the number of fishing vessels, and the size of the boats and their nets. They were especially preoccupied by the arrival of even larger, refrigerated vessels, which permit even longer trips, and create potential conflicts with the crews of the smaller boats.

The fishermen were favorable to the introduction of a closed season prohibiting fishing during the species' main breeding period, but were depending on the government to do this, which would give them access to financial compensation subsidies. They also expressed their concerns for the possible collapse of fishery stocks, given their lack of training or experience for other types of occupation.

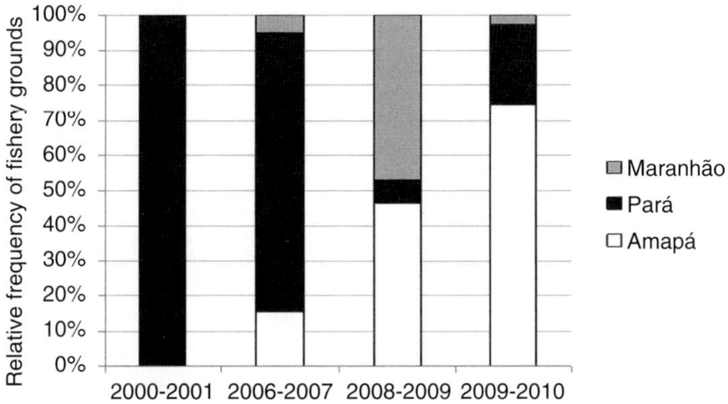


**Figure 4.5** Relative number of trips to fishing grounds in Amapá, Pará, and Maranhão undertaken by the Bragança *S. brasiliensis* fishing fleet between 2000 and 2010.

## 4.4 Discussion

The observed increase in catch per unit effort must be understood as a consequence of higher fishing power and the changes in fishing area. Therefore, CPUE should not be used as an indicator of the abundance of fish in the sea. These changes could be masking a steep

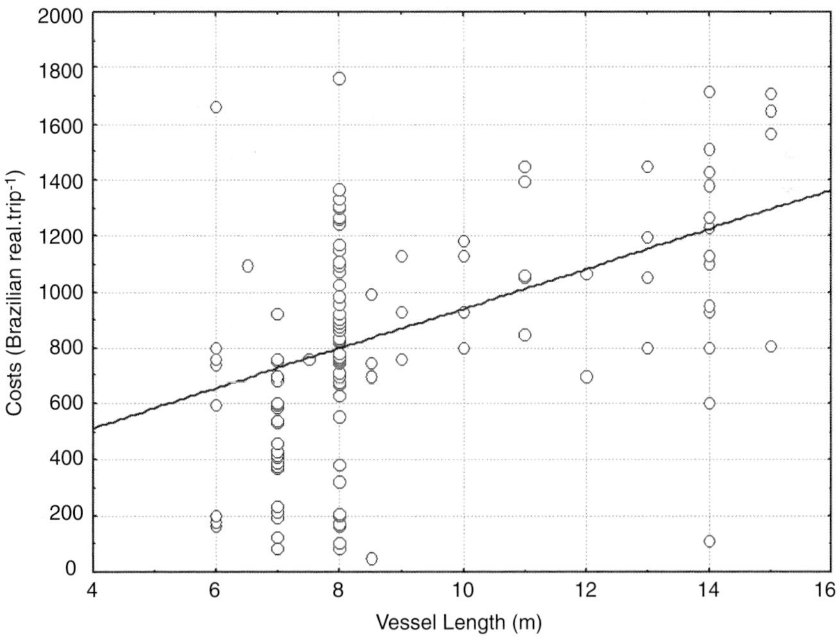


**Figure 4.6** Relationship between boat length (m) and the cost of fishing trips by vessels (BRL)) of the Bragança *S. brasiliensis* fishing fleet in 2009. One Brazilan Real = approximately US$0.60.

decline of fish stock (Thurstan et al., 2010). The discrepancy between the total catch from the government and the productivity indexes obtained in our scientifc collection of data indicate that there are sampling gaps in official statistics, which are considered unreliable (Isaac et al., 2006a, 2008).

Additionally, in contrast to other types of economic activity, *S. brasiliensis* fishery as well as other Amazonian small scale fisheries, do not generate economies of scale (Almeida et al., 2000). This refers to the cost advantages that any enterprise should obtain due to expansion of scale (Silvestre, 1987). In the case of the Serra Spanish mackerel fishery, the increased fishery power is accompanied by an increase in the cost of fishing operations.

The unregulated expansion of fishery productivity and catches may result in a major decline in stocks (Pauly et al., 2002). In the Northern Hemisphere diverse authors estimate that the biomass of predatory and large fish has strongly declined in the past century due to the impact of fisheries (Christensen et al., 2003; Myers and Worm, 2003; Jennings and Blanchard, 2004; Rosenberg et al., 2005). The increase of fishery intensity combined with state subsidies, has led in various fisheries to the over-exploitation of resources and over-capitalization (Sumaila et al., 2007). This global phenomenon may affect industrial and artisan fisheries alike, although in the latter it is often difficult to detect, given the frequent lack of reliable data (Anticamara et al., 2011).

The idea that natural resources are part of the common public domain is seen as a universally valid and natural law (Scudder and Connelly, 1985). From this same viewpoint, current Brazilian legislation guarantees the access of all citizens to natural resources, including fishery stocks. However, the state is also responsible for the establishment of the limits of sustainability for the exploitation of oceanic resources through the emission of licenses or the application of policies to avoid their depletion.

Reliable scientific data on fishery stocks or the social and economic characteristics of the artisan fishermen who exploit these stocks are extremely scarce in developing countries. A wide range of factors, such as the large number of fishing communities and landing ports, their distance from major urban centers, and their multi-gear or multi-species approaches, combine to make these fisheries "invisible" in the eyes of the state. This is despite the fact that these small-scale fisheries represent an important source of livelihood for a relatively large population (Pauly, 2006). In Brazil, almost 60% of the catch from marine/estuarine zones is produced by small-scale fisheries, and in the northern region this sector contributes 90% of total production (Isaac et al., 2006a). Studies in other parts of the world have shown that small-scale fisheries employ a larger number of fishermen for a given level of production, and also earmark a larger proportion of the catch for human consumption, are more fuel-efficient, and provide better enforcement of stock management policies, at both national and international levels (Sumaila et al., 2001).

In Brazil, however, public policies that regulate the fisheries focus primarily on the industrial sector, probably because it is the best organized and has the greatest political and economic power (Isaac et al., 2006a). The lack of reliable data and the political "invisibility" of small-scale fisheries have resulted in an almost complete lack of regulatory measures or management strategies for this sector.

Only two general items of federal legislation apply to the specific case of gill-net fishing in northern Brazil. One law introduced in 1988 (Nr 7679/88) prohibits the harvesting of any species, marine or freshwater, anywhere in the country during the breeding season. This law is normally impracticable, due not only to the lack of reliable data on the breeding seasonality of most species, but also because if this information were available, it would result in a continuous closed season, given the diversity of species exploited around the Brazilian coast and continental water. Off Pará alone, Isaac et al. (2006b) recorded the harvesting of 95 different species, including 61 bony fish, 20 cartilaginous fish, 11 crustaceans, and three mollusks. Although the fishermen interviewed in the present study were interested in the establishment of a closed season, the impact of this measure is difficult to evaluate due to lack of effective monitoring by the state. Thus, the effectiveness of closed seasons for the recuperation of stocks has yet to be reliably proven.

The second legal mechanism applicable to the *S. brasiliensis* fisheries of northern Brazil was published in 1998 (Nr 121/98) and establishes a country-wide limit of 2.5 km on the length of gill-nets. The measure is completely arbitrary, given the diversity and varying scale of gill-netting techniques practiced in different regions of the country, which defies the adequate definition of a universal net size. In any case, most fisheries use nets that are longer than this limit. Kotas et al. (2007) reported that bottom-fishing operations in southern Brazil use nets with a length of more than 25 km, while the mean length of nets in the Bragança fleet is now over 6 km. The control of the application of this legislation is almost impossible, given the dimensions of the Brazilian coast and the number of fishery fleets.

Overall, then, there are no effective legal mechanisms in force to regulate and limit the ongoing increase in the harvesting of *S. brasiliensis* in northern Brazil. This, together with the deficient official monitoring of the effort, may eventually result in the destabilization of stocks and, ultimately, the collapse of the fisheries. In the meantime, the authorities are providing incentives for the increase of productivity through cheap credit for the purchase of more powerful equipment and subsidies on fishing supplies. On the other hand, the subsidies can explain the technological changes observed in the fishery. This clearly contradicts both the few scientific data available showing the increase of fishing power, and the perception of the fishermen themselves, who are witnessing a decline in the productivity of the fishing grounds near the coast and in the mean size of fish.

Political incentives for the development of the fishery are not aimed specifically at *S. brasiliensis,* but they may mask cases of low productivity and result in over-capitalization and over-exploitation (Sumaila et al., 2010). While the stocks of most species, including *S. brasiliensis* are clearly approaching the limits of sustainability (Lucena-Fredou and Asano-Filho, 2006), the government maintains the public programs which finance the expansion of fleets and the purchase of fishing equipment. With the stocks of most species in decline, this policy has resulted in widespread disaster (Isaac, 2006), tying up public resources which otherwise could be applied to research or conservation of the species, or even in the education of fishermen.

Studies in northern Brazil have shown that the size at maturity, mean body size, of *S. brasiliensis* have all decreased in the past decade (Lucena-Fredou and Asano-Filho, 2006). The global conservation status of the species was recently assessed by the International Union for Conservation of Nature (IUCN) and while there are clear signs of decline in its populations, it was classified as "Least Concern" (Colette et al., 2011, Texeira Lessa et al., 2011). However, IUCN recommends an updated evaluation of the Brazilian stocks of the species, which were last evaluated in 2000, well before the recent trends observed in the present study.

In addition to the management of stocks, a number of potential strategies may support the development of the sector, including the improvement of the quality of the catch, the processing of sub-products (heads, skins, scales, etc.), or the avoidance of catches of undesirable or endangered species, such as many cartilaginous fish, in the case of this gill-net fishery. These practices may have many positive effects over the long term, and should not only be encouraged, but should substitute the current programs that support the development of fishery industries and the increase in productivity. The principle of precaution needs to be applied here to slow the growth of fishery power and to divert resources to the more systematic, long-term monitoring of fishing practices, productivity, and catches, and in particular, the reliable assessment of stocks and ecosystem impacts.

We conclude that there is an urgency to enforce traditional measures (effort monitoring and control and improvement of the quality of products) and/or to implement non-conventional approaches (participatory management and marine protected areas) which may contribute to the sustainability of the fisheries as well as to income and food security for the fishermen (Pauly et al., 2005).

## Acknowledgements

We gratefully acknowledge the Brazilian Council for Research and Technological Development-CNPq for the financial support for this research.

## References

Almeida, O.T., McGrath, D.G.E. and Ruffino, M.L. (2000) The commercial fisheries of the Lower Amazon: an economic analysis. *Fisheries Management and Ecology* **8**, 253–269.

Almeida, Z. (2008) Os recursos marinhos e estuarinos do Maranhão: Bioecologia, Tecnologia, Sócio-economia, Estado da Arte e Manejo. Ph.D. thesis (Doctoral Degree in Zoology), Universidade Federal do Pará, Belém. 293 pp.

Anticamara, J.A., Watson, R., Gelchu, A. and Pauly, D. (2011) Global fishing effort (1950–2010): Trends, gaps and implications. *Fisheries Research* **107**, 131–136.

Batista, V. and Fabré, N.N. (2001) Temporal and spatial patterns on serra *Scomberomorus brasiliensis* (Teleostei, Scombridae) catches from the fisheries on the Maranhão Coast, Brazil. *Brazilian Journal of Biology* **61**(4), 541–546.

Brito, C.F.S. and Furtado, I. Jr, (2010) Dinâmica sazonal da CPUE da serra, *Scomberomorus brasiliensis,* capturada com rede de emalhar do tipo serreira no estado do pará. *Arquivos de Ciências do Mar Brazil* **43**(1), 88–95.

Christensen, V., Guénette, S., Heymans, J.J., Walters, K.J., Watson, R., Zeller, D. and Pauly, D. (2003) Hundred-year decline of North Atlantic predatory fishes. *Fish and Fisheries* **4**, 1–24.

Collette, B.B., Carpenter, K.E., Polidoro, B.A., Juan-Jordá, M.J., Boustany, A., Die, D.J., Elfes, C., Fox, W., Graves, J., Harrison, L., McManus, L., Minte-Vera, C.V., Nelson, R., Restrepo, V., Schratwieser, J., Sun, C.L., Amorim, A., Brick-Peres, M., Canales, C., Cardenas, G., Chang, S.K., Chiang, W.C., De Oliveira Leite, N. Jr, Harwell, H., Lessa, R., Fredou, F.L., Oxenford, H.A., Serra, R., Shao, K.T., Sumaila, R., Wang, S.P., Watson, R. and Yáñez, E. (2011) High value and long life-double jeopardy for tunas and billfishes. *Science* **333**(6040), 291–292.

Collette, B.B., Russo, J.L. and Zavala-Camin, L.A. (1978) *Scomberomorus brasiliensis,* a new species of Spanish mackerel from the western Atlantic. *Fisheries Bulletin* **76**, 273–280.

Cunha, D.B., Vallinoto, M., Rego, P.S., Schneider, H., Sampaio, I. and Santa Brígida, E.L. (2008). Análise populacional de *Scomberomorus brasiliensis* (Scombridae, Perciformes) da costa brasileira através de sequências da alça D e do gene citocromo b. Anais do 54º Congresso Brasileiro de Genética. Salvador – Bahia. Available in: http://web2.sbg.org.br/congress/sbg2008/pdfs2008/24229.pdf (accessed 10 September 2012).

FAO (1995) Code of Conduct for Responsible Fisheries. Food and Agriculture Organization, Rome. 41 p.

Isaac, V.J. (2006) Explotação e manejo dos recursos pesqueiros do litoral amazonico: Um desafio para o futuro. *Ciência e Cultura, (SBPC), Brazil*, **58**(3), 33–36.

Isaac, V.J., Espírito-Santo, R.V., Bentes, B.S., Castro, E.R. and Sena, A.L. (2006b) Diagnóstico da pesca no litoral do Pará. In: Isaac, V.J., Martins, A., Haimovici, M. and Andrigueto, J.M. (eds). *A Pesca Marinha e Estuarina do Brasil no Início do Século XXI: Recursos, Tecnologias, Aspectos Socioeconômicos e Institucionais*. Ed. UFPA, Belém, pp. 11–40.

Isaac, V.J., Martins, A.S., Haimovici, M., Castelo, J.P. and Andrigueto, J.M. (2006a) Síntese do estado de conhecimento sobre a pesca marinha e estuarina do Brasil. In: Isaac, V.J., Martins, A., Haimovici, M. and Andrigueto, J.M. (eds). *A Pesca Marinha e Estuarina do Brasil no Início do Século XXI: Recursos, Tecnologias, Aspectos Socioeconômicos e Institucionais*. Ed. UFPA, Belém, pp. 181–186.

Isaac, V.J., Santo, R.V. and Nunes, J.L.G. (2008) A estatística pesqueira no litoral do Pará: resultados divergentes. *Pan-American Journal of Aquatic Sciences* **3**(3), 205–213.

Jennings, S. and Blanchard, J.L. (2004) Fish abundance with no fishing: predictions based on macroecological theory. *Journal of Animal Ecology* **73**, 632–642.

Kotas, J.E., Souza, N.L, Fiedler, F.N., Giffoni, B.B., Sales, G., Consulim, C.E.N. and Diegues, R.F. (2007). Diagnóstico das pescarias com rede de emalhe no sudeste/sul do Brasil, com frotas sediadas em Ubatuba (SP), Itajaí, Navegantes e Porto Belo (SC), entre os anos de 2005 e 2006. Resumo expandido. Congresso Latino Americano de Ciências do Mar – COLACMAR. Florianópolis, SC.

Lima, P.R.S. (2000) Dinâmica populacional do serra, *Scomberomorus brasiliensis* no litoral ocidental maranhense: estrutura da população, reprodução e nutrição (Osteichtyes; Scombridae). Monography (Biological Sciences), Universidade Federal do Maranhão, São Luís, 21 pp.

Lima, P.R.S. (2004) Dinâmica populacional do serra *S brasiliensis* no litoral ocidental do Maranhão. 2004. Master Dissertation, Universidade Federal Rural de Pernambuco. Recife, 63 pp.

Lima, P.R.S., Lessa, R.P., Castro, A.C.L. and Azevedo, J.WJ. (2009) Tamanho e idade de primeira maturação do serra *Scomberomorus brasiliensis* (Osteichthyes; Scombridae- Collette Russo and Zavala-Camin, 1978) no litoral ocidental do Maranhão – Brasil. *Boletim do Laboratório de Hidrobiologia (UFAMA)* **22**, 39–43.

Lucena-Frédou, F.L. and Asano-Filho, M. (2006) Recursos pesqueiros da região Norte. In:. Programa Revizee: Avaliação do potencial sustentável de recursos vivos na zona econômica exclusiva: Relatório Executivo, Ministério de Médio Ambiente, Secretaria de Qualidade Ambiental, Brasília, pp.121–152.

Mourão, K.R.M., Frédou, F.L., Espírito-Santo, R.V., Almeida, M.C., Silva, B.B., Frédou, T. and Isaac, V. (2009) Sistema dc produção pesqueira pescada amarela – *Cynoscion acoupa* Lacèpede (1802): um *estudo de caso no litoral nordeste do Pará – Brasil. Boletim do Instituto de Pesca, São Paulo* **35**(3), 497–511.

MPA (Ministério da Pesca e Aquicultura) (2010) Boletim estatístico da pesca e aqüicultura. Brasil 2008–2009. Available at: http://www.sepaq.pa.gov.br/files/u1/anuario_da_pesca_completo.pdf (accessed 10 September 2012).

Myers, R. and Worm, B. (2003) Rapid worldwide depletion of predatory fish communities. *Nature* **423**, 280–283.

Nobrega, M.F. and Lessa, R.P.T. (2009) Age and growth of Spanish mackerel (*Scomberomorus brasiliensis*) off the northeastern coast of Brazil. *Neotropical Ichthyology* **7**(4), 667–676.

Nobrega, M.F., Lessa, R.P.T., Lucena, F.M. and Hazin, F.H.V. (2009) Dinâmica de populações e avaliação dos estoques dos recursos pesqueiros do Nordeste do Brasil: espécies pelágicas,

*Scomberomorus brasiliensis* In: *Dinâmica de Populações e Avaliação dos Estoques dos Recursos Pesqueiros do Nordeste do Brasil.* Martins & Cordeiro Ltda. Eds., Fortaleza, pp. 64–74.

Pauly, D., Watson, R. and Alder, J. (2005) Global trends in world fisheries: impacts on marine ecosystems and food security. *Philosophical Transactions of the Royal Society London B Biological Science* **360**(1453), 5–12.

Pauly, D. (2006) Major trends in small-sclae marine fisheries, with emphasis on developing countries, and some implications for the social sciences. *Maritime Studies (MAST)* **4**(2):7–22.

Pauly, D., Christensen, V., Guénette, S., Pitcher, T.J., Sumaila, U.R., Walters, C.J., Watson, R. and Zeller, D. (2002) Towards sustainability in world fisheries. *Nature* **418**, 689–695.

Rosenberg, A.A., Bolster, W.J., Alexander, K.E., Leavenworth, B.W., Cooper, A.B. and McKenzie, M.G. (2005) The history of ocean resources: modelling cod biomass using historical records. *Frontiers in Ecology and the Environment* **3**(2), 84–90.

Scudder, T. and Connelly, T. (1985) Management systems for riverine fisheries. FAO, Fisheries Technical Paper 263, 85 pp.

Silva, G.C., Castro, A.C.L. and Gubiani, É.A. (2005) Estrutura populacional e indicadores reprodutivos de *Scomberomorus brasiliensis* Collette, Russo e Zavala-Camin, 1978 (Perciformes: Scombridae) no litoral ocidental maranhense. *Acta Scientiarum. Biological Sciences, Maringá* **27**(4), 383–389.

Silvestre, J. (1987) Economies and diseconomies of scale. In: Eatwell, J., Milgate, M. and Newman, P. *The New Palgrave: A Dictionary of Economics.* Palgrave Macmillan, London. pp. 80–84.

Sumaila, R.U., Khan, A.S., Dyck, A.J., Watson, R., Munro, G., Tydemers, P. and Pauly, D. (2010) A bottom-up re-estimation of global fisheries subsidies. *Journal of Bioeconomics* **12**, 201–225.

Sumaila, U.R., Khan, A., Watson, R., Munro, G., Zeller, D., Baron, N. and Pauly, D. (2007) The World Trade Organization and global fisheries sustainability. *Fisheries. Research* **88**, 1–4.

Sumaila, U.R., Liu, Y. and Tyedmers, P. (2001) Small versus large-scale fishing operations in the North Atlantic. In: Pitcher, T., Sumaila, U.R. and Pauly, D. (eds) *Fisheries impacts on North Atlantic ecosystems: Evaluations and policy exploration. Fisheries Centre Research Reports* **9**(5), 94 pp.

Teixeira Lessa, R.P., Juan Jorda, M., Carpenter, K.E., Di Natale, A., Fredou, F.L., Collette, B., Miyabe, N., Nelson, R., Fox, W., De Oliveira Leite, N. Jr, Travassos, P.E.P.F., Minte-Vera, C., Hazin, F.H.V., Boustany, A., Oxenford, H., Amorim, A.F. and Graves, J. (2011) *Scomberomorus brasiliensis*. In: *IUCN 2011. IUCN Red List of Threatened Species.* Available at http://www.iucnredlist.org/details/170335/0 (accessed 10 September 2012).

Thurstan, R.T., Brockington, S. and Roberts, C.M. (2010) The effects of 118 years of industrial fishing on UK bottom trawl fisheries. *Nature Communications* **1**(15), 1–6.

Chapter 5

# A Combination of Seagrass and Macroalgal Beds Enhances Survival of Young-of-the-Year Rockfish *Sebastes Cheni*

*Yasuhiro Kamimura and Jun Shoji*

**Abstract**

Japanese black rockfish, *Sebastes cheni,* is a dominant component of the fish fauna in vegetated habitats in temperate coastal waters of the western North Pacific. Juvenile rockfish migrate to macroalgal (*Sargassum* spp.) beds in early March when they are at a total length of about 20 mm and then inhabit seagrass (*Zostera* spp.) beds through summer in temperate coastal waters of the western North Pacific. We tested the hypothesis that seasonal changes in vegetation of coastal habitats affect survival of juvenile black rockfish in the Seto Inland Sea, south-western Japan, and propose conserving coastal areas with a combination of different habitats (macroalgal and seagrass beds) to improve the black rockfish population based on its life history and the phenology of the habitat vegetation. Physical and biological surveys were conducted from February to August in macroalgal and seagrass beds in central Seto Inland Sea, Japan, where macroalgae (mostly *Sargassum* spp.) grow from winter to spring and seagrass (*Zostera marina*) during summer. Mean daily cohort-specific mortality coefficient of black rockfish from 20 to 50 mm (macroalgae-dependent period, March to May: 0.06 day$^{-1}$), which was estimated by the use of otolith daily increments, was slightly higher than that from 50 to 80 mm (seagrass-dependent period, June to August: 0.03 day$^{-1}$). Mesocosm experiments showed a significant increase in juvenile predation rate in tanks without vegetation, indicating that vegetation contributed as predation refuge. A combination of the two different types of vegetation, macroalgae and seagrass, within a small spatial scale potentially increases the accumulative survival of young-of-the-year rockfish from the post-migration period (March) through summer in the Seto Inland Sea.

*Global Challenges in Integrated Coastal Zone Management*, First Edition. Edited by Erlend Moksness, Einar Dahl and Josianne Støttrup.

## 5.1 Introduction

Vegetated habitats such as seagrass and macroalgal beds in estuarine and coastal waters have been considered as important nurseries for fish because they have a variety of ecological functions such as feeding grounds and predation refuges for early life stages of fish (Adams, 1976; Azuma, 1981; Nagelkerken et al., 2000). To date, many studies have reported that vegetated habitats support larger numbers of fish compared to adjacent unvegetated habitats (Sogard, 1992; Heck et al., 2003). Although the role of these habitats as nurseries has been accepted by many scientists and the public, the nursery-role concept has rarely been verified. Quantitative analyses on growth, survival and movement (recruit to population) of young fish, as well as their abundance, are important components for accurate estimations of fish production (Beck et al., 2001). Many efforts have focused on species composition of fish assemblages and their feeding and growth analyses in vegetated habitats, while information on the quantitative fluctuations of cohort abundance, biomass and production are limited (Rooker et al., 1999; Kamimura et al., 2011).

Black rockfish *Sebastes cheni* is widely distributed in temperate and sub-arctic waters in the western North Pacific and is a commercially and recreationally important species in this area. Larvae and juveniles are one of the most dominant components of fish assemblages in macroalgal and seagrass beds during late winter through summer (Fuse, 1981; Nagasawa et al., 2000; Plaza et al., 2001). In the central Seto Inland Sea, Japan, early juvenile rockfish (*c*. 20 mm total length) migrate from offshore waters into macroalgal (*Sargassum* spp.) beds in early March and grow up to about 50 mm in total length by the end of May, when macroalgae disappears and seagrass (*Zostera marina*) becomes dominant in the vegetated habitat (Kamimura and Shoji, 2009).

A recent survey showed that annual production of black rockfish juveniles was 13 080–18 360 g $ha^{-1}$ $year^{-1}$ in a vegetated habitat in the central Seto Inland Sea, which was converted into an economic value (654 000–918 000 Y $ha^{-1}$ $year^{-1}$, where 100 Y = 1 \$; Kamimura et al., 2011). The total economic value of the ecosystem services of the vegetated habitat has been suggested to be much higher than estimates in previous studies (19 004 \$ $ha^{-1}$ $year^{-1}$ for seagrass and algal beds; Costanza et al., 1997) if the provisioning and regulating services which originate from these fish production are included in the previous estimates (Kamimura et al., 2011). Latitudinal comparison of abundance and growth of juvenile rockfish in seagrass beds has indicated that rockfish juvenile production in the Seto Inland Sea is among the highest in Japan, while there are spatio-temporal fluctuations in the abundance and growth rates (Shoji et al., 2011). A scheme of effective conservation of coastal habitats is indispensable for sustainable utilization of rockfish resources and its ecosystem services since reclamation of the shallow water ecosystems has been considered to have decreased fisheries resources in the Seto Inland Sea (Okaichi et al., 1996).

In the present study, we tested the hypothesis that a combination of different coastal habitats improves production of juvenile rockfish and proposed the importance of conservation of coastal habitats referring to the phenology of different types of vegetation (e.g. macroalgae and seagrass). In order to estimate the mortality of juvenile rockfish, quantitative samplings were conducted at a fine time scale (1- to 2- week intervals) in macroalgal and seagrass beds. Fish were divided into several same-extrusion-period cohorts by the use of the otolith daily rings. Using the mortality coefficients of juvenile black rockfish estimated in the field, the effect of vegetation coverage (seagrass or macroalgae) on the decrease in the cohort abundance was evaluated. In addition, mesocosm experiments were conducted to test if habitat complexity affects predation on the juveniles by piscivorous fish predators.

## 5.2 Methods

### *5.2.1 Field sampling*

Fish sampling and environmental surveys were conducted on a vegetated area (about 50 m wide, 500 m long) off the south-western coast of Aba Island, central Seto Inland Sea, Japan (Figure 5.1), at an interval of 1–2 weeks from February to August 2007. Aba Island is an uninhabited island with a coast of about 2 km on its south-western side. The vegetation is dominated by patches of seagrass during summer and macroalgae (mostly *Sargassum fusiforme, S. patens, S. piluliferum* and *Chorda filum*) during winter. Mean density of seagrass fluctuates between 20 and 160 shoots $m^{-2}$ and mean biomass (wet weight) of the macroalgae between 100 and 3500 g $m^{-2}$ (Kamimura and Shoji, 2009). The sea bottom of the vegetation area is comprised primarily of sand with occasional small stones (< 100 mm diameter), on which the macroalgae grow. Fish were collected using a round seine net (2 m height, 3 mm mesh aperture; Kamimura and Shoji, 2009) during a tidal level between 50 and 150 cm during daytime. The surveys were conducted at four separate locations within the vegetated area. Fish samples were preserved in 10% seawater formalin solution

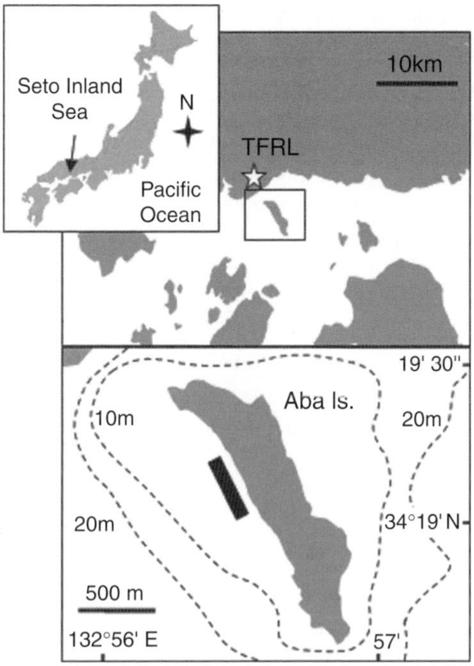


**Figure 5.1** Map of the sampling site off Aba Island, central Seto Inland Sea, Japan, where biological and physical surveys were conducted in order to estimate the mortality coefficient of juvenile rockfish.

and sub-samples of rockfish were preserved in 90% ethanol for otolith analysis. Water temperature and salinity were measured during each sampling.

In the laboratory, rockfish were identified according to Kai and Nakabo (2008) and total length (mm) and wet weight (g) were measured. Mean fish abundance was expressed as the number of fish 100 $m^{-2}$. Right-side lapilli (up to a maximum of $n = 50$ for each sampling date) were removed and daily increments (Plaza et al., 2001) were counted at 400–1000 × magnification using a compound light microscope connected to a monitor. The age–length relationship established for each sampling date was used to estimate the age of other specimens for which the otoliths were not analysed. Extrusion dates were used to separate the juveniles into specific cohorts, defined as individuals extruded within a 10-day period (mid December to late February: cohorts A to H). An exponential model was fitted to the decline in the abundance of each rockfish cohort. Mortality coefficients ($day^{-1}$) were estimated for the macroalgae-dependent period ($M_m$; March to May, 20–50 mm) and seagrass-dependent period ($M_s$; June to August, 50–80 mm). Part of fish abundance data published in the previous study (as total juvenile rockfish abundance data in Kamimura et al., 2011) was used for estimation of mortality coefficient during the macroalgae-dependent period. Effects of the existence of either seagrass or macroalgae on mortality of juvenile rockfish were accessed by simulating the decrease in the juvenile abundance under four conceptual vegetation conditions.

### *5.2.2 Mesocosm experiments*

Mesocosm experiments were conducted in order to test the hypothesis that habitat complexity reduces predation mortality of juvenile rockfish. The effect of physical complexity on juvenile survival was evaluated by comparing predation rates by piscivorous fish predators between vegetated and unvegetated tanks. The bottom of 1.0 $m^3$ black polycarbonate tanks was divided into four areas. Artificial, plastic seaweed was set on one of the four sections in the vegetated tanks at a density found in natural macroalgal beds around the surveyed area (Figure 5.1) in order to exclude possible effects of prey animals on juvenile behavior. There was no vegetation in one tank (unvegetated as control). Chinese sea bass *Lateolabrax* sp. were used as predators for the experiment. Fifty Chinese sea bass (average 260.3 mm) were acclimated in the tanks for 24 h prior to the experiment. Two predators together with 30 juveniles (average 28.2 mm) were introduced into each tank (six replicates for each condition). Predators and juveniles were not reused. The number of juveniles predated by the sea bass for 6 h was counted at the end of the experiment. The predation rate was expressed as the number of juveniles predated on by a predator per hour (no. fish predated $predator^{-1}$ $h^{-1}$) for each trial and was compared between the two conditions. The experimental protocols are also detailed in Shoji et al. (2007).

## 5.3 Results

Water temperature ranged between 11.4 °C (19 March) and 25.4 °C (10 August) and salinity between 31.9 (19 April) and 32.9 (10 August). A total of 7629 (15.4–66.7 mm) rockfish larvae and juveniles were collected in the macroalgal bed and 652 (46.2–78.3 mm) in the seagrass bed. Mean fish abundance was < 10.0 100 $m^{-2}$ in February then increased in to a maximum of 150.8 100 $m^{-2}$ on 24 March. Effects of sampling date on the abundance and fish total length were significant both in the macroalgal and seagrass beds ($p < 0.05$). The mean total length increased from about 20 mm in February to about 50 mm in May. Extrusion date of black rockfish ranged between 18 December and 22 February.

Abundance data from six (cohorts B to G) and five (cohorts B to F) black rockfish cohorts were available for estimation of the daily mortality coefficient in macroalgal ($M_m$) and seagrass beds ($M_s$). Data from one cohort were excluded from the analysis for $M_s$ because of the small sample size. Mean abundance of each cohort decreased from 68.9 ± 28.0 to 2.3 ± 1.8 100 $m^{-2}$ during the macroalgae-dependent period and then decreased to 1.5 ± 1.9 100 $m^{-2}$ during the seagrass-dependent period. The $M_m$ increased from 0.03 in cohort C to 0.11 for cohort G. The effect of season on the mortality coefficient was significant (Spearman's rank correlation coefficient, $p < 0.05$). The $M_s$ varied between 0.01 and 0.04 without effect of season ($p > 0.05$). Mean (SD) $M_m$ (0.06 ± 0.03) was significantly higher than that of $M_s$ (0.03 ± 0.02, $U$-test, $p < 0.05$).

Mesocosm experiments showed that mean predation rate in the unvegetated tank (0.27 ± 0.04 fish predated $predator^{-1}$ $h^{-1}$) was about sevenfold that in the vegetated tank (0.04 ± 0.05 fish predated $predator^{-1}$ $h^{-1}$). The effect of vegetation on the juvenile predation rate was significant ($U$-test, $p < 0.05$).

As shown in the conceptual illustration (Figure 5.2), survival rate of young-of-the-year rockfish is expected to be highest when the rockfish inhabit vegetation for 6 months

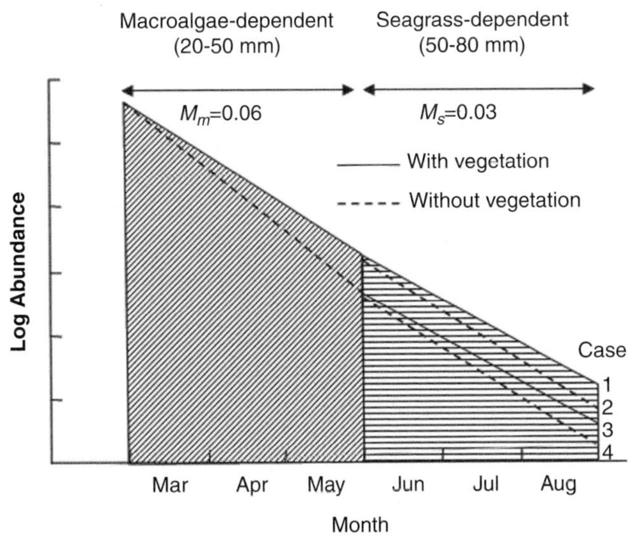


**Figure 5.2** A conceptual illustration of survivorship during the macroalgae-dependent and seagrass-dependent periods of a black rockfish *Sebastes cheni* cohort in the Seto Inland Sea. The young-of-the-year black rockfish inhabits macroalgae beds from March through May when 20–50 mm long and seagrass beds from June through August when 50–80 mm long. Predation is considered as the major factor affecting the survivorship. Mortality coefficient (under conditions with vegetation) was estimated in the field for each period. Condition of vegetation during the macroalgae-dependent and seagrass-dependent periods were expressed by the combination of with/without vegetation (macroalgae; seagrass) for four cases. Case 1: with; with, Case 2: with; without, Case 3: without; with, Case 4: without; without.

(macroalgal bed: February to May; seagrass bed: June to August: Case 1). On the contrary, survivorship should be minimal when no vegetation exists (Case 4). Since the daily mortality coefficient during the macloalgae-dependent period ($M_m$) is higher than that during the seagrass-dependent period ($M_s$), survival rate could be higher in Case 2 (with only macroalgae) than in Case 3 (with only seagrass).

## 5.4 Discussion

Generally, predation is one of the most important factors for fish mortality during the early life stages including egg, larval and juvenile periods, since it prevails throughout all the early life stages (Houde, 1987; Bailey and Houde, 1989). Even a small change in the mortality rate from predation can drastically alter the accumulated mortality during the pre- and post-recruit period and consequential the year-class strength (Houde, 1987). Vegetated habitats have been considered to serve as predation refuge for young fishes (Heck et al., 2003) since a physically complex habitat decreases the ability of predators to pursue and capture prey (Ryer, 1988; Rooker et al., 1999; Horinouchi, 2007; Shoji et al., 2007). However, the majority of the previous studies focused on the effects of only a single species (or type) of vegetation on the evaluated predation rate. In the Seto Inland Sea, black rockfish inhabit coastal areas for more than 6 months during the first year of life, during which the dominant vegetation switches from macroalgae to seagrass (Fuse, 1981; Kamimura and Shoji, 2009). Therefore, a combination of different vegetations with different phenology such as macroalgae and seagrass in an area should be considered when we evaluate how habitat complexity affects juvenile survival.

Results of the present field survey and mesocosm experiments show that survival of black rockfish is affected by the habitat complexity made by vegetation. In addition, since the growing season of the macroalgae (winter to spring) differs from that of the seagrass (summer; Fuse, 1981; Kamimura and Shoji, 2009), co-existence of these two different types of vegetation within a small spatial scale is suggested to be important to increase the cumulative survivorship of juvenile black rockfish during 6 months after migration into the coastal habitat.

The mortality coefficients estimated for juvenile black rockfish in the present study (0.03–0.06 day$^{-1}$) were far lower than those estimated for fish larvae in a variety of species and ecosystems (overall average: 0.222 day$^{-1}$; Houde and Zastrow, 1993) and were comparable with those for juvenile cohorts of red drum *Scianops ocellatus* (0.134–0.139 day$^{-1}$; Rooker et al., 1999) and Japanese sea bass *Lateolabrax japonicus* (0.017–0.041 day$^{-1}$: Shoji and Tanaka, 2007). Generally, juvenile mortality is lower than larval mortality as mortality decreases as fish body size increases (Bailey and Houde, 1989). In addition, juvenile fish mortality has been considered to be lower in vegetated habitats, which serve as a predation refuge for juveniles (Larkum et al., 2006). Recently, however, field surveys in seagrass beds suggest that the vulnerability of juvenile fish to predation is higher during the nighttime than in daytime because of the higher abundance and biomass of fish predators at nighttime (Hindell et al., 2000; Guest et al., 2003; Kinoshita et al., 2012). Since previous studies, including the present study, have evaluated coastal habitats focusing on fish community structure and ecological aspects during daytime. Accumulation of ecological information on fish abundance, biomass and predator–prey interactions during nighttime is needed for a more comprehensive evaluation and effective utilization of coastal habitats.

## Acknowledgments

We express our thanks to Sadaharu Iwasaki, Yuji Iwamoto, Kotaro Hirai and Takuma Morita for assistance in the field sampling and to Koji Hirakawa for providing experimental fishes. This research was partly supported by the Nippon Life Insurance Foundation.

## References

Adams, S.M. (1976) The ecology of eelgrass, *Zostera marina* (L.), fish communities. I. Structural analysis. *Journal of Experimental Marine Biology and Ecology* **22**, 269–291.

Azuma, M. (1981) Importance of *Zostera* beds as nursery ground for juvenile inshore fishes. In: Yatsuka, T., Miura, A., Saito, Y. and Syudo, T. (eds.) *Seaweed beds*. Koseisha-koseikaku, Tokyo, pp. 34–56.

Bailey, K.M. and Houde, E.D. (1989) Predation on eggs and larvae of marine fishes and the recruitment problem. *Advances in Marine Biology* **25**, 1–83.

Beck, M.W., Heck, K.L., Able, K. W., Childers, D. L., Eggleston, D.B., Gillanders, B. M., Halpern, B., Hays, C.G., Hoshino, K., Minello, T.J., Orth, R.J., Sheridan, P.F. and Weinstein, M.P. (2001) The identification, conservation, and management of estuarine and marine nurseries for fish and invertebrates. *BioScience* **51**, 633–641.

Costanza, R., d'Arge, R., de Groot, R., Farber, S., Grasso, M., Hannon, B., Limburg, K., Naeem, S., O'Neill, R.V., Paruelo, J., Raskin, R.G., Sutton, P. and van den Belt, M. (1997) The value of the world's ecosystem services and natural capital. *Nature* **387**, 253–260.

Fuse, S. (1981) Ecology of the black rockfish in *Zostera* and *Sargassum* beds. In: Yatsuka, T., Miura, A., Saito, Y. and Syudo, T. (eds.) *Seaweed Beds*. Koseisha-koseikaku, Tokyo, pp. 24–33.

Guest, M.A., Connolly, R.M. and Loneragan, N.R. (2003) Seine nets and beam trawls compared by day and night for sampling fish and crustaceans in shallow seagrass habitat. *Fisheries Research* **64**, 185–196.

Heck, K.L., Hays, G. and Orth, R.J. (2003) Critical evaluation of the nursery role hypothesis for seagrass meadows. *Marine Ecology Progress Series* **253**, 123–136.

Hindell, J.S., Jenkins, G.P. and Keough, M.J. (2000) Variability in abundances of fishes associated with seagrass habitats in relation to diets of predatory fishes. *Marine Biology* **136**, 725–737.

Horinouchi, M. (2007) Review of the effects of within-patch scale structural complexity on seagrass fishes. *Journal of Experimental Marine Biology and Ecology* **350**, 111–129.

Houde, E. D. (1987) Fish early life dynamics and recruitment variability. *American Fisheries Society Symposium* **2**, 17–29.

Houde, E.D. and Zastrow, C.E. (1993) Ecosystem- and taxon-specific dynamic and energetics properties of larval fish assemblages. *Bulletin of Marine Science* **53**, 290–335.

Kai, Y. and Nakabo, T. (2008) Taxonomic review of the *Sebastes inermis* species complex (Scorpaeniformes: Scorpaenidae). *Ichthyological Research* **55**, 238–259.

Kamimura, Y. and Shoji, J. (2009) Seasonal changes in the fish assemblages in a mixed vegetation area of seagrass and macroalgae in the central Seto Inland Sea. *Aquaculture Science* **57**, 233–241.

Kamimura, Y., Kasai, A. and Shoji, J. (2011) Production and prey source of juvenile black rockfish *Sebastes cheni* in a seagrass and macroalgal bed in the Seto Inland Sea, Japan: estimation of the economic value of a nursery. *Aquatic Ecology* **45**, 367–376.

Kinoshita, H., Kamimura, Y., Hirai, K., Mizuno, K., Iwamoto, Y. and Shoji, J. (2012) Vulnerability of juvenile fish to piscivorous fish predators increases during nighttime in a seagrass bed in the central Seto Inland Sea, Japan. *Bulletin of Japanese Society of Fisheries Oceanography* **76**, 24–29.

Larkum, A.W.D., Orth, R.J. and Duarte, C.M. (2006) *Seagrasses: Biology, Ecology and Conservation*. Springer, Dordrecht, pp. 1–691.

Nagasawa, T., Yamashita, Y. and Yamada, H. (2000) Early life history of mebaru, *Sebastes inermis* (Scorpaenidae), in Sendai Bay, Japan. *Japanese Journal of Ichthyology* **41**, 231–241.

Nagelkerken, I., van der Velde, G., Gorissen, M.W., Meijer, G.J., van't Hof, T. and den Hartog, C. (2000) Importance of mangroves, seagrass beds and the shallow coral reef as a nursery for important coral reef fishes, using a visual census technique. *Estuarine, Coastal and Shelf Science* **51**, 31–44.

Okaichi, T., Komori, S. and Nakanishi, H. (1996) *Biological Resources and Environments of the Seto Inland Sea*. Koseisha-koseikaku, Tokyo. pp. 1–272.

Plaza, G., Katayama, S., and Omori, M. (2001) Otolith microstructure of the black rockfish, *Sebastes inermis*. *Marine Biology* **139**, 797–805.

Rooker, J.R., Holt, S.A., Holt, G.J. and Fuiman, L.A. (1999) Spatial and temporal variability in growth, mortality, and recruitment potential of postsettlement red drum, *Sciaenops ocellatus*, in a subtropical estuary. *Fishery Bulletin* **97**, 581–590.

Ryer, C.H. (1988) Pipefish foraging: effects of fish size, prey size and altered habitat complexity. *Marine Ecology Progress Series* **48**, 37–45.

Shoji, J., Sakiyama, K., Hori, M., Yoshida, G. and Hamaguchi, M. (2007) Seagrass habitat reduces vulnerability of red sea bream *Pagrus major* juveniles to piscivorous fish predator. *Fisheries Science* **73**, 1281–1285.

Shoji, J. and Tanaka, M. (2007) Density-dependence in post-recruit Japanese seaperch *Lateolabrax japonicus* in the Chikugo River, Japan. *Marine Ecology Progress Series* **334**, 255–262.

Shoji, J., Toshito, S., Mizuno, K., Kamimura, Y., Hori, M. and Hirakawa, K. (2011) Possible effects of global warming on fish recruitment: shifts in spawning season and latitudinal distribution can alter growth of fish early life stages through the changes in daytime. *ICES Journal of Marine Science* **68**, 1165–1169.

Sogard, S.M. (1992) Variability in growth rates of juvenile fishes in different estuarine habitats. *Marine Ecology Progress Series* **85**, 35–53.

Section 2
# Adaptation/Mitigation to Change in Coastal Systems

Chapter 6

# Natural Resource Potential of Macroalgae Harvesting in the Baltic Sea—Case Study Trelleborg, Sweden

*Emma Risén, Joseph Santhi Pechsiri, Maria E. Malmström, Nils Brandt and Fredrik Gröndahl*

**Abstract**

The interest in harvesting biomass from the Baltic Sea has increased in recent years. However, there is a lack of available data on macroalgae biomass and of cost-effective methods for site-specific quantification of macroalgae. In this study, macroalgae biomass has been quantified in Trelleborg and thus the nutrient reduction that could be achieved by harvesting on a regional scale. The biomass was estimated on the basis of existing inventories of macroalgae, photic zone distribution and bottom substrata. An independent model for estimating the potential of macroalgae growth was applied where factors affecting the growth of macroalgae, for example nutrients, light and temperature, were considered. The estimated summer stock of macroalgae biomass along the 58 km coastal stretch in Trelleborg amounts to 19 000 tonnes dry weight (dwt) red filamentous algae. If 10–30% of this summer stock were to be harvested, a nutrient reduction of 50–150 t of nitrogen could be achieved. The model for estimating biomass proved promising and worthy of further investigation.

## 6.1 Introduction

Eutrophication of the Baltic Sea increased significantly from the 1940s to the 1980s (Kautsky et al., 1986; Elmgren, 1989). As a consequence, fast growing red filamentous macroalgae within the Baltic Sea has become increasingly common during the past three decades (Elmgren, 1989; Kautsky, 1994; Isaeus et al., 2004; Swedish Institute for the Marine Environment, 2011). Filamentous macroalgae cause major problems when they detach and drift by creating anoxic conditions and, in addition, the biomass causes huge problems for the fishing and tourist industries (Eklund et al., 2005). In 1992, drifting algae caused an emergency stoppage in a nuclear power plant in Finland (Kiirikki et al., 1998).

---

*Global Challenges in Integrated Coastal Zone Management*, First Edition. Edited by Erlend Moksness, Einar Dahl and Josianne Støttrup.

Furthermore, the algae leak toxic effluents during decomposition that can be harmful to marine organisms (Eklund et al., 2005).

In the sea around Trelleborg, which is part of the central Baltic proper and situated in southern Sweden, red filamentous macroalgae dominate the macroalgae community (Kautsky and Kautsky, 1994). Today, large amounts of biomass are removed from the shoreline at Trelleborg to facilitate recreational activities and tourism. The retrieved biomass is dumped back into the Baltic Sea at other locations and therefore no nutrient retention is achieved. However, the 'Trelleborg concept' has been devised as a method for nutrient reduction. This concept involves harvesting wild, free-floating macroalgae in shallow waters and using the biomass for biogas production and the digestate as fertiliser (Gröndahl et al., 2009). This may be a suitable nutrient reduction method and the additional benefits, such as biomass removal from the shoreline and biogas production, are important added values to the concept.

Recently, several similar projects aimed at harvesting and using maritime biomass from the Baltic Sea have been initiated (Müller, 2010; Aldentun, 2011). Therefore, it is essential to address the sustainability of harvesting wild macroalgae before a number of permanent harvesting regimes are established in the Baltic Sea. One important sustainability aspect of harvesting is the natural resource potential. Therefore, the overall aim of the present study was to quantify the natural resource potential of red macroalgae along a Baltic Sea coastal stretch, using Trelleborg as a case study.

## 6.2 Methods

### *6.2.1 Case study area, Trelleborg*

Trelleborg has a 58 km long shallow, coastal stretch (Figure 6.1). The coastal stretch off Trelleborg has suitable growth conditions for ephemeral macroalgae, with large annual nutrient rich freshwater runoff and a long shallow coastline with warm waters during summer. Most of the coastline is classified as moderately exposed or exposed to wave action (Müller et al., 2007). The coastal stretch has sandy bottoms with boulders, stones and rocks, as well as hard rock areas (Müller, 2008). Two of the dominant algal species are *Polysiphonia fucoides* and *Ceramium tenuicorne* (Lundgren et al., 2009).

The total quantity of filamentous biomass at two monitoring stations (Figure 6.1) has been assessed in a regional monitoring programme performed by scuba divers. The salinity along the Trelleborg coastline fluctuates from 6.5 to 9.5 PSU and macroalgae growth is considered to be nitrogen-limited rather than phosphorus-limited during summer (Lundgren et al., 2009). Data from the regional monitoring programme (see Lundgren et al., 2009) from a depth of 5 m were used as input in the biomass prediction model in this work. Input data is based on monthly average values on temperature and nutrients, publicly available (Lundgren et al., 2009). Daily average values on surface irradiance were used as input for the model and were provided by the Swedish Meteorological and Hydrological Institute (Swedish Meteorological and Hydrological Institute, 2011).

The photic depth at Trelleborg is typically 14 m during summer (Lundgren et al., 2009). However, scuba diving within the area indicated that algae occurrence decreases rapidly

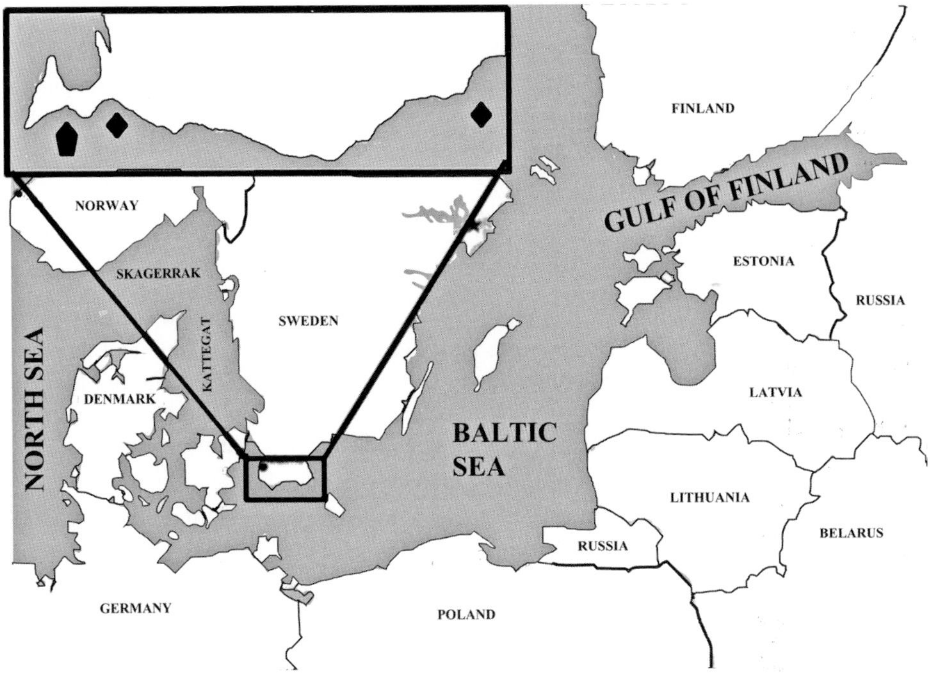


**Figure 6.1** Map of the Baltic Sea with an enlargement of the case study area in Trelleborg. The filled rhombs in the inserted magnification indicate the two stations within the regional monitoring programme, Kämpinge (left) and Abbekås (right). The larger rhomb indicates Falsterbo station, where salinity, temperature and nutrient data are recorded.

after 10 m (Gröndahl, 2010). Therefore, as a conservative estimate, this study assumed that the suitable depth range for red filamentous algae growth is 0–10 m.

For the shallow area of 0–3 m depth, the red algae occur as drifting algae mats and is therefore not dependent on the bottom substrata (Gröndahl, 2010). The average coverage of drifting algae at 0–3 m depth during August 2009 was 30% (Gröndahl, 2010). At a depth of 3–10 m, both attached and unattached algae are abundant (Lundgren et al., 2009).

## 6.2.2 Biomass quantification

The total natural resource potential in the case study area, $m_{tot}$ [tonnes dry weight] (tonnes dwt), was calculated by summing the natural resources in shallow waters, $m_s$ [tonnes dwt], and deeper waters, $m_d$ [tonnes dwt], according to Equations 6.1–6.3.

$$m_{tot} = m_s + m_d \tag{6.1}$$

The natural resource potential for shallow depths (0–3 m) in the study area, where drifting red macroalgae completely dominate, was estimated using Equation 6.2. The red filamentous biomass at shallow depths, $m_s$ [tonnes dwt], was estimated from the total area in the depth interval, $A_s$ [km$^2$], the percentage of the total bottom area covered with drifting algae, $\phi_s^{field}$ [%], and the average density of drifting algae mats, $c_s^{field}$ [gram dry weight m$^{-2}$] (g dwt m$^{-2}$). Both $\phi_s^{field}$ and $c_s^{field}$ were calculated as an average value from 12 transect investigations in Trelleborg during August 2009 reported by Gröndahl (2010). All parameters used are listed in Table 6.1.

$$m_s = A_s * \left(\frac{\phi_s^{field}}{100}\right) * c_s^{filed} * 10^{-12} \tag{6.2}$$

The red filamentous biomass in the depth interval 3–10 m was estimated using Equation 6.3. In this depth interval, there were hard bottom substrata available and therefore both unattached and attached macroalgae were assumed to be present. The red filamentous biomass in deeper waters (3–10 m) was estimated from the bottom area in the depth interval, $A_d$ [km$^2$], the percentage of the total area covered with hard substrata, $\phi_d^{est}$ [%], and the average coverage of total red filamentous algae (attached and unattached), $c_d^{field}$ [g dwt m$^{-2}$]. An average value of $c_d^{field}$ at 4 and 6 m depth during summer in 2002–2009 was calculated from monitoring data reported by Lundgren et al. (1999).

$$m_d = A_d * \left(\frac{\phi_d^{est}}{100}\right) * c_d^{field} * 10^{-12} \tag{6.3}$$

$A_d$ and $A_s$ were estimated from a nautical map (scale 1 : 5 000) using a publicly available digital area-estimate tool provided by the Swedish Maritime Administration (Swedish Maritime Administration, 2011). Estimation of $\phi_d^{est}$ was based on the assumption that a stony or rocky shoreline indicates an area with stony/rocky bottom substrata. Data on the amount of shoreline covered with stones and boulders or rock were therefore taken from maps of beach types along the Swedish coastline reported by the Geological Survey of Sweden (Müller et al., 2007). This led to the assumption that 50% of the sea bottom under 3–10 m deep waters in Trelleborg is covered with stones, boulders or rock and is thus suitable for growth of attached algae.

The uncertainty in $m_d$ was assessed and set to represent the uncertainty of $m_{tot}$, since the biomass at deeper waters completely dominates the total biomass amount, as will be shown. The uncertainty of the biomass quantification was estimated through error propagation assuming uncorrelated uncertainties. The relative standard deviation of $A_d$ was evaluated by estimating the same area twice and the relative standard deviation of $c_s^{field}$ were represented by average deviations of five replicates for each field sampling occasion. The error in $\phi_d^{est}$ was conservatively estimated. The uncertainties of each parameter and for the biomass estimate on deeper waters are shown in Table 6.1.

### 6.2.3 Biomass coverage from growth modelling

When monitoring data on the average algae coverage are lacking for Equations 6.2 and 6.3, model predictions can be used to predict biomass per square metre. The model presented here is based on hydrographical data commonly available from regional and national monitoring programmes. Using the equation from Kiirikki et al. (1998), the biomass net growth per square metre of red macroalgae was calculated with the following equation:

$$\Delta c = G_{gross}c - L(c - c_{mim})\Delta t \tag{6.4}$$

The accumulation rate of biomass, $\Delta c$ [gram dry weight m$^{-2}$ day$^{-1}$] depends on the gross growth rate constant, $G_{gross}$ [day$^{-1}$], biomass per square metre $c$ [g dwt m$^{-2}$], and loss rate constant per day, $L$ [d$^{-1}$]. The overwintering biomass is set by $c_{min}$ [g dwt m$^{-2}$]. The gross growth function depends on the theoretical maximum growth rate, $\mu_{max}$ [day$^{-1}$], which is limited by restriction functions shown in Equation 6.5. Here, the maximum gross growth, $\mu_{max}$, is set to be 75% lower from September on, since shortened day length and increased water temperature is thought to induce settlement and senescence for spring and summer macroalgae (Pato et al., 2011). Equations 6.5–6.7 are taken from Martins and Marques (2001).

$$G_{gross} = \mu_{max} \times f(K) \times f(I) \times f(T) \times f_{ext}(NP) \tag{6.5}$$

The gross growth of algae biomass, $A$ [g dwt m$^{-2}$], assumes logistic growth up to a maximum biomass per square metre, $K$ [g dwt m$^{-2}$], as displayed in Equation 6.6 that quantifies $f(K)$. Furthermore, the temperature restriction $f(T)$of growth is described by Equation 6.7, where $T$ [°C] is temperature and $T_{opt\,growth}$[°C] the optimal temperature for growth. The parameter $T_x$ [°C] indicates if the temperature is below optimum for growth and is equal to 0 if the temperature $T$ is $< T_{opt\,growth}$ and equal to 25 if $T_{opt\,growth} \geq T_{opt\,growth}$. Nutrient limitations of growth, $f_{ext}(NP)$, is described by Equation 6.8, where $\Phi_{NH_4}$ [$\mu$g l$^{-1}$] and $\Phi_{NO_3}$ [$\mu$g l$^{-1}$] are ammonium and nitrate concentrations, respectively, in the water column. In addition, nitrogen limitation is dependent on the nitrogen half-saturation constant, $K_N^{MA}$ [$\mu$g l$^{-1}$]. This limiting function is altered from Martins and Marques (2001) so that the effect of internal nutrient concentrations is disregarded.

$$f(K) = 1 - \left(\frac{c}{K}\right) \tag{6.6}$$

$$f(T) = exp\left[-2.3\,\mathrm{x}\left(\frac{T - T_{opt\,growth}}{T_x - T_{opt\,growth}}\right)^2\right] \tag{6.7}$$

$$f_{ext}(NP) = \frac{\Phi_{NH_4} + \Phi_{NO_3}}{K_N^{MA} + \Phi_{NH_4} + \Phi_{NO_3}} \tag{6.8}$$

The light limiting function, Equation 6.9, assumes logistic growth up to an optimal irradiance, $I_{opt}$ [$\mu$mol photon m$^{-2}s^{-1}$] and was described by Steele (1965). Light limitation is determined by light $I_Z$ [$\mu$mol photon m$^{-2}$s$^{-1}$] at the depth z [m]. The depth dependence of light is determined with Equation 6.10, where $k = \frac{1.7}{Z_{SD}}$.

$$f(I) = \frac{I_Z}{I_{opt}} \times e^{\left(1-\frac{I_Z}{I_{opt}}\right)} \tag{6.9}$$

$$I_Z = I_0 e^{-kz} \tag{6.10}$$

Equation 6.11 describes the loss of algae biomass and was taken from Kiirikki et al. (1998). The temperature dependence of the loss is identical to Equation 6.7, but with the optimal temperature $T_{opt\ loss}$ set to 20 °C. In Table 6.1, all parameters included in the growth model are shown.

$$L = L_{max} * f(T)_{loss} \tag{6.11}$$

#### 6.2.3.1 *Uncertainties in growth model*

The uncertainties of each parameter were assessed and the propagated error (assuming uncorrelated uncertainties) of the processes gross growth rate and the loss rate of biomass were estimated (Table 6.1) (Mandel, 1984). During January to early summer, the dominant process is in general gross growth and during summer and autumn the loss process is generally dominating. The dominant process was set to govern the overall uncertainty for model predictions for the respective time intervals.

## 6.3 Results

### *6.3.1 Quantification of biomass in Trelleborg*

The results of the biomass quantification using field observations of algae coverage are shown in Table 6.2. Using Equations 6.2 and 6.3, it was estimated that a total of about 19 000 tonnes dwt of filamentous algae is present as summer stock in Trelleborg. Most of the stock is found in the depth interval 3–10 m ($m_d$). As indicated in Table 6.2, the uncertainty within $m_d$ is estimated to ± 7 000 tonnes dwt.

For comparison with previous studies, Table 6.3 shows estimates made by Kautsky (1994) of the red filamentous standing stock based on transect investigations of coverage per square metre at different depths in the Kalmar Sound area in the northern part of the Swedish Baltic proper. The area was estimated from digital maps of bottom depths using the same technique as in this study. However, Kautsky (1994) took shoreline slope into account; for simplicity this was not done in our study, which may result in a slight underestimate of the biomass amount.

### *6.3.2 Estimation of harvesting potential in Trelleborg*

The harvesting method considered in our case study area is designed to remove floating algae mats from the water column at shallow depths and from the shoreline. The method is described in detail in Trelleborg Municipality (2011). In this study, annual harvesting of 10–30% of the total summer stock is used as an example, resulting in an annual harvesting potential in Trelleborg of an estimated 2 000–6 000 tonnes dwt of red filamentous biomass. Harvesting of 10–30% of the stock is chosen as an example since Gröndahl (2010) showed that correspondingly 30% of the summer stock estimate in this study was found as beach cast within the case study area in August 2009. It is therefore plausible that this amount could be harvested in the shoreline and at shallow depths.

**Table 6.1** Parameters for biomass prediction.

| Parameter | Description | Values | Units | Estimated uncertainty for the case study area (%) | Source |
|---|---|---|---|---|---|
| $c_{min}$ | Overwintering biomass | 0.5 | g dwt $m^{-2}$ | 30% | Estimated by Kiirikki et al. (1998) for spring species of filemantous macroalgae in Gulf of Finland, Baltic Sea. |
| $K_{TB}$ | Maximum density in Trelleborg | 400 | g dwt $m^{-2}$ | 20% | Derived from monitoring data in Trelleborg of red filamentous macroalgae (Lundgren et al., 2009). |
| $L_{max}$ | Maximum loss rate constant | 0.15 | $d^{-1}$ | 20% | Estimated based on Kiirikki et al. (1998) in the Gulf of Finland for *Ceramium sp.*, Baltic Sea. |
| $\mu_{max}$ | Maximum growth rate constant at optimal temperature | 0.78 (0.195 autumn and winter) | $d^{-1}$ | 30% | *Ceramium sp.* in Gulf of Finland, Baltic Sea (Kiirikki et al., 1998).<br>Literature range 0.1–0.78, (Bowie et al., 1985; Biber et al., 2004). |
| $K_N^{MA}$ | Macroalgal Nitrogen half saturation constant | 19 | $\mu g\, l^{-1}$ | 30% | Kiirikki et al. (1998) for red filamentous algae (e.g., *Ceramium)* in the Gulf of Finland, Baltic Sea. |
| $I_{opt}$ | Optimal PPFD for photosynthesis | 100 | $\mu$ mol photo $m^{-2}\, sec^{-1}$ | 20% | Literature range 90-300 (Fralick and Mathieson, 1975; Bowie et al., 1985; Biber et al., 2004). |
| $T_{opt\,growth}$ | Optimal temperature where growth rate is maximum | 15 | °C | 10% | Kiirikki et al. (1998) Red filamentous algae (e.g., *Ceramium*) in the Baltic Sea. |
| $T_{opt\,loss}$ | Optimal temperature for loss | 20 | °C | 10% | Kiirikki et al. (1998) for filamentous algae (e.g., *Ceramium* ) in Gulf of Finland, Baltic Sea. |
| $T_{min\,growth}$ | Lower temperature limit where growth ceases | 0 | °C | 10% | *Polysiphonia urceolata* in the North Sea (Fortes and Luning, 1980). |
| $T_{max\,growth}$ | Upper temperature limit where growth ceases | 25 | °C | 10% | *Polysiphonia urceolata* in the North Sea (Fortes and Luning, 1980). |

**Table 6.2** Estimated summer stock of red filamentous algae in Trelleborg and parameters used in the quantification (equation 6.1) with uncertainties within parentheses.

| Assessed area [depth, m] | $c^{field}$ [g dwt m$^{-2}$] (uncertainty) | $A$ [km$^2$] (uncertainty) | Φ [%] (uncertainty) | Filamentous biomass [tonnes] (uncertainty) |
|---|---|---|---|---|
| $m_s$ (0–3) | 84 | 18 | 30 | 1 500 |
| $m_d$ (3–10) | 280 (± 45) | 124 (± 1) | 50 (± 20) | 17 000 (± 7000) |
| $m_{tot}$ | | 142 | | 19 000 |

Based on the average elemental composition of algae, this harvesting would involve a nutrient reduction of 50–150 tonnes nitrogen and 4–13 tonnes phosphorus, as shown in Table 6.4. The estimated nitrogen reduction potential corresponds to approximately 5–15% of the freshwater runoff input of nitrogen to the area as reported by Müller et al. (2007). As shown in Table 6.4, it is theoretically feasible to heat 170–500 houses annually on the methane produced from 10–30% of the summer stock. This calculation is based on data on biogas potential and methane content reported by Biswas (2009) from co-digestion of red filamentous macroalgae with manure, as detailed in the footnote to Table 6.4.

### *6.3.3 Modelled standing stock of red filamentous algae in Trelleborg*

The results of the biomass estimation model are shown in Figure 6.2 as biomass coverage [g dwt m$^{-2}$] over the years: 2005–2009. The height and timing of the biomass peaks in the model are determined by environmental conditions, that is light, temperature and

**Table 6.3** Estimation of algae stock for different coastal stretches of the Baltic Sea.

| Coastal stretch | Filamentous biomass [tonnes dwt] | Coastline [km] | Biomass per km coast [tonnes km$^{-1}$] | Bottom area [km$^2$] | Source |
|---|---|---|---|---|---|
| Swedish southern coast, (attached stock) | 25 200[b] | 2 192 | 12[a] | NA | Kautsky, 1994; Kautsky and Kautsky, 1994 |
| Swedish Baltic proper coast, (attached stock) | 100 000[b] | 10 640 | 9[a] | NA | Kautsky, 1994; Kautsky and Kautsky, 1994 |
| Estonia, Kassari Bay (unattached stock) | 28 000[c] | NA | NA | 120 | Paalme et al., 2004; Martin et al., 2006; Kersen and Martin, 2007 |

NA, Not analysed

[a] $\delta = \frac{m}{D}$, where $\delta$ is biomass per km coastline, m (tonnes) is the amount of macroalgae and D (km) is the length of coastline examined.

[b] Assuming a carbon content of 24% based on Atkinson and Smith (1983).

[c] From reported wet weight with an assumed dry matter content of 20% (Greenwell et al., 1984).

**Table 6.4** Harvesting, nutrient reduction and biogas potential in Trelleborg.

| Biomass [tonnes dwt][c] | N reduction[a] [tonnes dwt] | P reduction[b] [tonnes dwt] | N reduction as fraction of annual freshwater runoff in Trelleborg (%) | Houses that can be heated (energy produced kWh/year)[d] |
|---|---|---|---|---|
| 2000 (10% of tot. pot.) | 50 | 4 | 5% | Around 170 houses (3 371 MWh/year) |
| 6000 (30% of tot. pot.) | 150 | 13 | 15% | Around 500 houses (10 113 MWh/year) |

[a]Assuming a nitrogen content of 2.5% of dry weight (Gröndahl et al., 2009).
[b]Assuming a phosphorus content of 0.21% of dry weight (Gröndahl et al., 2009).
[c]With an average dry matter content of 15% (Dahl et al., 2009).
[d]$\theta \frac{\varepsilon \cdot 6 \cdot \omega \cdot m \cdot \tau}{\upsilon}$ where $\theta$ is the number of houses that can be heated; $\in$ (0.49) [$m^3$ $kg^{-1}$ VS] based on (Biswas, 2009) is the biogas potential; 6 (65%) is the methane content [% $CH_4$] (Biswas, 2009); $\omega$ (0.54) [kg VS (kg dwt algae)$^{-1}$] (Gröndahl et al., 2009) is the volatile solids content of the algae; m (2 000 000) [kg dwt] is the amount of algae; $\tau$ (9.8) [kWh $m^{-3}$ $CH_4$] I energy content of 1 $m^3$ methane (Edström and Nordberg, 2004); and $\upsilon$ (20 000) [kWh $year^{-1}$] is the average annual energy demand for a house in Sweden.

availability of nutrients in the water column. As shown in Figure 6.2, the main growth period for the red filamentous species is in spring and early summer according to the model.

The uncertainties of main processes due to parameterisation were assessed as described in Section 6.2.3.1 and are presented in Table 6.1. The propagated uncertainty of the gross growth, $G_{gross}$, the dominant process during spring, is 47% and the propagated uncertainty of the loss process, $L$, (dominant process during summer) is 37%. These uncertainties are based on conservative estimates.

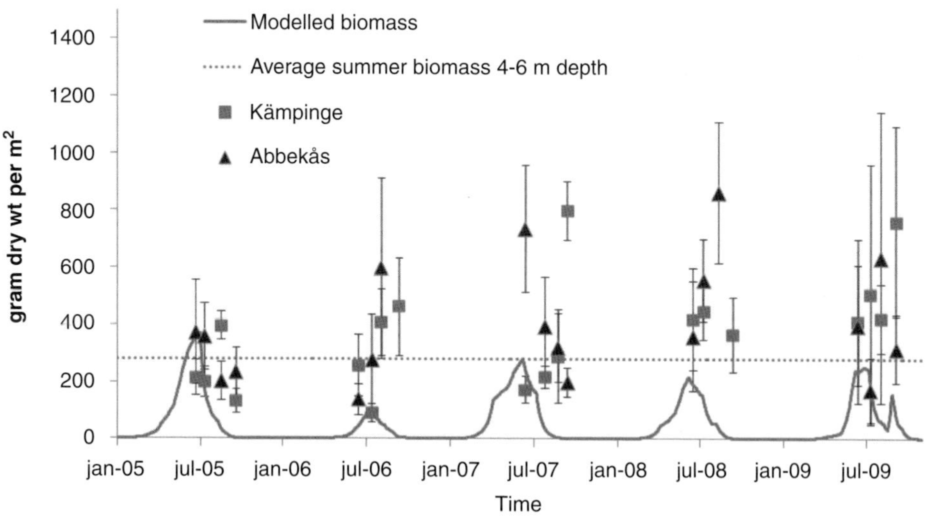


**Figure 6.2** Total filamentous biomass per square metre at 4 m depth. Field data from two monitoring stations (squares and triangles), with standard deviations (error bars) and average summer biomass (dotted line); continuous line indicates independent model predictions.

## 6.4 Discussion

### 6.4.1 *Quantification of biomass in Trelleborg*

As shown, the estimates in the present study are high in comparison with those reported by Kautsky (1994) for Kalmar Sound regarding tonnes per km shoreline. This difference was expected, since Trelleborg has a very shallow coastline, with larger water volumes within the photic zone and higher salinity than the northern parts of the Baltic proper, within which Kalmar Sound is situated. Furthermore, the Trelleborg region has a large summer stock of red filamentous algae, since the environmental and hydrodynamic conditions support both unattached and attached seasonal stocks. However, the estimated stock in Trelleborg is only around half of that reported for Kassari Bay in Estonia, although the Trelleborg study area is larger. This indicates a larger natural resource potential in Kassari Bay in comparison with Trelleborg. The reason for this may be that Kassari Bay is a relatively sheltered and shallow area with suitable growth conditions. Furthermore, Kassari Bay has the largest known stock of unattached red macroalgae reported for the Baltic Sea (Kersen and Martin, 2007).

The uncertainties of each parameter used for the quantification in our study are presented. These uncertainties are consistent with previous assessments of biomass potential in natural systems (Kiirikki et al., 1998; Öberg, 2006). Furthermore, there is a conceptual uncertainty embodied in the parameter $A_d$, associated with the division into different growth regimes based on water depth. Nevertheless, our quantification results indicate that our method can provide important first-hand indicators of stock size when field monitoring data on algae biomass is unavailable.

### 6.4.2 *Estimation of harvesting potential in Trelleborg*

In this study, harvesting of 10–30% of the stock was used as an example. However, the amount of the summer stock to be harvested is not a fixed number but rather a decision that has to be made based on regional ecosystem resilience within the site-specific context. In addition, harvesting potential is of course highly dependent on harvesting technique and harvesting intensity. Nevertheless, the quantification of the total summer stock sets the upper limit for macroalgae harvesting. If, from a theoretical perspective, 100% of the summer stock in Trelleborg were harvested, the nitrogen removal would amount to 50% of annual nitrogen freshwater runoff within the case study area.

Furthermore, the harvested biomass can be utilized for energy production. However, the potential energy output from biomass processing must be set in proportion to the energy input required in the system for harvesting, handling and processing the biomass. This aspect will be addressed in future work.

### 6.4.3 *Modelled standing stock of red filamentous algae in Trelleborg*

Without calibration, the model results agree within about a factor of five with field data for Trelleborg during summer months at 4 m depth. These results are also in agreement with previous studies of red filamentous algae in the Baltic Sea (Kiirikki and Lehvo, 1997;

Martin et al., 2006). The estimations of stock showed a decline in biomass during August, when warmer waters lead to increased grazing and decomposition, for all years except 2009, when there was a second biomass peak during the end of August.

As shown, field observations used as validation data are associated with large uncertainties; the five replicates for each sampling occasion differ by as much as a factor of 2 within each station, and between the stations the variations are within a factor of 4. Considering these large uncertainties in validation data, our model could present satisfactory, independent, first-hand estimates of biomass stocks that correlates to validation data within about a factor of five for our case study area. Furthermore, the propagated parameter uncertainty of the gross growth, and the loss processes of 47% and 37%, respectively, indicate that parameterisation is important for model predictions. However, site-specific information on parameters is often not readily available and the predictive capacity of our model is, as mentioned, satisfactory for first-hand estimations of biomass growth.

There are other quantification methods for biomass, such as remote sensing techniques. These techniques can assess abundance over a vast area and assess the amount of surface floating algae (e.g. Vahtmäe et al., 2006; Nezlin et al., 2007). However, since our aim was to assess biomass quantity from readily available public data for our case study area as an example of a site with typical data availability, remote sensing was not considered. Furthermore, the benefit of a model based on irradiance, temperature, nutrients and light is the possibility of predicting biomass when environmental conditions are altered or predicting biomass stock response to harvesting events; this will be further investigated in future work.

## 6.5 Conclusions

Despite the current increasing interest in harvesting macroalgae in the Baltic Sea, few previous studies have estimated the amount of macroalgae biomass available and there is a lack of cost-effective methods for site-specific quantification of macroalgae. This study has quantified macroalgae biomass in the case study area of Trelleborg, Sweden, and tested an algal biomass prediction model on the area.

Our assessment of the coast of Trelleborg indicates a considerable summer stock of red filamentous algae of 19 000 tonnes dwt. If, as an example, 10–30% of this estimated stock was harvested, the nutrient reduction potential would be considerable, in the magnitude of 50–150 tonnes nitrogen annually which corresponds to 5–15% of the annual freshwater input of nitrogen to the area. An added value of harvesting algae is the potential to produce methane from the biomass. According to our estimates, harvesting of 10–30% of the summer stock could supply heat to 170–300 houses annually. The proposed, independent biomass model using readily available data gives an acceptable level of agreement between predictions and field data given. The model for estimating natural resource potential is thus promising and worthy of further investigation for other sites.

## Acknowledgements

We are very grateful to two anonymous referees and the editors for valuable comments on the manuscript. Furthermore, we thank Dr Per Olsson, Toxicon, Sweden, for providing data from Sydkustens Vattenvårdsförbund. We also thank Dr Eva Blidberg, Industrial Ecology, KTH, Sweden, for comments on draft versions of this manuscript.

STRÅNG data used here are from the Swedish Meteorological and Hydrological Institute, and were produced with support from the Swedish Radiation Protection Authority and the Swedish Environmental Agency. This study was funded by The Swedish Research Council Formas; *Committed to Excellence in Research for Sustainable Development.* Grant number 229-2009-468.

## References

Aldentun, Y. (2011) Verksamhetsplan 2011 Nya substrat från havet. Report **2011/014–404**, Regionförbundet Kalmar län, Kalmar, Sweden [In Swedish].

Atkinson, M. J. and Smith, V. S. (1983). C:N:P ratios of benthic marine plants. *Limnologogy and Oceanography* **28**, 568–574.

Biber, P. D., Harwell, M. A., Cropper, J. and Wendell, P. (2004). Modeling the dynamics of three functional groups of macroalgae in tropical seagrass habitats. *Ecological Modelling* **175**, 25–54.

Biswas, R. (2009) *Biomethanation of Red Algae from the Eutrophied Baltic Sea.* Master Thesis Environmental Science Program, Faculty of Arts and Sciences, LiU-TEMA/ES-A-09/002-SE, Linköping University, Linköping, Sweden.

Bowie, G. L., Mills, W. B., Porcella, D. B., Campbell, C. L., Pagenkopf, J. R., Rupp, G. L., Johnson, K. M., Chan, P. W. H. and Gherini, S. A. (1985) Rates, constants and kinetic formulations in surface water quality modeling. Report **EPA/600/3-85/040**, Second Edition, US Environmental Protection Agency, USA.

Edström, M. and Nordberg, Å. (2004) Producera biogas på gården; gödsel, avfall och energigrödor blir värme och el. Report **104**, JTI Institutet för jordbruks- och miljöteknik. Sweden [in Swedish].

Eklund, B., Svensson, A. P., Jonsson, C. and Malm, T. (2005) Toxic effects of decomposing red algae on littoral organisms. *Eustarine, Coastal and Shelf Sience* **62**, 621–626.

Elmgren, R. (1989) Manś impact on the ecosystem of the Baltic Sea: energy flows today and at the turn of the century. *Ambio* **18**, 326–332.

Fortes, M.-D. and Luning, K. (1980) Growth rates of North Sea macroalgae in relation to temperature, irradiance and photoperiod. *Helgoland Marine Research* **34**, 15–29.

Fralick, R. A. and Mathieson, A. C. (1975) Physiological ecology of four *Polysiphonia* species (Rhodophyta, Ceramiales). *Marine Biology* **29**, 29–36.

Greenwell, M., Bird, C. J. and McLachlan, J. (1984) Depth-related variation in gross chemical composition of several seaweeds. *Aquatic Botany* **20**, 297–305.

Gröndahl, F. (2010) Marin inventering utförd vid Trelleborgs kommuns kustremsa den 16–30 augusti 2009. Report **11**. Trelleborgs kommun, Trelleborg, Sweden [In Swedish, short version in English].

Gröndahl, F., Brandt, N., Karlsson, S. and Malmström, M. E. (2009) Sustainable use of Baltic Sea natural resources based on ecological engineering and biogas production. In: Brebbia, C. A. & Tiezzi, E., (Eds.) Ecosystems and Sustainable Development VII – the Proceedings of the Seventh International Conference on Ecosystems and Sustainable Development (ECOSUD), 2009 Chianciano Terme, Italy. WIT Press, Southampton, UK, pp. 153–161.

Isaeus, M., Malm, T., Persson, S. and Svensson, A. (2004) Effects of filamentous algae and sediment on recruitment and survival of *Fucus serratus* (Phaeophyceae) juveniles in the eutrophic Baltic Sea *European Journal of Phycology* **39**, 301–307.

Kautsky, H. (1994) Quantitative distribution of sublittoral plant and animal communities along the Baltic Sea gradient. In Eleftheriot, A., Ansell, A. D. and Smith, C. J. (Eds.) *Biology and Ecology of Shallow Coastal Waters, 28th European Marine Biology Symposium,* **28**.

Kautsky, N., Kautsky, H., Kautsky, U. and Waern, M. (1986) Decreased depth penetration of *Fucus vesicolosus* (L.) since the 1940s indicates eutrophication of the Baltic Sea. *Marine Ecology Progress Series* **28**, 1–8.

Kautsky, U. and Kautsky, H. (1994) Coastal Productivity in the Baltic Sea. *In:* Eleftheriot, A., Ansell, A. D. and Smith, C. J. (Eds.) *Biology and Ecology of Shallow Coastal Waters 28th European Marine Biology Symposium,* **28**.

Kersen, P. and Martin, G. (2007) Annual biomass loss of the loose-lying red algal community via macroalgal beach casts in the Väinameri area, NE Baltic Sea. *Proceedings of the Estonian Academy of Sciences Biology and Ecology* **56**, 278–289.

Kiirikki, M., Haapama, J., Koponen, J., Ruuskanen, A. and Sarkkula, J. (1998) Linking the growth of filamentous algea to the 3D-ecohydrodynamic model of the Gulf of Finland. *Environmental Modelling & Software* **13**, 503–309.

Kiirikki, M. and Lehvo, A. (1997) Life strategies of filamentous algae in the northern Baltic Proper. *Sarsia* **82**, 259–267.

Lundgren, F., Olsson, P., Sjölin, A. and Nylander, W. (1999) Undersökning av fintrådiga alger i Öresund och längs Sydkysten – en metodikstudie. Report 98/99 Toxicon AB, Trelleborg, Sweden [In Swedish].

Lundgren, F., Olsson, P., Sjölin, A. and Nylander, W. (2009) Sydkustens Vattenvårdsförbund Undersökningar längs Sydkusten Annual Reports 98/99 and 2002–2009. Toxicon AB, Trelleborg, Sweden [In Swedish].

Mandel, J. (1984) *The Statistical Analysis of Experimental Data*, Dover Edition John Wiley & Sons Ltd, New York, pp. 403.

Martin, G., Paalme, T. and Torn, K. (2006) Growth and production rates of loose-lying and attached forms of red algae *Furcellaria lumbricalis* and *Coccotylus truncatus* in Kassari Bay, the West Estonian Archipelago Sea. *Hydrobiologica* **554**, 107–115.

Martins, I. and Marques, J. C. (2001) A model for the growth of opportunistic macroalgae (*Enteromorpha* sp.) in tidal eustaries. *Eustarine, Coastal and Shelf Sience* **55**, 247–257.

Müller, M. (2008) Fallstudie Trelleborg – Integrerad kustzonsförvaltning på lokal nivå i Östersjön genom geografisk kartläggning och scenarioanalys. Trelleborg Municipality, Trelleborg, Sweden [In Swedish, short version in English].

Müller, M. (2010) Miljökonsekvensbeskrivning för insamling av alger längs Trelleborgs kust. Report **9**, Trelleborg Municipality, Trelleborg, Sweden [In Swedish].

Müller, M., Hansson, A. and Hügert, J. (2007) Trelleborgs kommun for en hållbar utveckling – Hav i balans för en levande kust. Report **070704**, Trelleborgs Municipality, Trelleborg, Sweden [In Swedish].

Nezlin, N. P., Kamer, K. and Stein, E. D. (2007) Application of color infrared aerial photography to assess macroalgal distribution in an eutrophic eustary, Upper Newport Bay, California. *Eustaries and Coasts* **30**, 855–868.

Paalme, T., Martin, G., Kotta, J., Kukk, H. and Kaljurand, K. (2004) Distribution and dynamics of drifting algae mats in Estonian coastal waters during 1995–2003. In Aben, H. & Pärnapuu, I. (Eds.) *Proceedings of the Estionian Academy of Sciences Biology and Ecology* **53**, 260–268.

Pato, L., Martinez, B. and Rico, J. M. (2011) Environmental control of the annual erect phase of *Nemalion hemlinthoides* (Rhodophyta) in the field. *Scientia Marina* **75**, 263–271.

Steele, J. H. (1965) Notes on some theoretical problems in production ecology. In Goldman, C. R., (Ed.) *Primary Production in Aquatic Environments*. University of California Press, Berkley, California.

Swedish Institute for the Marine Environment (2011) Havet Om miljötillståndet i svenska havsområden. [Swedish Institute for the Marine Environment, Annual Report.] Växsjö, Sweden [In Swedish].

Swedish Maritime Administration (2011) *Nautical Maps by Sjöfartsverket* [Online]. Available at http://kartor.eniro.se/ (accessed 10 September 2012).

Swedish Meteorological and Hydrological Institute (2011) *Strång Model* [Online]. Available: http://strang.smhi.se/extraction/index.php?data=chrt&lev=0 (accessed 10 September 2012).

Trelleborg Municipality (2011) Evaluation of machines for the collection of algae. Wetlands, Algae Biogas – A southern Baltic Sea Eutrophication Counteract Project, South Baltic Program [Online]. Available: http://wabproject.pl/files/evaluation%20of%20algae%20machines.pdf (accessed 10 September 2012).

Vahtmäe, E., Kutser, T., Martin, G. and Kotta, J. (2006) Feasability of hyperspectral remote sensing for mapping benthic macroalgal cover in turbid waters – A Baltic Sea case study. *Remote Sensing of Environment* **101**, 342–351.

Öberg, J. (2006) Primary production by macroalgae in Kattegat, estimation from monitoring data, seafloor properties and model simulations. *Continental and Shelf Research* **26**, 2415–2432.

Chapter 7

# Conserving Mangrove Ecosystem for Climate Change Adaptation in the Ganges Basin

*M. Shahadat Hossain*

**Abstract**

The Ganges-Brahmaputra-Meghna river systems contain fluvial and tidal geomorphological deposits that create the world's largest delta in the form of coastal landscape and islands, which develop a complex mangrove ecosystem. Landsat TM image of the study area was analyzed to assess the spatial distribution of mangrove forest and accreted zone. Participatory field observations, transect walks, focus group discussions and key informants interviews were applied for the collection and examination of past and present records on ecosystems, livelihood and disaster associated with climatic variation and also to explore the goods and services of the mangrove ecosystem. A total of 27 014 ha of mangrove forest and 60 000 ha accreted marshy stable land were identified with their geospatial distribution in the Ganges basin. The mangrove ecosystem protects the shoreline and accumulates sediment to build new land along the islands. Due to the stabilization of new land, mangroves move seaward leaving the new land to be colonized by other plants. Mangrove trees dissipate wind-wave driven energy to enhance the natural barrier against natural disasters. Conservation of existing mangrove with further plantation in the accreted marshy stable lands can act as a bio-shield to climate change extreme events with diversified goods and services to sustain the livelihoods of the islanders.

## 7.1 Introduction

Vegetation maps are essential to provide information on the spatial distribution of existing mangroves and planning for plantation in the accreted stable zones of the Ganges basin in Bangladesh. Remotely sensed data have been popularly used in mangrove forest mapping for inventory and monitoring purpose in many parts of the world (Green et al., 1998; Trisurat et al., 2000; Hossain et al., 2003; Kovacs et al., 2008). The increasing use of remote sensing techniques in forest mapping is indeed made possible because of the high reflectance values

*Global Challenges in Integrated Coastal Zone Management*, First Edition. Edited by Erlend Moksness, Einar Dahl and Josianne Støttrup.

from forested areas in the near-infrared, moderate reflectance in the middle-infrared and low reflectance in the red spectral regions (Trisurat et al., 2000). Landsat TM data can for instance be used to identify the primary forest (Sader et al., 1990), and different succession stages (Mausel et al., 1993).

Mangroves play an essential role in maintaining a healthy coastal environment by providing protection for a myriad of juvenile aquatic species, functioning as a habitat for a variety of terrestrial fauna, operating in improving coastal protection and acting as a source of nutrients that sustains many complex food chains (Kovacs, 1999). Island communities such as those in our research area, the Ganges basin in the central coastal zone of Bangladesh, are mangrove civilizations. Their economic, social and cultural life is closely interwoven with the surrounding flora and fauna, its tidal and seasonal cycles and their associated fish, shrimp, crab and other floral and faunal reproduction seasons. However, these vital tropical ecosystems in the coastal intertidal zones, covering about 170 000 km$^2$ (Rollet, 1984; Khan and Hossain, 1996), continue to be under immense threat from a variety of human actions. Over the past 20 years, approximately 35% of the world's mangrove forest area has been lost (Valiela et al., 2001). One of the greatest limitations to their protection is the lack of proper inventory and monitoring. Traditional techniques of *in situ* field measurements of these forested wetlands are extremely tedious and labour intensive given the typical inaccessibility of these systems, as well as limited mobility resulting from the maze of roots and stems; thick, unconsolidated substrate; and tidal flooding (Kovacs et al., 2008). Consequently, there has been a recent interest in the use of remotely sensed imagery, which can be acquired periodically and over very large geographical areas, for mapping and monitoring these often vast and remote wetlands (Green et al., 1998). However, the accuracy of the final map is affected by the ability of the classification procedure to discriminate the vegetation. The ability to do this is partly a function of the sensors' resolution, and partly a function of the image processing method or classification procedure adopted (Trisurat et al., 2000).

The coastal zone of Bangladesh represents an area of 47 211 km$^2$ inhabiting 35 million people who live in 'high risk areas' along the coast and islands. The physiology, morphology and other natural conditions have made the area vulnerable to cyclones, storms and floods which are very devastating causing immense suffering and damage to people, property and the environment (Hossain, 2001). Approximately 53 damaging cyclones were reported in the coastal area of Bangladesh from 1793 to May 2009 causing about 1 000 000 human casualties, innumerable animal deaths, destroying hundreds of millions of houses, standing crops, coastal embankments and mangrove forest (Hossain, 2009). Therefore it is important to assess coastal community resilience and scaling up peoples' capacity to adapt to the climate extreme events. The objectives of the study were to assess the spatial distribution of mangrove with accreted zone for further plantation and to explore the goods and services of the mangrove ecosystem.

## 7.2 Methods

### *7.2.1 Study area*

The study area located in the lower Ganges-Brahmaputra-Meghna (GBM) basin consists of about 80 near-shore islands of Bangladesh between latitudes 21°50′ and 23°05′N and longitude 90°30′ and 91°35′E (Figure 7.1). The dominant soil characteristics are muddy

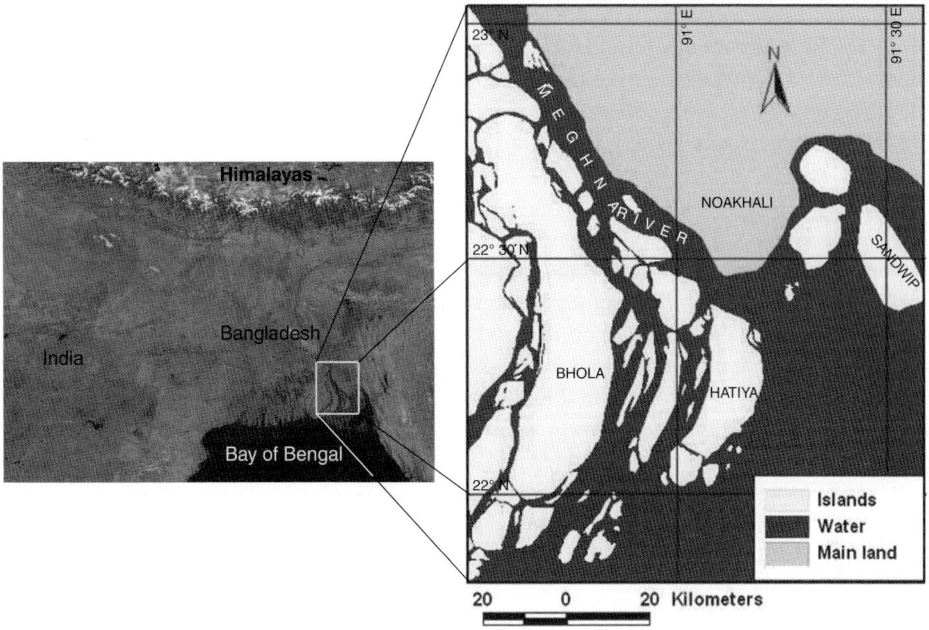


**Figure 7.1** MODIS image of December 2010 shows Ganges-Brahmaputra-Meghna river systems in Bangladesh. Classified Landsat TM image (right enlarged) shows geographical location of the islands in the Ganges basin of Bangladesh. (see Plate 7.1 for colour version)

soil and sandy-clay loam texture. The international rivers Ganges and Brahmaputra enter Bangladesh from the western and northern sides respectively, meet inside the country and then flow downstream as the Meghna River and finally empty into the Bay of Bengal. Since prehistoric times, the region has been one of the areas of most active sedimentation in the world (Khan et al., 1998). The tidal range at the Meghna River near the Noakhali coast ranges from 0.4 m at neap tide to 4.4 m at spring tide (Hossain, 2008). The mean annual rainfall is 2547 mm, of which more than 95% occurs during the monsoon season. The highest precipitation falls during May–September while the lowest is in October–November and March–April, and there is no rainfall during December–February. The mean annual minimum and maximum temperature is 23 and 31 °C with peaks of ~33 °C during March–June. Winter temperature drops down to 14–16 °C in December–January. The relative humidity is high, varying from 72% in March to 87% in July.

## 7.2.2 *Satellite image*

Landsat TM image was acquired from the Space Research and Remote Sensing Organization of Bangladesh (SPARRSO) on 4 March 2007. The pixel resolution of the image was 30 m. Bands 1, 5 and 7 were used in the analysis because these are considered the most useful for vegetation mapping (Manson et al., 2001). Geometric correction was performed with bilinear transformation (Research Systems Inc., 2000a) and the root mean square error (0.38) was controlled within less than one pixel (30 m) for 82 ground control points (GCPs).

Road joints, level crossings of road, prominent features and buildings were selected as most of the GCPs in the image. Ground data were obtained by field investigation of about 105 plots. All image processing was performed using ENVI (version 4.5) developed by Research Systems Inc., USA. The GIS software used in this study was ArcGIS for windows (version 9.3) developed by Environmental Systems Research Institute Inc., USA.

### 7.2.3 Image analysis

An unsupervised classification of the image using the ISODATA algorithm was operated for three series of spectral grouping of the same set of Landsat image, for comparison of the classified thematic maps. Minimum and maximum number of classes of 5–10, 10–15 and 15–20 were assigned for each analysis. Supervised classifications were carried out on the basis of region of interest (ROIs), where the training areas (selected during field investigation) were regions of terrain with known properties or characteristics. Groups of pixels of image were randomly selected and clustered into classes using standardized ENVI polygon drawing routines (Research Systems Inc., 2000b). Parallelepiped and maximum likelihood classification strategy was most useful for discriminating the category of interest (Trisurat et al., 2000; Hossain et al., 2007, 2008; Hossain, 2008).

### 7.2.4 Accuracy assessment

Classified Landsat TM image verification was carried out by making comparisons between predicted and actual land covers. A stratified simple random sampling from different land cover areas was followed to identify 146 sites for subsequent visit and assessment. The location of each field site was determined using Garmin map76CSx GPS. Classification accuracy was carried out by making comparisons between classified image maps and existing land cover. On-the-ground verification is the most reliable and also the most time consuming. Such an approach is appropriate to verify individual sites after the GIS has been employed to identify the mangrove forest cover. Consequently, the accuracy report is generated to calculate statistics of the percentages of accuracy based on the results of the error matrix using SPSS software (version 11.5). To assess the classification accuracy, a standard error matrix was determined, using data from the output map as the rows, and the reference data (ground truth points) as the columns in the matrix. To measure the improvement of the classification over a random assignment of pixels, the Kappa and Tau coefficients and their 95% confidence intervals were calculated for testing significant differences between rows and columns (Ma and Redmond, 1995; Congalton and Green, 1999; Hossain et al., 2007).

### 7.2.5 Field survey

Participatory assessment was carried out using field observations and community level group meetings with different stakeholder groups to gather primary information following the approaches of Pido (1995), IIRR (1998) and Hossain et al. (2004). Between April 2008 and February 2009, the research team carried out a series of 60 focus group discussions, 20 field observations, 5 transect walks, and 10 key informant interviews organized in co-operation with village leaders, followed by 18 neighbourhood group meetings and interviews relating to 1500 household heads including 30% women. All the efforts were for the collection and examination of past and present records on ecosystems, livelihood and disaster associated with climatic variation and also for exploring the goods and services of the mangrove ecosystem. Using a series of photographs, informants were asked to identify the species of mangroves and to provide a rough estimate of their relative abundance in their local area. Individuals were then asked to evaluate the goods and services of mangrove forest (Kovacs, 1999).

### *7.2.6 Resilience assessment*

Community resilience is the capacity of a community to adapt to and influence the course of environmental, social and economic change (Hossain, 2009). Human, physical, financial, natural and social assets of the islands were weighted and scored in terms of significance for the islanders' livelihood. The human, physical, financial, natural and social asset data were collected from four villages of Hatiya Island with the shadow of mangrove and embankment on the seaward side. Data were gathered through structured interviews from 60 fishermen; individuals with the greatest experience and respect within each village were deliberately sought. Each interviewee was asked to compare a series of elements (attributes and sub-attributes) that would best explain the island community resilience (ICR). In the context of this study, a simple three-level hierarchical structure was developed (Figure 7.2). The top level in the hierarchy is the ultimate goal of the decision-making analysis process. The lowest level in the hierarchy contains the evaluation objects.

The resilience assessment focused on 23 basic criteria, namely human asset (skilled fishermen, agriculturist, livestock raiser, traders), physical asset (fishing craft and gear, landing centre, ice mill, house, dike/road), financial asset (cash income, remittances, saving, liquid assets, credit), natural asset (fisheries, cropland, grassland, mangroves) and social asset (fishermen association, marketing network, social harmony, good governance, education). Ratings are systematically scored according to Saaty (1977) on a 17-point continuous scale from 1/9 (least important) to 9 (most important). Weights have been given according to the effectiveness of the criteria. The weight for each criterion was determined by pair-wise comparisons in the context of a decision-making process known as the analytical hierarchy process (Saaty, 1977, 1990), which was also recommended by Kovacs et al. (2004) and Hossain et al. (2009a). The individual weights and consistency measures were computed and the results summarized by computing the overall weights (Table 7.1). The consistency ratios (CR) of 0.0087 to 0.0817 for the table was within the ratio of equal to or less than 0.10 recommended by Saaty (1977), signifying a small probability that the weights were developed by chance.

## 7.3 Results

### *7.3.1 Spatial distribution of mangrove forest*

The thematic classes derived from maximum likelihood, parallelepiped and isodata classification produced 90%, 88% and 85% accuracy, respectively (Table 7.2). The accuracy of the image classification error matrix output the values of 0.87 and 0.96 for the Kappa

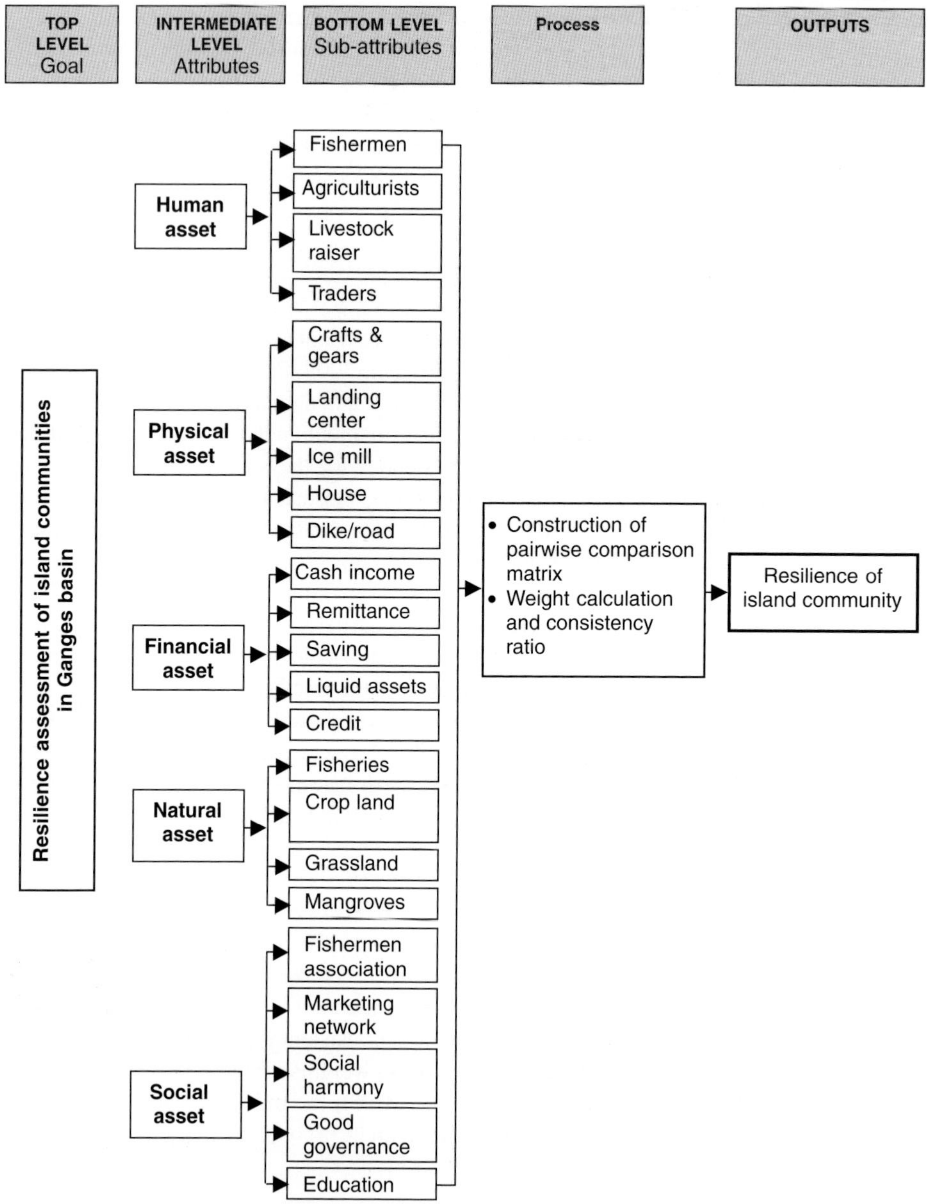


**Figure 7.2** The adapted Analytical Hierarchy Process (AHP) approach for resilience assessment of island communities in the Ganges basin.

**Table 7.1** A pair-wise comparison matrix for assessing the relative importance of different criteria for ICR in the Ganges basin (numbers show the rating of the row factor relative to the column factor).

| | Fishermen | Agriculturists | Livestock raiser | Traders | | Weight |
|---|---|---|---|---|---|---|
| *Human asset* | | | | | | |
| Fishermen | 1 | 1/2 | 1/5 | 1/7 | | 0.063 |
| Agriculturists | 2 | 1 | 1/2 | 1/4 | | 0.131 |
| Livestock raiser | 5 | 5/2 | 1 | 3/4 | | 0.322 |
| Traders | 8 | 4 | 3/2 | 1 | | 0.484 |
| $\lambda_{max} = 4.1240$, CI $= 0.0413$, CR $= 0.0459$ | | | | | | |

| | Crafts & gears | Landing centre | Ice mill | House | Dike/road | Weight |
|---|---|---|---|---|---|---|
| *Physical asset* | | | | | | |
| Crafts & gears | 1 | 1/2 | 1/4 | 2 | 3/4 | 0.112 |
| Landing centre | 2 | 1 | 1/2 | 4 | 3/2 | 0.224 |
| Ice mill | 4 | 2 | 1 | 8 | 3 | 0.448 |
| House | 1/2 | 1/4 | 1/7 | 1 | 1/2 | 0.061 |
| Dike/road | 3/2 | 3/4 | 1/3 | 5/2 | 1 | 0.155 |
| $\lambda_{max} = 5.1242$, CI $= 0.0311$, CR $= 0.0277$ | | | | | | |

| | Cash income | Remittance | Saving | Liquid asset | Credit | Weight |
|---|---|---|---|---|---|---|
| *Financial asset* | | | | | | |
| Cash income | 1 | 1/5 | 1/4 | 1/2 | 3/2 | 0.078 |
| Remittance | 5 | 1 | 5/4 | 5/2 | 8 | 0.394 |
| Saving | 4 | 3/4 | 1 | 2 | 6 | 0.307 |
| Liquid asset | 2 | 1/2 | 1/2 | 1 | 3 | 0.163 |
| Credit | 3/4 | 1/9 | 1/5 | 1/2 | 1 | 0.058 |
| $\lambda_{max} = 5.1823$, CI $= 0.0456$, CR $= 0.0407$ | | | | | | |

| | Fisheries | Agriculture land | Grassland | Mangroves | | Weight |
|---|---|---|---|---|---|---|
| *Natural asset* | | | | | | |
| Fisheries | 1 | 1/2 | 1/2 | 3 | | 0.185 |
| Agriculture land | 2 | 1 | 1 | 6 | | 0.370 |
| Grassland | 2 | 1 | 1 | 6 | | 0.370 |
| Mangroves | 1/2 | 1/5 | 1/5 | 1 | | 0.075 |
| $\lambda_{max} = 4.2205$, CI $= 0.0735$, CR $= 0.0817$ | | | | | | |

| | Fishermen association | Marketing network | Social harmony | Good governance | Education | Weight |
|---|---|---|---|---|---|---|
| *Social asset* | | | | | | |
| Fishermen association | 1 | 1/2 | 5/2 | 3/2 | 1/2 | 0.164 |
| Marketing network | 2 | 1 | 5 | 3 | 1 | 0.327 |
| Social harmony | 1/3 | 1/5 | 1 | 1/2 | 1/5 | 0.061 |
| Good governance | 3/4 | 1/4 | 2 | 1 | 1/2 | 0.121 |
| Education | 2 | 1 | 5 | 3 | 1 | 0.327 |
| $\lambda_{max} = 5.0401$, CI $= 0.0100$, CR $= 0.0089$ | | | | | | |

**Table 7.1** (*Continued*)

| | Human | Physical | Financial | Natural | Social | Weight |
|---|---|---|---|---|---|---|
| *Overall* | | | | | | |
| Human asset | 1 | 1/4 | 3/2 | 2 | 3/2 | 0.145 |
| Physical asset | 4 | 1 | 6 | 8 | 6 | 0.582 |
| Financial asset | 2/3 | 1/5 | 1 | 4/3 | 1 | 0.101 |
| Natural asset | 1/4 | 1/9 | 3/4 | 1 | 3/4 | 0.063 |
| Social asset | 3/4 | 1/6 | 3/2 | 4/3 | 1 | 0.109 |
| $\lambda_{max} = 5.0390$, CI $= 0.0098$, CR $= 0.0087$ | | | | | | |

and Tau coefficient, respectively at 95% confidence. A total of 27 014 ha mangrove forest was identified and the spatial mapping has clearly indicated the location and extent of the mangrove forest in the islands (Figure 7.3). Most of the mangrove forests in the islands are distributed in the intertidal zone, with the range of elevation 0–1 m during low tide.

The predominant land cover classes were identified as agriculture land, grassland, mangrove forest, secondary succession, bare substratum, and canal/creek. After validation of the classification, the area covered by each of the selected categories was estimated (Table 7.3). The selected islands of study are mainly characterized by agriculture, representing almost 68% of the total study area. The second and third largest land coverage was bare substratum (15%) and mangrove forest (7%) respectively, with the remaining coverage's amounting 2–6% each. In 131 sites, the classification map provides accurate figures of land cover according to field verification, whereas only 15 sites were found to have improper land cover. Thus, 90% of the classification output matched with the field verification. User accuracy (UA) and producer accuracy (PA) for each of the classes showed that Bas (bare substratum) had the lowest PA (0.73), although its UA was as high as 0.94. Agriculture was well discriminated from the rest of the classes (PA = 1.00 and UA = 0.94). The mangrove forest was highly represented in the sampling (42/146) and UA reached 0.95, and the PA had values above 0.90 (Table 7.4). A value of 1 indicates that the two outputs are in perfect agreement (no change has occurred), whereas if the two outputs are completely different from one another, then the Kappa value is −1. The Kappa (K) and Tau (T) coefficient had the values of 0.87 and 0.96 respectively at 95% confidence. We concluded that about 90% of the pixels were classified correctly, better than would be expected by a completely random classification.

Interviewees readily distinguished the three major taxa occupying in the mangrove forest common to these islands. *Sonneratia, Avicennia* and *Excoecaria* were all locally identified as 'keora', 'baen', and 'gewa' respectively. All interviewees identified *Sonneratia* and *Avicennia* as the two most abundant genera in the coastal islands with an average

**Table 7.2** Summary of overall accuracy of Landsat TM image classification for mangrove vegetation.

| Band combination | Unsupervised (%) | Supervised (%) | |
|---|---|---|---|
| | Isodata | Parallelepiped | Maximum likelihood |
| 2,3,4 | 84.25 | 87.29 | 89.43 |
| 3,4,5 | 85.01 | 88.11 | 89.73 |
| 7,3,4 | 83.71 | 87.04 | 88.76 |
| 7,3,5 | 86.52 | 89.52 | 90.92 |
| Overall | 84.87 | 87.99 | 89.71 |

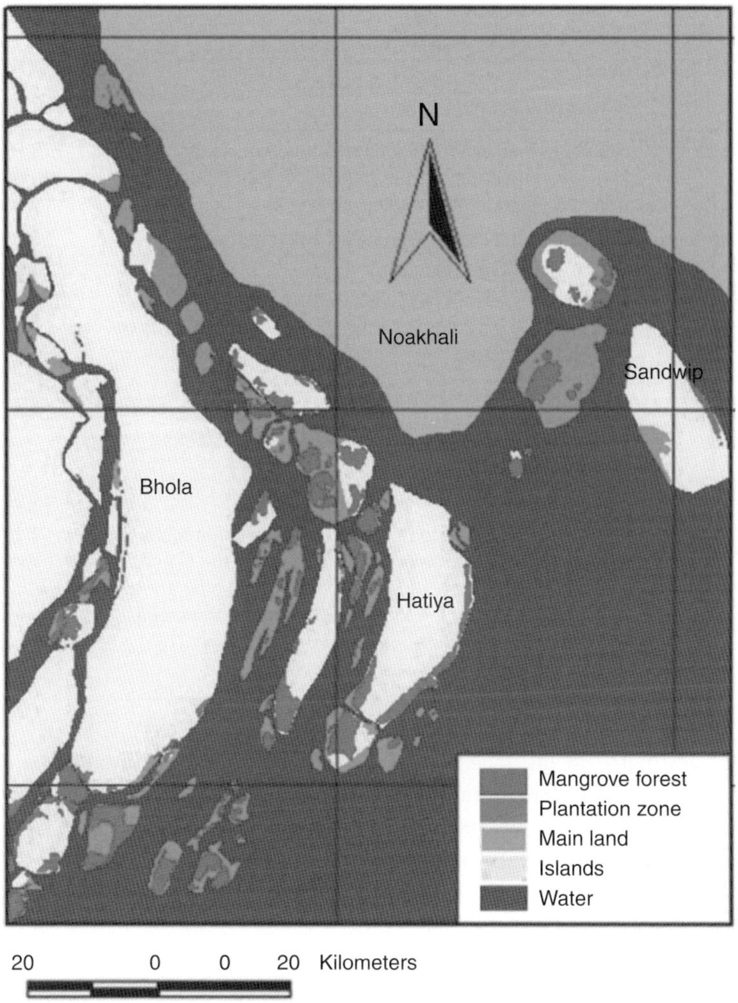


**Figure 7.3** Spatial distribution of mangrove forest and accreted stable zone in the Ganges basin from the supervised classification of the 2007 Landsat TM image. (see Plate 7.3 for colour version)

proportional estimate of 94% and 5%, respectively. 'These species grow well on new accretion areas with regular inundation' they added. The spatial mapping indicates the location and extent of about 60 000 ha accreted and marshy bare substratum (Figure 7.3) that can be planted with mangrove.

## *7.3.2 Resilience of island community*

The use of five attributes (human, physical, financial, natural and social assets) provides an integrated approach for island community resilience assessment (ICRA). Specifically, findings of pairwise comparison matrix (Table 7.1) provide informative examples of how islanders perceive that weight of particular sub-attributes impact on their capacity to build

**Table 7.3** Land cover categories used in the Landsat TM image supervised classification and resulting area (total and %).

| Land cover categories | Description | Area (ha) | % |
|---|---|---|---|
| Agriculture land | Soils with intensive agricultural production. Paddy cultivation during monsoon months. Part of the land used for vegetables in winter months | 276 973 | 68 |
| Bare substratum | Low vegetated and un-vegetated salt marshes and newly accreted land | 59 231 | 15 |
| Mangrove forest | Hydrophilic and halophytic vegetation, normally homogeneous, composed mainly by *Sonneratia* species and *Avicennia* species, located along the coastal belt and tidal flats | 27 014 | 7 |
| Secondary succession | Non-arboreous vegetation, halophytic species, and shrubs in areas further away from the canal/creeks, growing mainly on non-agriculture areas | 23 518 | 6 |
| Grassland | Natural grasslands and wet meadows | 11 923 | 3 |
| Canal/creek | Canals or creeks systems surface, navigational route of the islanders. Waterways which criss-cross the islands, and drain water from the surrounding lands, play a vital role in carrying runoff from agriculture and homestead sources | 8 136 | 2 |
| | **Total** | 406 795 | 100 |

and manage resilience in the face of climate change. The sub-attributes of mangrove, fishermen, house, credit and social harmony with 59, 55, 44, 44 and 42% weight respectively, have the highest role in ICRA at Ganges basin (Figure 7.4).

The interviewees ranked the mangroves as the most important natural asset. Common statements of the interviewees were 'the mangrove preserves life in the islands, we are all fishermen and get our food from there, it is our money-tree and life-tree'. Important

**Table 7.4** Error matrix corresponding to accuracy assessment for a supervised classification of Landsat TM image from the Ganges basin.

| Classification data | Reference data: Agr | Gra | Maf | Ses | Bas | Cac | Row totals | Producer's accuracy | User's accuracy |
|---|---|---|---|---|---|---|---|---|---|
| Agr | **33** | 2 | 0 | 0 | 0 | 0 | 35 | 1.00 | 0.94 |
| Gra | 0 | **15** | 4 | 0 | 0 | 0 | 19 | 0.88 | 0.79 |
| Maf | 0 | 0 | **40** | 2 | 0 | 0 | 42 | 0.91 | 0.95 |
| Ses | 0 | 0 | 0 | **16** | 6 | 0 | 22 | 0.89 | 0.73 |
| Bas | 0 | 0 | 0 | 0 | **16** | 1 | 17 | 0.73 | 0.94 |
| Cac | 0 | 0 | 0 | 0 | 0 | **11** | 11 | 0.92 | 1.00 |
| Column total | 33 | 17 | 44 | 18 | 22 | 12 | 146 | | |

Diagonal sum (bold) = 131; Overall accuracy = 0.897.
Kappa agreement = 0.872; Kendall's tau = 0.962.
Agr, agriculture land; Gra, grassland; Maf, mangrove forest; Ses, secondary succession; Bas, bare substratum; Cac, canal/creek

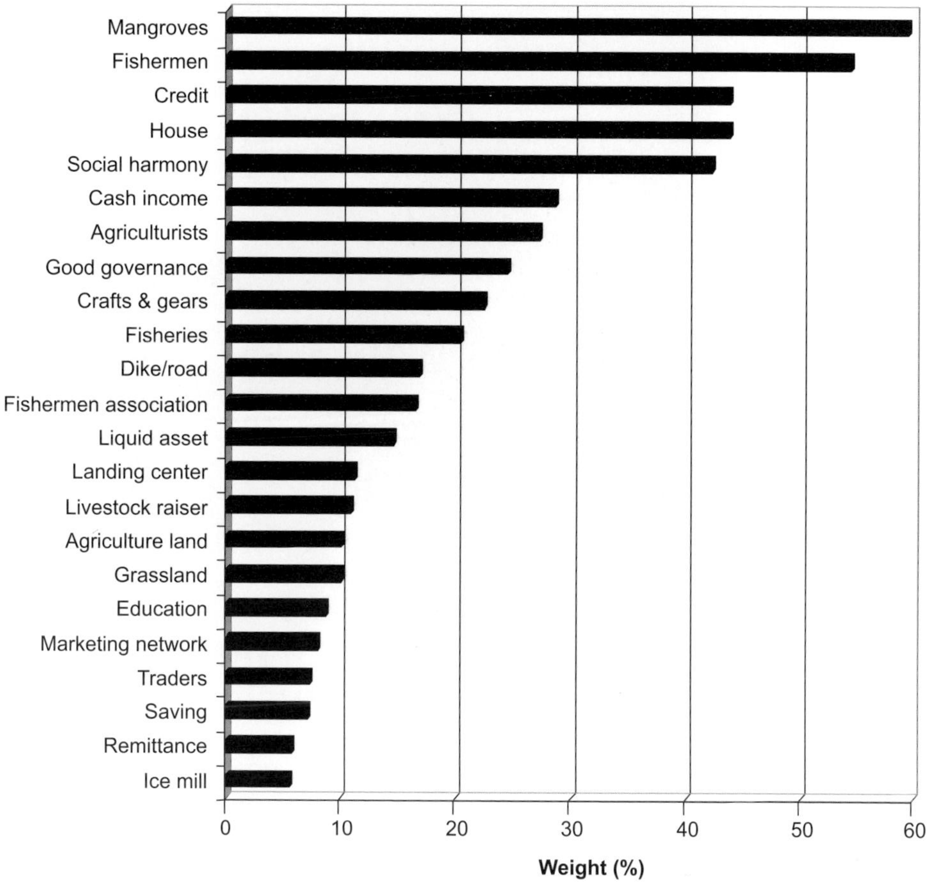


**Figure 7.4** Community resilience assessment scale shows the effectiveness of 23 sub-attributes in Ganges basin.

goods and services of mangroves upon which the local community can ensure their daily demand are shown in Figure 7.5. Fishing, shrimping and crabbing in the mangrove forest and adjacent area is an important source of income and livelihood option in the islands of Ganges basin. Some elderly fishermen reported that the structure of the mangrove trees enables them to withstand wave impacts and help to dissipate wave action from severe storms. Agriculture and livestock are the main land-use patterns on the islands. Cow and buffalo are kept for milk and meat, where milk indigenously preserves for sour card, ghol (indigenous drink) and ghee (clear butter for cooking) preparation to supply in city markets.

Considering the overall weights of all attributes, the results show that the natural asset is most significant to the islanders, and the vector of weights indicates that the physical asset is the least significant. The interviewees mentioned that most of the physical and natural assets are susceptible to cyclonic storm, that is mangroves uprooted, grassland covered with thick layer of sands, salinity intrusion in the crop land, loss of fishing crafts and gear,

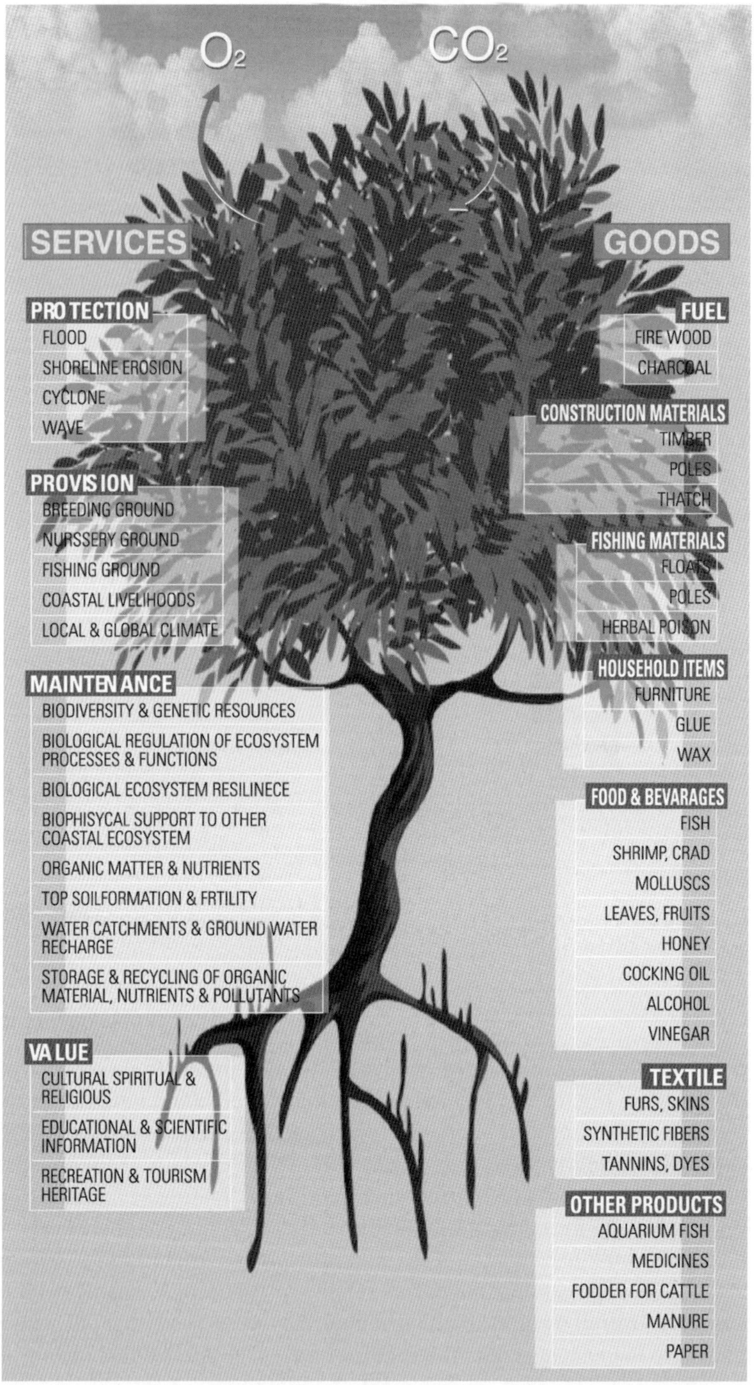


**Figure 7.5** The interviewees mentioned the productive and protective role of mangrove forest as goods and services in the Ganges Basin, Bangladesh.

and severe damage of house, road and dike. The equal weights associated with the social and financial assets of this model would suggest a similar significance to the islanders. The human asset on the islands with sufficient indigenous knowledge show professional performance in using other assets to maintain income and food security.

## 7.4 Discussion

In this study, mapping the mangrove forest in the coastal islands of Ganges basin was explored using both unsupervised and supervised classification techniques of Landsat TM data. There was a possible correlation between image processing effort and accuracy. The unsupervised image processing method is relatively rapid, requiring the operator to do little more than edit the final classes, but is the least accurate. The supervised classification procedure requires greater effort from the operator during the process of signature editing. As a result supervised classification requires the most effort, it is computationally intensive, but it is also cost-effective as the extra investment of time produces a significant increase in accuracy. The results reveal that higher classification accuracy is achieved with the supervised classification method which is in accordance with Trisurat et al. (2000) and Hossain et al. (2003). This may be due to the analyst's increased control in defining signatures for the classification decision rule (Joria and Jorgenson, 1996).

The identification of *Sonneratia apetala, Avicennia officinalis* and *Excoecaria agallocha* in Hatiya, Sandwip, Bhola and Noakhali corresponds with the finding of Siddiqi (2001). New alluvial lands are flat and found along tidal rivers, estuaries and coast, which are the ideal sites for *Sonneratia apetala.* The species prefers a higher degree of tidal inundation and grows over a wide range of salinity (Hossain and Lin, 2001). *Avicennia officinalis* seems to have preference for higher salinity, somewhat coarser soils and a lower degree of tidal inundation. The sign of stability of the newly accreted mudflat is the growth of natural grasses such as *Oryza coarctata* and *Myrostachya wightiana* along with some halophytic plants (*Suaeda nudiflora*) (Hossain et al., 2003).

The protective benefits of mangrove forest against tropical cyclone and wave action is important and well-recognized (Hossain et al., 2003, 2008; Barbier et al., 2008). The importance of mangroves as nursery grounds for the larval and juvenile stages of fin fishes, shrimps, crabs, and cockles has been highlighted by many researchers around the world (Khan and Hossain, 1996; Hossain, 2001; Glaser, 2003; Lee, 2004; Hossain et al., 2009b). The annual economic value of mangroves, using the cost of the products and services they provide, has been estimated to be \$200 000—900 000 per ha (Wells, 2006). To maintain a balance between social and economic aspects, an integrated resource management approach, for example mangrove-nursery-livestock or mangrove-livestock-pond-agriculture, may be examined (Hossain and Das, 2010). Within the field of climate change research, resilience studies have made an important contribution. The coastal zone and islands are regularly visited by cyclone, tidal surge and salinity intrusion (Hossain et al., 2009b). Each of the 23 sub-attributes can contribute to the overall resilience index, and those contributions can change over time as different scenarios unfold over the course of this century at different locations. The resilience assessment scale can serve as a powerful tool for the comparative analysis of livelihood dependency of the island community.

## 7.5 Conclusion

It is essential to keep the mangrove forest undisturbed to attract the mangrove-dependent species to enhance biodiversity. Newly accreted lands have to be planted through a community-based, production-sharing approach. The establishment of a well-developed mangrove forest zone along the coast can ensure different goods and services for the local community. Ensuring the balance between the proper assessment of products and the dependency of the local community, the mangrove forest can develop with extra income generating options. The health of the mangrove forest also affects the health of other economically and biologically important ecosystems. Moreover, the well-developed mangrove forest can act as bio-shield to mitigate the effect of climate change disasters. In this way, the local community may get diversified benefit through the conservation of mangrove forest as well as plantation in the newly accreted stable coastal lands.

## References


Barbier, E.B., Koch, E.W., Silliman, B.R., Hacker, S.D., Wolanski, E., Primavera, J., Granek, E.E., Polasky, S., Aswani, S., Cramer, L.A., Stoms, D.M., Kennedy, C.J., Bael, D., Kappel, C.V., Perillo, G.M.E. and Reed, D.J. (2008) Coastal ecosystem-based management with nonlinear ecological functions and values. *Science* **3**, 321–323.

Congalton, R.G. and Green, K. (1999) *Assessing the Accuracy of Remotely Sensed Data: Principles and Practices*. Boca Raton: Lewis Publishers.

Glaser, M. (2003) Interrelations between mangrove ecosystem, local economy and social sustainability in Caeta Estuary, North Brazil. *Wetlands Ecology and Management* **11**, 265–272.

Green, E.P., Mumby, P.J., Edwards, A.J. and Ur-Clark, C.D. (1996) A review of remote sensing for tropical coastal resources assessment and management. *Coastal Management* **24**, 1–40.

Green, E.P., Clark, C.D., Mumby, P.J., Edwards, A.J. and Ellis, A.C. (1998) Remote sensing techniques for mangrove mapping. *International Journal of Remote Sensing* **19**, 935–956.

Hossain, M.S. (2009) *Climate Change Resilience by Mangrove Ecosystem*. PRDI, Dhaka, Bangladesh, 33 pp.

Hossain, M.S. (2008) Land suitability analysis for sustainable aquaculture development in Noakhali coast. PhD dissertation, University of Chittagong, Bangladesh.

Hossain, M.S. (2001). Biological aspects of the coastal and marine environment of Bangladesh. *Ocean & Coastal Management* **44**, 261–282.

Hossain, M.S. and Das, N.G. (2010) GIS-based multi-criteria evaluation to land suitability modelling for giant prawn (*Macrobrachium rosenbergii*) farming in Companigonj Upazila of Noakhali, Bangladesh. *Computers and Electronics in Agriculture* **70**, 172–186.

Hossain, M.S. and Lin, C.K. (2001) Land Use Zoning for Integrated Coastal Zone Management: Remote Sensing, GIS and RRA Approach in Cox's Bazar Coast, Bangladesh. ITCZM Publication Series, No. 3, Integrated Tropical Coastal Zone Management, Asian Institute of Technology, Bangkok, Thailand, 25 pp.

Hossain, M.S., Lin, C.K., Tokunaga, M. and Hussain, M.Z. (2003) Remote sensing and GIS application for suitable mangrove afforestation area selection in the coastal zone of Bangladesh. *Geocarto International* **18**, 61–65.

Hossain, M.S., Khan, Y.S.A., Chowdhury, S.R., Kashem, M.B. and Jabbar, S.M.A. (2004) Environmental and socio-economic aspects: a community based approach from Chittagong Coast, Bangladesh. *Jahangirnagar University Journal of Science* **27**, 155–176.

Hossain, M.S., Chowdhury, S.R., Das, N.G. and Rahaman, M.M. (2007) Multi criteria evaluation approach to GIS-based land suitability classification for tilapia farming in Bangladesh. *Aquaculture International* **1**, 425–443.

Hossain, M.S., Sam, W. and Shamsuddoha, M. (2008) Care mangrove forest care coastal people. Institute of Marine Sciences and Fisheries, Chittagong University, Bangladesh.

Hossain, M.S., Chowdhury, S.R., Das, N.G., Sharifuzzaman, S.M. and Sultana, A. (2009a) Integration of GIS and multicriteria decision analysis for urban aquaculture development in Bangladesh. *Landscape and Urban Planning* **90**, 119–133.

Hossain, M.S., Wong, S., Chowdhury, M.Z.R. and Shamsuddoha, M. (2009b) Remote sensing and GIS application to mangrove forest mapping in the Meghna Deltaic Islands of Bangladesh. *Bangladesh Journal of Marine Sciences and Fisheries* **1**, 81–96.

IIRR (1998) Participatory methods in community-based coastal resource management. International Institute of Rural Reconstruction, Silang, Cavite, Philippines.

Joria, P.E. and Jorgenson, J.C. (1996) Comparison of three methods for mapping Tundra with Landsat digital data. *Photogrammetric Engineering and Remote Sensing* **61**, 321–327.

Khan, Y.S.A. and Hossain, M.S. (1996) Impact of shrimp culture on the coastal environment of Bangladesh. *International Journal of Ecology and Environmental Sciences* **22**, 145–158.

Khan, Y.S.A., Hossain, M.S., Hossain, S.M.G.A. and Halimuzzaman, A.H.M. (1998) An environmental assessment of trace metals in the Ganges-Brahmaputra-Meghna estuary. *Journal of Remote Sensing and Environment* **2**, 103–117.

Kovacs, J.M. (1999) Assessing mangrove use at the local scale. *Landscape and Urban Planning* **43**, 201–208.

Kovacs, J.M., Vandenberg, C.V., Wang, J. and Flores-Verdugo, F. (2008) The use of multipolarized spaceborne SAR backscatter for monitoring the health of a degraded mangrove forest. *Journal of Coastal Research* **24**, 248–254.

Kovacs, J.M., Malczewski, J. and Flores-Verdugo, F. (2004) Examining local ecological knowledge of hurricane impacts in a mangrove forest using an analytical hierarchy process (AHP) approach. *Journal of Coastal Research* **20**, 792–800.

Lee, S.Y. (2004) Relationship between mangrove abundance and tropical prawn production: a re-evaluation. *Marine Biology* **145**, 943–949.

Ma, Z. and Redmond, R.L. (1995) Tau coefficients for accuracy assessment of classification of remote sensing data. *Photogrammetric Engineering and Remote Sensing* **61**, 435–439.

Manson, F.J., Loneragan, N.R., McLeod, I.M. and Kenyon, R.A. (2001) Assessing techniques for estimating the extent of mangroves: topographic maps, aerial photographs and Landsat TM images. *Marine & Freshwater Research* **52**, 787–792.

Mausel, P., Wu, Y., Li, Y., Moran, E.F. and Brondizio, E.S. (1993) Spectral identification of successional stages following deforestation in the Amazon. *Geocarto International* **4**, 61–71.

Pido, M.D. (1995) The application of rapid rural appraisal techniques in coastal resources planning: experience in Malampaya Sound, Philippines. *Ocean and Coastal Management* **26**, 57–72.

Research Systems Inc. (2000a) ENVI user's guide. ENVI version 34, Research Systems Inc., USA.

Research Systems Inc. (2000b) *Exploring ENVI, Training Course Manual*. Better Solutions Consulting Limited, Liability Company, USA.

Rollet, B. (1984) La ecologia de los manglares con referen cia especial a la base biologica para la ordencion sostenida, forestal y pesca. Food and Agriculture Organization of the United Nations, Rome.

Saaty, T.L. (1977) A scaling method for priorities in hierarchical structure. *Journal of Mathematical Psychology* **15**, 234–281.

Saaty, T.L. (1990) *The Analytic Hierarchy Process: Planning, Priority Setting, Resource Allocation*. RWS Publications, Pittsburgh.

Sader, S.A., Stone, T.A. and Joyce, A.T. (1990) Remote sensing of tropical forests: an overview of research and applications using non-photographic sensors. *Photogrammetric Engineering and Remote Sensing* **56**, 1343–1351.

Siddiqi, N.A. (2001) *Mangrove Forestry in Bangladesh.* Institute of Forestry and Environmental Sciences, University of Chittagong, Chittagong, Bangladesh.

Trisurat, Y., Eiumnoh, A., Murai, S., Hussain, M.Z. and Shrestha, R.P. (2000) Improvement of tropical vegetation mapping using a remote sensing technique: a case of Khao Yai National Park, Thailand. *International Journal of Remote Sensing* **21**, 2031–2042.

Valiela, I., Bowen, J.L. and York, J.K. (2001) Mangrove forests: One of the Worlds threatened major tropical environments. *BioScience* **51**, 807–815.

Wells, S. (2006). In the front line: Shoreline protection and other ecosystem services from mangroves and coral reefs. UNEP-WCMC, Biodiversity series 24.

Chapter 8

# Monitoring Humpback Whale (*Megaptera novaeangliae*) Behaviour in a Highly Urbanised Coastline: Gold Coast, Australia

*Jan-Olaf Meynecke, Silje Vindenes and Daniella Teixeira*

**Abstract**

The east coast of Australia experiences one of the world's largest annual humpback whale (*Megaptera novaeangliae*) migration, with an estimated 14 000 individuals in 2010. However, increasing coastal development is accelerating the environmental pressure on migrating marine megafauna. Consequently, solutions to better manage humpback whale presence in urbanised waters are required. We have developed a novel survey method that can be applied to operating whale watch vessels, better integrating the tourism industry into research and ultimately coastal management in urbanised coastal waters. Preliminary results from the first season of observation (May–November 2010) in the Gold Coast bay showed a successful survey return of over 500 individuals that included 14 286 behavioural state observations. The data were analysed in terms of most commonly observed behaviours, movement, pod size and composition. The numbers of mothers with calves were highest in September and October and both resting and feeding behaviours were documented, indicating the importance of the bay for these individuals. Our pilot study demonstrated that the benefits of whale watch, boat-based data collection can outweigh its limitations when strategically deployed and carefully analysed.

## 8.1 Introduction

Eastern Australian humpback whales migrate annually from their summer feeding grounds (December–March) in Antarctica to their tropical breeding grounds in the Coral Sea in winter (May–October). These large-scale migration patterns are well known and well documented (e.g. Dawbin, 1966; Garrigue et al., 2000; Jenner et al., 2001; Noad et al., 2008). However, the movement patterns along this migratory route, the degree of interchange,

*Global Challenges in Integrated Coastal Zone Management*, First Edition. Edited by Erlend Moksness, Einar Dahl and Josianne Støttrup.

smaller-scale habitat use and the factors that influence the choice of these areas are poorly understood (Castro et al., 2008). It is likely that these are driven by physical and biological factors that can be impacted by high-density coastal development as is experienced on the Gold Coast in south-east Australia. The humpback whale is listed as 'vulnerable' under the Australian *Environment Protection and Biodiversity Conservation Act 1999* (EPBC Act). The estimated worldwide abundance is believed to be greater than 60 000 individuals (Reilly et al., 2008) and the east Australian population is estimated to be 7000 $\pm$ 660 for 2004 with an annual rate of increase of 10.6 $\pm$ 0.5% for 1987–2004 (Noad et al., 2008), which would suggest about 14 000 individuals in 2010.

The shape of the Gold Coast inshore area functions similar to a bay in which the current reverses, resulting in calm waters particularly during October each year (Castelle et al., 2007). This makes the Gold Coast a potential resting spot, as documented by anecdotal reports. In 2009, an estimated 50 000 people partook in the whale watching season on the Gold Coast (Gold Coast, 2010). South-east Queensland is among the most populated areas in Australia with more than 2 million people and a yearly increase of up to 14% (ABS, 2010), making it one of the fastest growing regions in the world (Baum and O'Connor, 2005). This development consequently increases the pressure upon the adjoining marine environment through habitat loss, constructions, sand mining and pumping, shipping, noise and water pollution. This is increasing the rate of loss of biodiversity and abundance of near-shore marine species (Jackson et al., 2001).

The stretch of coastline within the Gold Coast region is one of the few major developed coastlines in the world where dugongs, whales and dolphins are still present (Chilvers et al., 2005). Despite this, the impacts of such development on marine megafauna, including whales, are poorly understood. As humpback whales frequent the shallow near-shore waters, it is likely that their susceptibility to disturbances from anthropogenic sources is high. Migrating humpback whales rely on stores for energy and thus the whales are particularly vulnerable to the energetic costs of continued disturbance. In order to improve the management of humpback whales in near-shore waters around Australia, information on the behavioural activities of humpback whales at aggregation sites is required in combination with collected long-term point observations. Three aggregation areas have been identified: (1) the southern end of the Great Barrier Reef, (2) Hervey Bay and (3) the Gold Coast region (DEH, 2005). The southern end of the Great Barrier Reef, in particular around Swine Reef, is a suspected calving ground (Chaloupka et al., 1999). Hervey Bay lies approximately 300 km north of the Gold Coast. Over 20 years of studies in Hervey Bay have revealed that 30–50% of the humpback whales divert from their southward migration and move into the bay from August to October (Paterson, 1991; Chaloupka et al., 1999). This enclosed bay formed by Fraser Island on the east site is a confirmed resting area, in particular for mothers with calves, most likely due to the calm and shallow waters (Franklin et al., 2010).

Here we aim to investigate the behaviour, pod sizes and compositions of 518 humpback whales observed in the Gold Coast bay during their northern and southern migration in 2010 and to compare the findings with the aggregation site in Hervey Bay. Our objective was to provide a first assessment of humpback whale behaviour in a recognised aggregation area along their migration route. It is hypothesised that the bay is used as a resting area for mothers with calves predominantly during the southern migration.

## 8.2 Methods

### *8.2.1 Study area*

The study area comprises the Gold Coast inshore waters from Point Lookout, North Stradbroke Island, in the north to Tweed Heads in the south and 15 nautical miles (nm) east of the foreshore including state and federal coastal waters (Figure 8.1). The Gold Coast inshore area can be described as a sickle/sense shaped coastal bay facing east to the Coral Sea. This region is wave dominated with an average deep-water significant wave height of approximately 1 m (0.8–1.4 m) (Mirfenderesk and Tomlinson, 2007, 2008). The Gold Coast climate is sub-tropical with a hot humid wet season (November–April) and a mild dry season (May–October). The region is influenced by easterly to south-easterly trade winds coming from the Pacific Ocean bringing moist, warm air and constant winds of varying degrees. Generally, windier conditions occur during the summer months and calmer conditions during the winter.

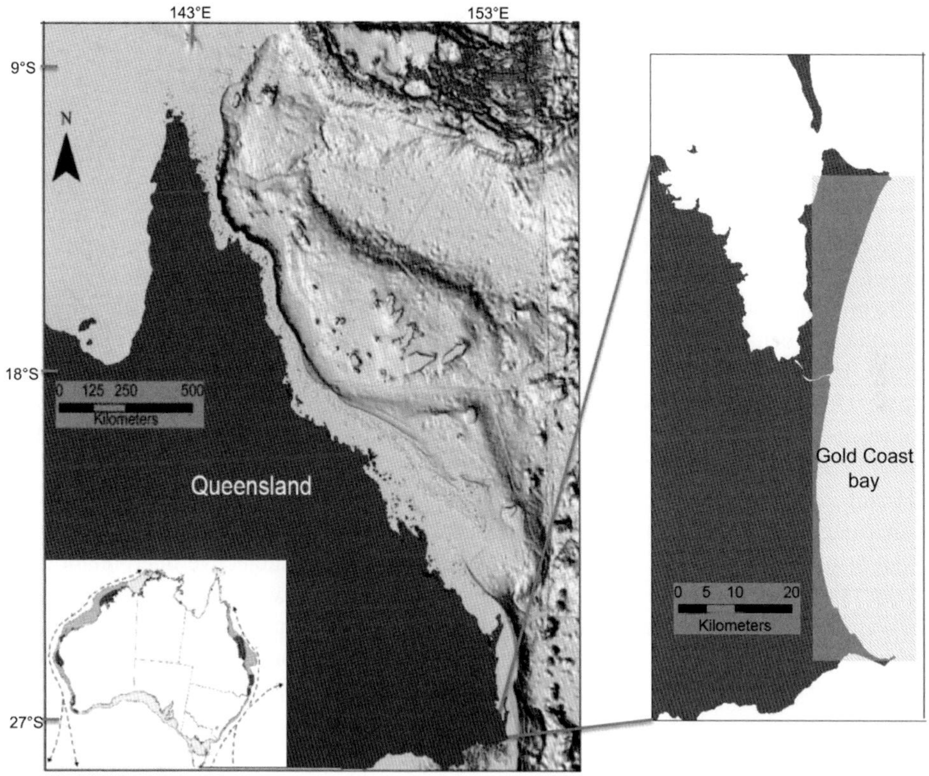


**Figure 8.1** Location of the Gold Coast bay on the eastern coast of Australia. The extent of the observation area is indicated by the grey box. The inserted map of Australia is showing migratory pathways of Australina humpback whale populations. Recognised aggreagtion sites are indicated in dark grey along the Australia coast (DEH, 2005).

### 8.2.2 Data collection

Surveys were conducted on board the commercial whale watch vessel '*Spirit of Migloo*', operated by Sea World Whale Watch. This catamaran is 24 m long, and 9 m wide with a 7 m high viewing platform. In accordance with the Australian EPBC Act, the vessel had to keep a distance of 100 m from whales and 300 m if three or more boats were present at the same whales or if the vessel was moving at a speed higher than 6 knots. When calves were present, the vessel had to stay a minimum of 300 m from the calf unless the calf moved such that the vessel was within 300 m, in which case the vessel had to be disengaged or moved out of the caution zone (300 m).

Surveys were undertaken for 5 days a week on two separate trips each day with duration of 2.5 h each. During these trips, the date, time, coordinates, bearing, pod size, calf presence and displayed behaviour/s were recorded. Maximum distance of the vessel from the shore was 15 nm. Locations of the humpback whale pods encountered were recorded by using the vessel's GPS system. The direction in terms of north, south, west and east of pod movement was assessed at the first sighting of a pod. The travel speed of the whales was estimated from the boat speed as an average value while the vessel was positioned alongside the whales as they travelled and changes recorded when the speed changed by more than 1 knot. Pod refers to a singleton and two or more humpback whales swimming side-by-side (Clapham, 1993; Corkeron et al., 1994). 'Adults' describes the number of whales in a pod that were not calves: however, this classification does not imply sexual maturity. Humpback whales under 10 m total length were considered juveniles and whales under 6 m long were considered calves when accompanied by a significantly larger whale with whom they maintained a constant and close relationship (Tyack and Whitehead, 1983), with the adult being assumed to be the mother (Clapham et al., 1999). The size of the whale was also compared to a known distance alongside the vessel.

Behavioural data were collected using a standardised field data sheet to ensure that the recording of behaviours was consistent between survey trips and observers (Mann et al., 2000). This sheet listed 21 different behavioural states frequently observed in humpback whales (Kaufman and Forestell, 1986; Isaacs and Dalton, 1992; Mann et al., 2000) and also collected information on blow per surfacing, travel speed (knots), dive time and resting time (Table 8.1).

Using a behavioural key, which described each of these behaviours, observed behaviours were categorically assigned as one of these behavioural states. Additionally, video footage of all displayed behaviours was recorded, which allowed for subjective interpretation between observers. Classifying any given observed behaviour was discussed between observers when necessary. Behaviours were observed and assigned categories by the same two observers who jointly undertook all field surveys throughout the season.

### 8.2.3 Statistical methods

To examine variation in pod size and composition, the whole sample frequencies and percentages of pod size were reported. In addition, we reported the pod size categories by 'calves present' or 'no calves present'. To further analyse the behavioural data, we reported the frequency of occurrence of each behaviour and investigated the differences between northern and southern migrations and between pods with and without calves

**Table 8.1** Description of 21 behavioural stages recorded during humpback whale observations.

| Common behaviour (> 100 counts) | Behaviour description | Less common behaviour (< 100 counts) | Behaviour description |
|---|---|---|---|
| Slip under | Whale submerges without rounding out and/or arching the peduncle | Mugging | Whale or whale pod stay within a few meters of a vessel |
| Peduncle arch | Peduncle arch appears at the surface (e.g. in attempt to dive deeper); | Pec wave | Whale extends one pectoral straight up while lying on its side at the surface or both pectoral fins are waved in the air while the whale lies on its back |
| Round out | Diving descent by arching its body (after the last inhalation); | Head slap | Whale propels half its body out of the water in a nearly perpendicular direction and hits the water surface with a pound |
| Fluke up dive | Tail flukes brought straight up into the air, exposing the entire ventral surface | Head rise | Head is brought up above the surface at a 45-90° angle (eye is generally not exposed) |
| Pec slap | Pectoral fins are slapped at the surface while whale rolls or lies on its side | Bubbling | Release of continuous, controlled amounts of air |
| Tail slap | Tail stock and fluke smacked forcefully on water surface | Tail slash | Whale strikes its tail in a side to side, slashing movement |
| Breach | Whale propels itself out of the water (clearing the surface with two-thirds of its body or more) | Tail cocking | Whale is lying upright in the water, caudal peduncle bent and slightly arched at its posterior extremity (flukes curled down) |
| Peduncle slap | Rear portion of the body, including the caudal peduncle and the flukes thrown up out of the water and then brought down sideways | Spy hop | Head is positioned vertically above water with eyes exposed |
| Fluke down dive | Flukes are brought clear of the water (ventral surface not exposed) | Tail extension | Flukes and caudal peduncle extended straight into the air |
| Head lunge | Head brought above surface while whale lunges forward fast | Motorboating | Whale swims rapidly at the surface with head above body parallel to the surface |
| | | Snaking | S-shaped postural display anterior portion of the head is angled out of the water (dorsal fin above the surface, peduncle arched) |

present. A one-way analysis of variance (ANOVA) using SPSS 19.0 was used to determine significant differences between observed behaviours between pods with calves present and pods without calves present.

## 8.3 Results

We recorded 518 individual humpback whales, and 14 286 behavioural state observations, between May and November 2010 in the Gold Coast bay. The first and last recorded whale sightings were on the 21 May 2010 and 4 November 2010, respectively. A total of 101 survey days were conducted for 5 days per week. Data were obtained on all survey trips. The total survey time was 400 h and observations of humpback whales were undertaken for a total of 142 h. The average rate of survey time (time spent on the water looking for humpback whales) per week was 16 h and the average observation time was about 6 h throughout the study period. Survey hours per week ranged from 2.5 to 35 h depending on weather and season (Figure 8.2).

Of the 21 recognised humpback whale behavioural states, 'slip under' was the most common accounting for nearly 50% of all observations. The next most commonly observed behavioural state was 'round out' which accounted for 10% of all observations. Eleven of the 21 behaviours were observed at a frequency of less than 1% of total recorded observations (Figure 8.3).

About 40% of all movement directions of pods recorded during the main northern migration (May–July) where in a northern direction. Movements during the main southern migration (September–November) were less directed, with 20% swimming directly towards the south and the rest in other directions.

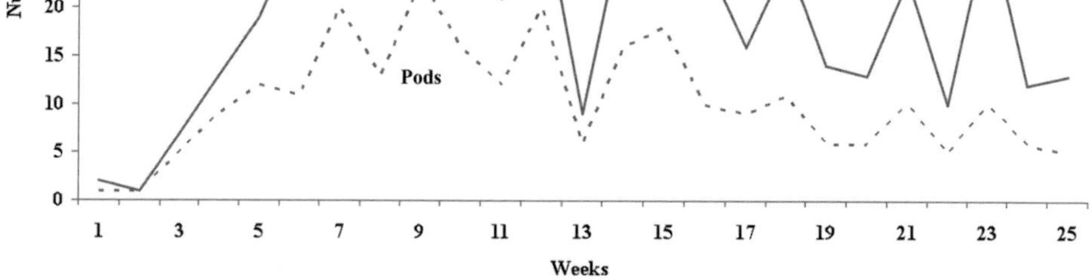


**Figure 8.2** Weekly observations of humpback whale pods, individual whales (a) and observation hours (b) from May to November 2010 at the Gold Coast bay. (*Continued*)

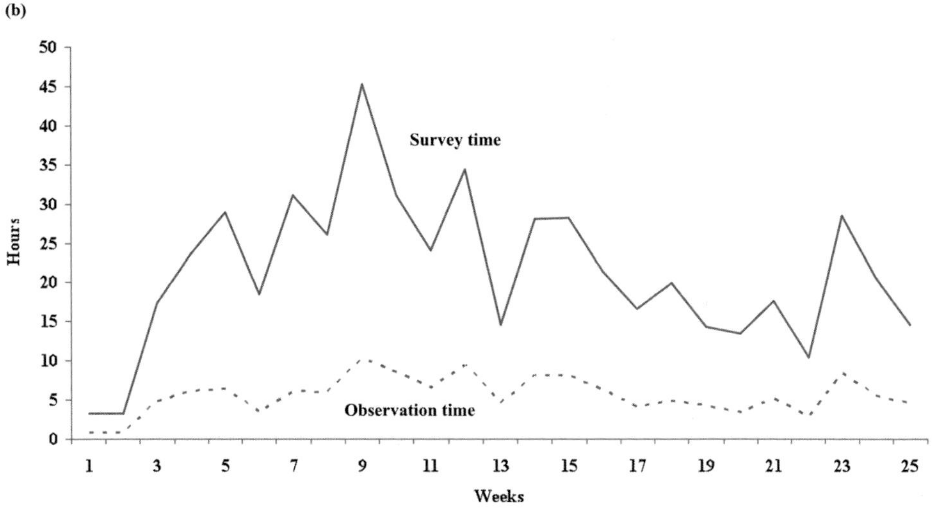


**Figure 8.2** (*Continued*)

## *8.3.1 Pod size and composition*

A total of 262 pods were observed over the duration of the migration season. Pod size ranged from one to six individuals, with a mean of 2.42. Approximately one-third of the pods were singletons (33%) and nearly half were pairs (45%), one-third of which were mother–calf pairs. Nearly a quarter of the pods (22%) consisted of three or more individuals (Figure 8.4,

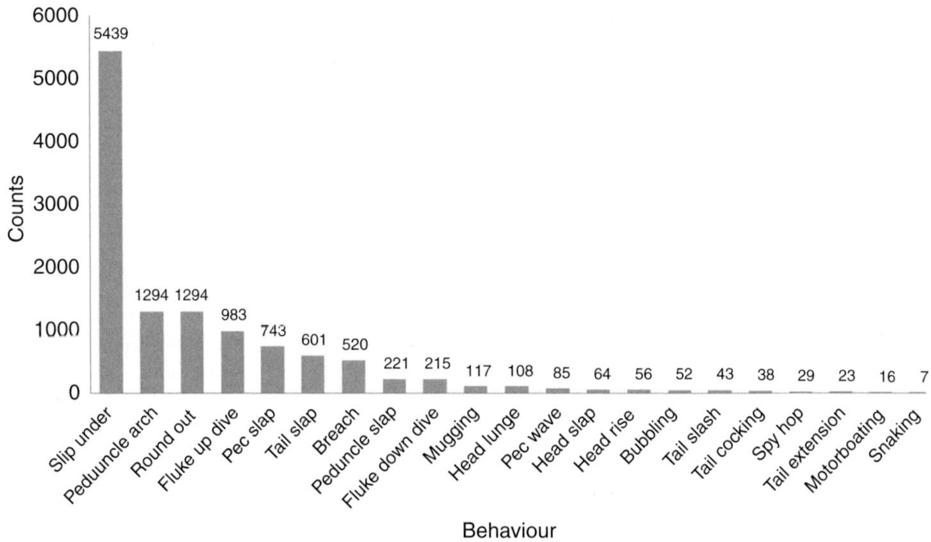


**Figure 8.3** Frequencies of 21 observed behavioural states.

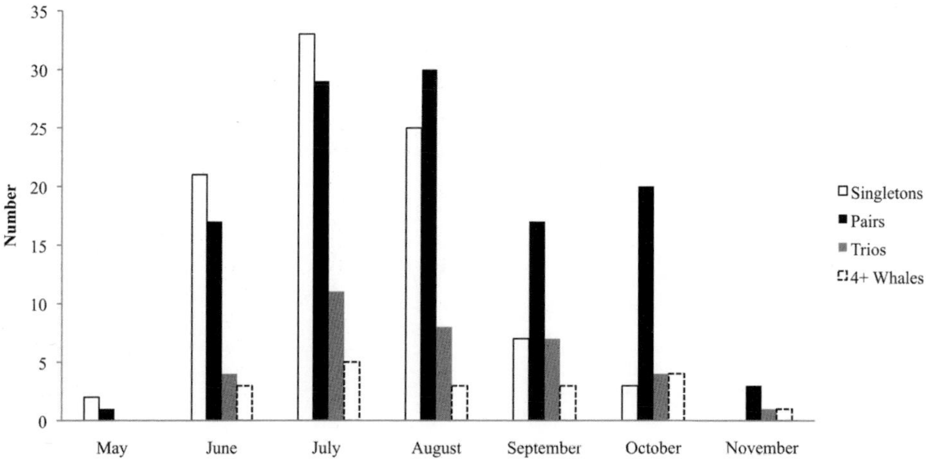


**Figure 8.4** Proportions of pod size over months.

Table 8.2). A total of 40 pods had calves (15%) and these pods predominantly occurred as mother–calf pairs. However, on some occasions an assumed escort was present and a single observation documented two mothers and two calves in a single pod.

## *8.3.2 Mothers with calves*

The first mother–calf pair was observed in late June (26/6/10). This was an early observation for a newborn calf and the sighting suggested that the calf may have been born during the northern migration, possibly only days prior to the sighting in the Gold Coast bay. The greatest number of calves was observed in October (38%), followed by August (21%) and September (19%). The behavioural state 'slip under' occurred significantly more frequently in mother–calf pairs than in pods without calves ($F(1,74) = 4.42$, $p < 0.01$). Breaching occurred significantly more frequently in mother-calf pods than in pods without calves ($F(1,501) = 6.52$, $p < 0.01$) and occurred significantly more frequently in calves than in mothers ($F(1,74) = 5.83$, $p < 0.02$).

The mean dive time of mothers and calves was 3.6 min and the mean travel was 2.35 knots. In contrast, the mean travel speed in pods without calves was 3.5 knots with a

**Table 8.2** Summary of number of whales in pods between May and November 2010 in the Gold Coast bay and number of whales in pods with no calves present and calves present.

| | Pods with no calves present | | Pods with calves present | |
|---|---|---|---|---|
| Individuals per pods | *n* | % | *n* | % |
| 1 | 91 | 40.5 | 0 | 0 |
| 2 | 90 | 40.0 | 27 | 73.0 |
| 3 | 28 | 12.4 | 7 | 18.9 |
| 4+ | 16 | 7.1 | 3 | 8.1 |
| Total | 225 | 100.0 | 37 | 100.0 |

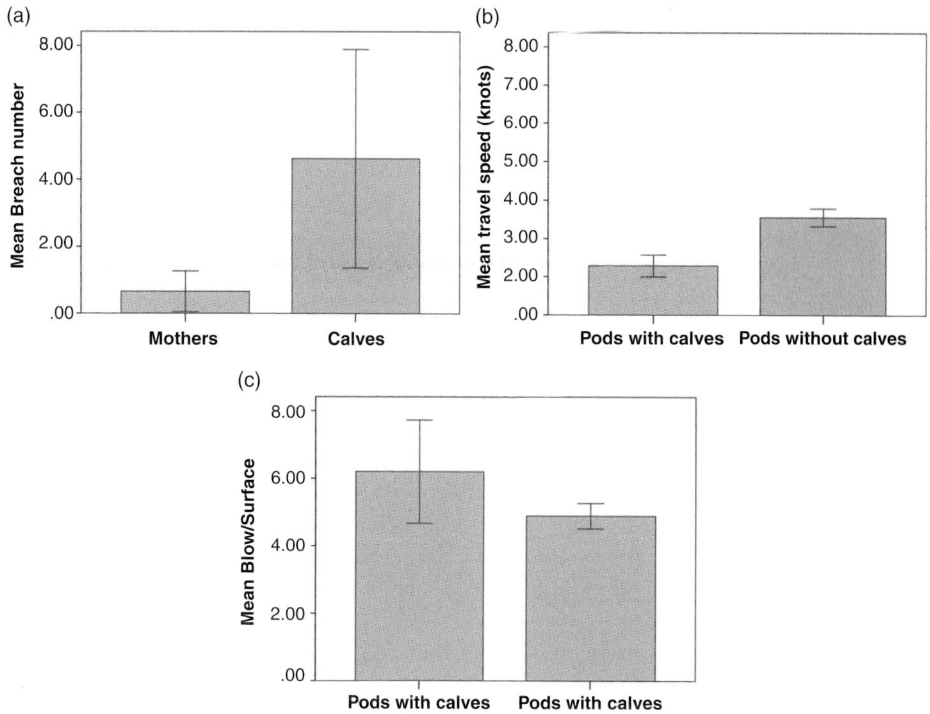


**Figure 8.5** ANOVA results for (a) comparison of breaching between mother and calves: comparison between pods with and without calves in regards to (b) mean travel speed; and (c) number of blows per surfacing.

similar average dive time. As expected, the mean blow/surfacing was significantly higher in mother–calf pods, at 6.2, than in pods without calves ($F(1{,}458) = 5.83$, $p < 0.02$), where the mean was 4.89. These observations are consistent with a significant decrease in swim speed in post-August (southern migration) observations ($F(1{,}360) = 27.08, p < 0.01$) (Figure 8.5).

## 8.4 Discussion

The analyses of the first year of continuous observations of humpback whales in the Gold Coast bay showed a large number of sightings (518 in 101 observation days), confirming the region as an aggregation area (DEH, 2005). Mother–calf pairs were mainly sighted from August until November inside the Gold Coast bay indicating that the bay-like environment of the Gold Coast waters may function as a resting spot. Additionally, we determined that calves frequently used the area for activities such as breaching. The observed pods were significantly slower when calves were present (Figure 8.5) with the general swimming speed being related to the direction of migration – faster during northern migration and slower during southern migration. This was also confirmed by the finding that the whales were moving on more direct paths during northern migration than during the southern migration.

The findings of this study are coherent with those long-term studies in Hervey Bay (Corkeron et al., 1994; Paterson et al., 1994; Dawbin, 1997; Chaloupka et al., 1999; Franklin et al., 2010). This bay is located at 25°S, 153°E. It is a wide, shallow, often sandy bay ($<$ 18 m deep) (Corkeron et al., 1994) and approximately 175 nm north of the Gold Coast bay. Hervey Bay has been recognised as a resting ground and previously as a breeding ground for humpback whales (Corkeron et al., 1994). The area is frequently visited by humpback whales during the southern migration for resting and socialising.

The average number of whales per pod sighted in Hervey Bay during an intensive survey period of 13 years was 2.26 (Franklin et al., 2010). This is in line with our first year of observation at the Gold Coast bay returning an average of 2.4 whales per pod. Pods with two whales present (pairs) were the most frequently observed at 45% (Hervey Bay, 57%) (Franklin et al., 2010) and one-third of all pods observed in the Gold Coast bay had one or more calves present (Hervey Bay, 40%) (Franklin et al., 2010). Proportionately, calves were present less frequently in Gold Coast pods than those in Hervey Bay. This may indicate lower numbers of mother–calf pairs in the Gold Coast bay. However, mother–calf pairs may particularly use the 3 nm zone (shallow $<$ 20 m and sandy bottom) for resting and this area was not covered by our surveys. It has been suggested that females with calves prefer shallower waters close to shore to minimize predation by sharks and/or to avoid harassment by males (Whitehead and Moore, 1982; Glockner and Venus, 1983; Mattila et al., 1989; Smultea, 1994), or as a function of social organisation (Ersts and Rosenbaum, 2003).

Similar to Hervey Bay, the proportion of pods with calves present increased and the number of singletons decreased towards the end of the year. High numbers of singletons were sighted in July, at which time the whales are finding their partners (Garrigue et al., 2001). The first calves observed in the Gold Coast bay occurred end of June. Therefore, calves accompanied by mothers may be between a few weeks to 3 months of age (Chittleborough, 1953; Chittleborough, 1958), supporting the hypothesis that the Gold Coast bay is not a calving ground, but rather a stopover for mothers with calves during the southern migration.

There are benefits of using whale watch operators as platforms of opportunity for observations. Data from whale watch vessels can provide information on distribution, without expensive survey effort (e.g. Weinrich et al., 2000). The frequency and intensity of survey opportunities for certain areas is often high and enables fine-scale analyses (Higham and Lück, 2008). Behavioural observations from land would allow more observers but also limit the time of observation of an individual whale. The whale watch boats often drift alongside the whales for 30–60 min with their engines turned off, providing an excellent opportunity to observe individual behaviour. However, the use of whale watch boats for behavioural studies is confounded by some factors. Observations of behaviours may not reflect the entire range of behaviours displayed in an area. When using boats as observation platforms, physical factors such as changing water properties that may influence whale presence and behaviour as well as the influence of wave height (visibility) on naked eye observation can be confounding factors. A detailed investigation into the effect of physical factors that may influence behaviour was not possible due to limited data points. It is also difficult to collect distribution and abundance information for a larger area (e.g. $>$ 100 km$^2$) from whale watch boats. There is a tendency of whale watch boats to continuously go to the same areas and they are restricted to remaining outside the 3 nm zone in Queensland, thereby providing biased information. For more detailed studies, it is proposed to use a combination of land and boat surveys and increase the number of observers as well as spatial coverage.

## 8.5 Conclusions

Information on humpback whale movement patterns and core activities in urbanised coastal waters can be collected with the assistance of the tourism industry. This is useful for estimating trends and to investigate areas of further research and may ultimately help to improve the management of humpback whales in urbanised coastlines. The whale watching industry can make valuable contributions to the understanding of cetacean populations and the number of whale watch operations actively collecting data should increase. Involving whale watch tour operators also allows for raising environmental awareness amongst the passengers (Higham and Carra, 2003). Our data on pod characteristics of humpback whales in the Gold Coast bay indicated that the bay provides an important habitat for whales and is frequently used by mother–calf pairs, as a temporary stopover during their southern migration to Antarctic feeding grounds. Similar behaviours were observed compared to the aggregation site in Hervey Bay confirming the importance of the Gold Coast bay as a habitat for humpback whales. However, site fidelity and total number of individuals using the Gold Coast bay require further investigation. Future studies in the Gold Coast bay should use measures of individual identification to estimate time of residence (Katona et al., 1979; Kniest et al., 2010) and undertake abundance and distribution estimations (Vigness-Raposa et al., 2010).

## Acknowledgements

The authors thank Whale Watch Australia/Sea World Whale Watch for supporting the data collection and the numerous people who made this field work possible, including David Robertson, Adrian Nelson and Nik Beezum. We also thank friends and colleagues and two anonymous reviewers for comments.

## References

ABS (2010) *Year Book Australia Statistics*. Australian Bureau of Statistics, Canberra.

Baum, S. and O'Connor, K. (2005) Regional Population and employment change in Australia 1991–2001: inertia in the face of rapid change? *GeoJournal* **62**, 85–94.

Castelle, B., Bourget, J., Molnar, N., Strauss, D., Deschamps, S. and Tomlinson, R. (2007) Dynamics of a wave-dominated tidal inlet and influence on adjacent beaches, Currumbin Creek, Gold Coast, Australia. *Coastal Engineering* **54**, 77–90.

Castro, C., Aceveco, J., Allen, J., Dalla Rosa, L., Florez-Gomzalez, L., Aguaya-Lobo, A., Rasmussen, K., Llano, M., Garita, F., Forestell, P., Secchi, E.R., Godos, I.G., Ferrina, D., Kaufman, G., Scheidat, M. and Pastene, L.A. (2008) Migratory movements of humpback whales (*Megaptera novaeangliae*) between Machalilla National Park, Ecuador and Southeast Pacific. International Whaling Commission report SC/60/SH23 pp. 6.

Chaloupka, M., Osmond, M. and Kaufman, G. (1999) Estimating seasonal abundance trends and survival probabilities of humpback whales in Hervey Bay (east coast Australia). *Marine Ecology-Progress Series* **184**, 291–301.

Chilvers, B.L., Lawler, I.R., Macknight, F., Marsh, H., Noad, M. and Paterson, R. (2005) Moreton Bay, Queensland, Australia: an example of the co-existence of significant marine mammal populations and large-scale coastal development. *Biological Conservation* **122**, 559–571.

Chittleborough, R.G. (1953) Aerial observations on the humpback whale, *Megaptera nodosa* (Bonnaterre), with notes on other species. *Australian Journal of Marine and Freshwater Research* **4**, 219–228.

Chittleborough, R.G. (1958) The breeding cycle of the female humpback whale, *Megaptera nodosa* (Bonnaterre). *Australian Journal of Marine and Freshwater Research* **9**, 1–20.

Clapham, P.J. (1993) Social organization of humpback whales on a North Atlantic feeding ground. *Symposia of the Zoological Society* **66**, 131–145.

Clapham, P. J., Wetmore, S.E., Smith, T.D. and Mead, J.G. (1999) Length at birth and at independence in humpback whales. *Journal of Cetacean Research and Management* **1**, 141–146.

Corkeron, P.J., Brown, M., Slade, R.W. and Bryden, M.M. (1994) Humpback whales, *Megaptera-novaeangliae* (Cetacea, Balaenopteridae), in Hervey Bay, Queensland. *Wildlife Research* **21**, 293–305.

Dawbin, W.H. (1966) The seasonal migratory cycle of humpback whales. In: Norris, K.S. (Ed.). *Whales, Dolphins and Porpoises*. University of California Press, Berkeley, 145–170.

Dawbin, W.H. (1997) Temporal segregation of humpback whales during migration in Southern Hemisphere waters. *Memoirs of the Queensland Museum* **42**, 105.

DEH (2005) *Humpback Whale Recovery Plan*. Department of Environment, Commonwealth of Australia, Canberra, 11 pp.

Ersts, P.J. and Rosenbaum, H.C. (2003) Habitat preference reflects social organization of humpback whales (*Megaptera novaeangliae*) on a wintering ground. *Journal of Zoology* **260**, 337–345.

Franklin, T., Franklin, W., Brooks, L., Harrison, P., Baverstock, P. and Clapham, P.J. (2010) Seasonal changes in pod characteristics of eastern Australian humpback whales *(Megaptera novaeangliae)*, Hervey Bay 1992–2005. *Marine Mammal Science* **27**, 134–152.

Garrigue, C., Forestell, P.H., Greaves, J., Gill, P., Naessig, P.J., Patenaude, N.M. and Baker, C.S. (2000) Migratory movements of humpback whales (*Megaptera nevaeangliae*) between New Caledonia, East Australia and New Zealand. *Journal of Cetacean Research and Management* **2**, 111–115.

Garrigue, C., Greaves, J. and Chambellant, M. (2001) Characteristics of the New Caledonian humpback whale population. *Memoirs of Queenland Museum* **47**, 539–546.

Glockner, D.A. and Venus, S.C. (1983) Identification, growth rate and behavior of humpback whale (*Megaptera novaeangliae*) cows and calves in the waters off Maui, Hawaii, 1977–1979. In: Payne, R.S. (Ed.). *Communication and Behavior of Whales*. Westview Press, Boulder, CO, 223–258.

Gold Coast (2010) Whale Watching at the Gold Coast. Available at http://www.goldcoast.com.au/article/2009/06/02/84415_gold-coast-lead-story.html (accessed 10 September 2012).

Higham, J.E.S. and Carra, A.M. (2003) Sustainable wildlife tourism in New Zealand: An analysis of visitor experiences. *Human Dimensions of Wildlife* **8**, 25–36.

Higham, J.E.S. and Lück, M. (2008) Marine Wildlife and Tourism Management: In search of a scientific approach to sustainability. In: Higham, J.E.S. and Lück, M. (Eds), *Marine Wildlife and Tourism Management*. Biddles Ltd, King's Lynn, 1–19.

Isaacs, R. and Dalton, T. (1992) *The Australian Guide To Whale Watching*. Weldon, Sydney, 96 pp.

Jackson, R.B., Carpenter, S.R., Dahm, C.N., McKnight, D.M., Naiman, R.J., Postel, S.L. and Running, S.W. (2001) Water in a changing world. *Ecological Applications* **11**, 1027–1045.

Jenner, K.C., Jenner, M.-N.M. and McCabe, K.A. (2001) Geographical and temporal movements of humpback whales in western Australian waters. *Australian Petroleum Production & Exploration Association Journal* **38**, 692–707.

Kaufman, G.D. and Forestell, P.H. (1986) *Hawaii's Humpback Whales: A Complete Whale Watchers Guide*. Pacific Whale Foundation Press, Hawaii, 176 pp.

Katona, S., Baxter, B., Brazier, O., Kraus, S., Perkins, J. and Whitehead, H. (1979) Identification of humpback whales by fluke photographs. In: Winn, H.E. and Olla, B.L. (Eds), *Cetaceans Behavior of Marine Animals: Current Perspectives in Research*. Plenum, New York, **3**, 33–44.

Kniest, E., Burns, D. and Harrison, P., (2010) Fluke matcher: A computer-aided matching system for humpback whale (*Megaptera novaeangliae*) flukes. *Marine Mammal Science* **26**, 744–756.

Mann, J., Connor, R.C., Tyack, P.L. and Whitehead, H. (2000) *Cetacean Societies Field Studies of Dolphin and Whales*. University of Chicago Press, London, 435 pp.

Mattila, D.K., Clapham, P.J., Katona, S.K. and Stone, G.S. (1989) Population composition of humpback whales, *Megaptera-novaeangliae,* on Silver Bank, 1984. *Canadian Journal of Zoology-Revue Canadienne De Zoologie* **67**, 281–285.

Mirfenderesk, H. and Tomlinson, R. (2007) Numerical modelling of tidal dynamic and water circulation at the Gold Coast Broadwater, Australia. Proceedings of the 9th International Coastal Symposium. *Journal of Coastal Research* **50**, 277–281.

Mirfenderesk, H. and Tomlinson, R. (2008) Observation and analysis of hydrodynamic parameters in tidal inlets in a predominantly semidiurnal regime. *Journal of Coastal Research* **24**, 1229–1239.

Noad, M.J., Dunlop, R.A., Paton, D. and Cato, D.H. (2008) An update of the east Australian humpback whale population (E1) rate of increase. International Whaling commission report SC/60/SH31.

Paterson, R., Paterson, P. and Cato, D.H. (1994) The status of humpback whales *Megaptera-novaeangliae* in East Australia 30 years after whaling. *Biological Conservation* **70**, 135–142.

Reilly, S.B., Bannister, J.L., Best, P.B., Brown, M., Brownell Jr., R.L., Butterworth, D.S., Clapham, P.J., Cooke, J., Donovan, G.P., Urbán, J. and Zerbini, A.N. (2008) *Megaptera novaeangliae.* In: IUCN 2011. IUCN Red List of Threatened Species. Version 2011.2. www.iucnredlist.org (accessed 10 September 2012).

Smultea, M.A. (1994) Segregation by humpback whale (*Megaptera-novaeangliae*) cows with a calf in coastal habitat near the Island of Hawaii. *Canadian Journal of Zoology-Revue Canadienne De Zoologie* **72**, 805–811.

Tyack, P. and Whitehead, H. (1983) Male competition in large groups of wintering humpback whales. *Behaviour* **83**, 132–154.

Vigness-Raposa, K.J., Kenney, R.D., Gonzalez, M.L. and August, P.V. (2010) Spatial patterns of humpback whale (*Megaptera novaeangliae*) sightings and survey effort: Insight into North Atlantic population structure. *Marine Mammal Science* **26**, 161–175.

Weinrich, M.T., Kenney, R.D. and Hamilton, P.K. (2000) Right whales (*Eubalaena glacialis*) on Jeffreys Ledge: A habitat of unrecognized importance? *Marine Mammal Science* **16**, 326–37.

Whitehead, H. and Moore, M.J. (1982) Distribution and movements of west-indian humpback whales in winter. *Canadian Journal of Zoology-Revue Canadienne De Zoologie* **60**, 2203–2211.

Section 3

# Coastal Governance

Chapter 9

# An Overview of the Implementation of SAF Methodology Regarding the Stakeholder Response, in the Mussel-Farming Area of Chalastra, Thermaikos Gulf

*Zoi I. Konstantinou and Yannis N. Krestenitis*

**Abstract**

The topic of this work is the interaction between the scientific team and the stakeholders participating in the Systems Approach Framework implementation, regarding the sustainable management of mussel-farming activity in Chalastra, Thermaikos Gulf, Greece. Results from the stakeholders' involvement will be examined to identify if, at the end of the project, the experiment had succeeded in its initial goals of bringing science closer to policy and using existing knowledge to produce a management tool to support deliberations. More particularly, this paper focuses on the manner that the key governance issues were addressed regarding what concerns (a) the stakeholders and "policy-makers" engagement, (b) the developed management tool, and (c) the structuring of the relationship between science and policy. The primary goal is to provide an insight to the initially unforeseen obstacles and particularities that the management attempt confronted, the efforts to overcome them and the lessons learned from the procedure.

## 9.1 Introduction

Nowadays, the importance of stakeholder involvement in policy-making has been a matter of endless societal discussions (van Asselt and Rijkens-Klomp, 2002). Regarding environmental issues, almost every assessment incorporates some reference to the involvement of stakeholders in the process (Voinov and Bousquet, 2009). The Systems Approach Framework (SAF; Hopkins et al., 2011) is a methodology combining general systems theory and

*Global Challenges in Integrated Coastal Zone Management*, First Edition. Edited by Erlend Moksness, Einar Dahl and Josianne Støttrup.

soft systems methodology, targeted in addressing coastal zone management sustainability issues (Tett et al., 2011a) and committed to including the stakeholders in the decision-making process (Tett et al., 2011b). The argument is that the prospect of bridging the gap between science and policy is greatest when the relevance of stakeholder's interests is acknowledged by scientists and policy-makers (Mette, 2011). Although there are already developed stakeholders' structures in most European countries, in Greece the concept of stakeholders' engagement and participation in environmental management issues is at its infancy. This paper discusses the implementation of the SAF in the mussel-farming area of Chalastra, Thermaikos Gulf, regarding the stakeholders' response, in an effort to reveal the particularities of the situation, the ways that those were addressed and finally to discuss the possible added value created from the implementation.

## 9.2 Study area and policy-issue

The coastal area of Chalastra (Figure 9.1), located at NW side of the inner Thermaikos Gulf, 20 km NW of Thessaloniki, at the delta of Axios, Loudias and Aliakmon rivers, is a small area of 1.35 km$^2$, protected under the Ramsar Convention as part of the deltaic system.

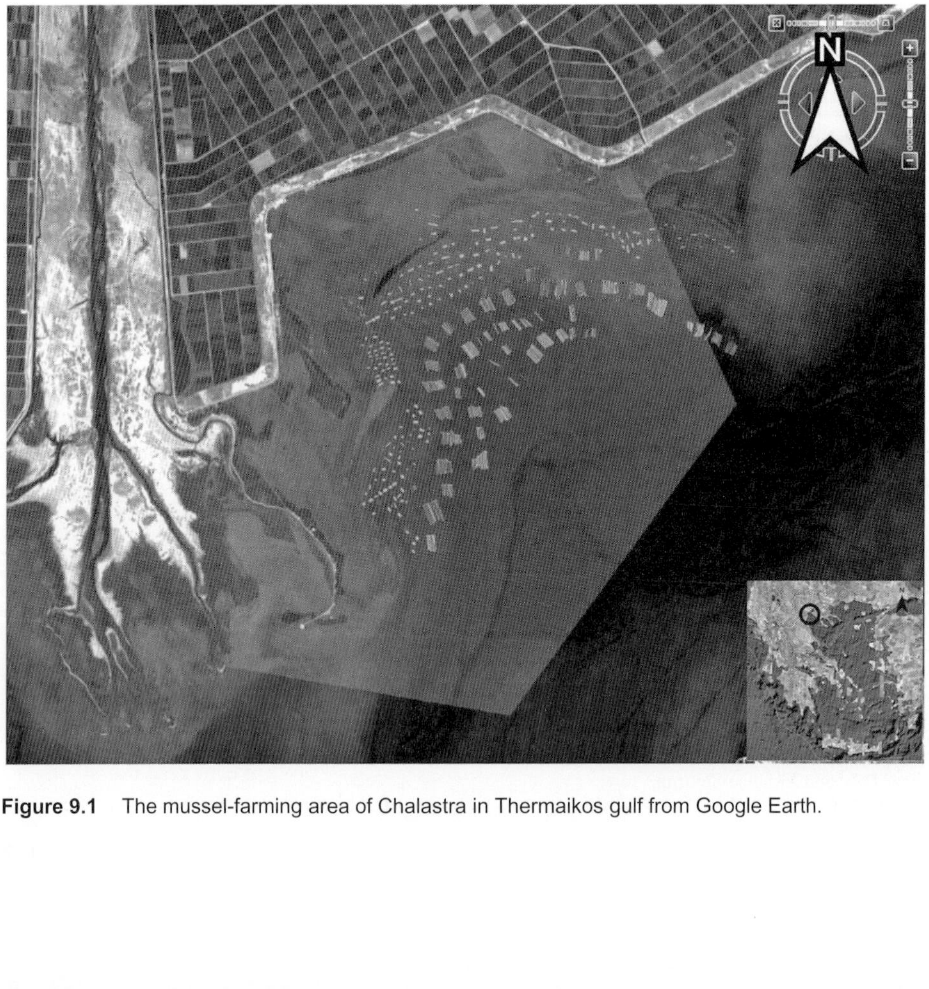


**Figure 9.1** The mussel-farming area of Chalastra in Thermaikos gulf from Google Earth.

The main marine activity in Chalastra is intensive farming of the Mediterranean mussel *Mytilus galloprovincialis*. Local mussel production is very important, producing 30% of the total Greek national production. The activity is family-based, providing employment to a significant part of the local population, both directly and through associated activities (NCMR, 2001). Two cultivation systems are used: (1) large area, long-line type cultivation, and (2) small area, pole type cultivation (NCMR, 2001; ATEITH, 2007). The local farmers use the pole cultivation mainly to support the long-line.

During the past decade, a significant reduction of the quality and quantity of the mussel production is observed in the area. Additionally a significant increase in the duration of harmful algal bloom events is causing severe selling restrictions and thus amplifying the pressure on the activity (Konstantinou et al., 2012).

At the beginning of the SAF implementation in Chalastra, the focus was placed on the environmental aspects of the activity. The policy-issue (McFadden and Priest, 2011) seemed straightforward: an area with high mussel production faces reduction in quantitative and qualitative terms. Yet, following the SAF and through the interaction with the stakeholders (McFadden et al., 2010), it was revealed that the problems of the area were related to the deficiencies in governance and regulation, transforming it to an issue of jurisdictional conflict (McCreary et al., 2001). Besides the challenge of engaging stakeholders in the participation processes, there was the additional challenge of incorporating the governance context into the management tool.

## 9.3 Methods

The mussel-farming area of Chalastra has been under occasional monitoring in the past (NCMR, 2001; ATEITH, 2007) and there has been a former effort of engaging local mussel-farmers in the principles of participatory management through open discussion (Zanou et al., 2005). Yet, inevitably, that effort faced funds limitation and stopped. One of the goals of the science–policy interface through the SAF was to create a stakeholders' dynamic that could exist independently from any project.

The process of stakeholders' engagement was structured, adapting to the circumstantial and cultural particularities of the area (UNEP, 1999; Treby and Clark, 2004; Zanou et al., 2005; Vanderlinder et al., 2011). The main components of this procedure were: (a) two rounds of private interviews, (b) two stakeholder meetings, and (c) two meetings of a stakeholders' collaboration group.

### *9.3.1 Private interviews*

The first round of investigative and informative interviews (Vanderlinder et al., 2011) was conducted during January–March 2008. The responsible public representatives provided information about: (a) the operational framework and (b) the current problems of the activity. Additionally they were informed about the goals of the SAF implementation in the area. Table 9.1 lists the authorities interviewed and their role in the activity's management. The representatives of the mussel-farming associations gave their opinion on the operational framework of the activity and details about the activity–community linkages.

**Table 9.1** The main public authorities implicated in the management of mussel-farming activity in Chalastra and their responsibilities.

| Public Authority | Responsibilities related to mussel-farming activity |
|---|---|
| Ministry of Environment | Regulating aquaculture nationally |
| Regional Office for Environment | Issues and renews operational licenses |
| Public Land Service | Holding and shares areas' property rights |
| Municipality of Chalastra | Collects contributory benefits and provides consulting for the issuing of operational licenses |
| Prefectural Office for Aquaculture | Monitors the production and provides consulting for the issuing of operational licenses |
| Prefectural Veterinarian Office | Monitors HAB's, imposes selling restrictions and provides consulting for the issuing of operational licenses |
| Prefectural Office for Environment | Provides consulting for the issuing of operational licenses and for changes in regulation |
| Regulatory Organization of Thessaloniki | Provides consulting for the issuing of operational licenses and for changes in regulation |
| Regional Office of Agricultural Development | Provides consulting for the issuing of operational licenses and for changes in regulation |
| Authority for Thermaikos Gulf Management | Provides consulting for the issuing of operational licenses and for changes in regulation |
| Authority for the Protection and Management of Axios – Loudias – Aliakmon Delta | Provides consulting for the issuing of operational licenses and for changes in regulation |
| Prefectural Office for Productive Activities Development | Provides consulting for the issuing of operational licenses and for changes in regulation |
| Prefectural Office for Productive Activities Management | Provides consulting for the issuing of operational licenses and for changes in regulation |
| Local Coastguard Authority | Imposes fines to farms not owning operational license after combination. |

Personal interviews were selected against questionnaires to provide flexibility and space for communication.

The second round of interviews aimed at extracting any available data to support the management tool. The public representatives were asked to provide data related to the mussel production or to indicate alternative sources they knew about. In parallel, a questionnaire survey was conducted among the mussel-farmers to obtain data for the economic characteristics of the activity, such as the related costs, revenue and profits (Konstantinou et al., 2012).

To maintain the communication base that was created through the interviewing process, frequent contact was established. This communication was necessary to build the trust of stakeholders towards the scientific team (Mette, 2011).

## 9.3.2 *Stakeholders' meetings*

The first stakeholders' meeting was organized in April 2010. Its goals were: (a) to present the draft management tool, (b) to present and discuss the results of specific management scenarios, and (c) to initiate a deliberation over the area's problems. The meeting was targeted exclusively towards the authorities and the representatives of the mussel-farmers in an effort to gain an official base. A neutral facilitator was used to moderate the meeting and to "interpret" the scientific language, assisting with the science–stakeholder interface (Mette et al., 2011).

In June 2010, in co-operation with Municipality of Chalastra and the Authority for the Protection and Management of the Estuaries, the second joint stakeholders' meeting was organized. The co-organization of the meeting was pursued to reinforce the engagement of crucial actors to the participatory procedures and to increase the validity of the process. This meeting was open to all stakeholders, thus promoting the importance of public participation in management. Its goals were: (a) to provide time for deliberation over the policy-issue and (b) to propose and support the creation of a stakeholders' collaboration group ("multi-stakeholder task force"; McCreary et al., 2001) for the area.

### *9.3.3 Stakeholders' collaboration group meetings*

The stakeholders' collaboration group comprises of the leaders of the mussel-farmers' associations, three scientists from different disciplines and three representatives of key public authorities. The purpose of a stakeholders' collaboration group is to promote self-disclosure and communication among the participants (Kreuger and Casey, 2000). The creation of such a structure can prove a flexible way to develop a communication habit that will survive after the end of the SAF implementation in the area. Thus it can be the base for development of the stakeholders' attitude regarding participatory approaches and conflict resolution.

Meetings of the group were organised for July 2010 and April 2011. The goal of both meetings was to introduce the participants to the concept of exchanging information and to discuss the solution of issues that could lead in alterations of the activity's operational framework. The first meeting was devoted to discussing funding opportunities for the development of the activity, whereas the key issue of the second meeting was the significant changes in the new national legal framework for aquaculture and the way that it will affect Chalastra's mussel-farming area.

## 9.4 Results

### *9.4.1 Operational framework of mussel-farming*

The outcomes of the first round of interviews enabled the comprehension of the events that lead to current institutional situation in the area. The development of mussel-farming in Chalastra was initiated in the early 1980s. The activity proved to be successful, hence strong investment ensued, resulting in a doubling of the operational licenses by the 1990s. Initially, there was an effort for regulation that proved superficial due to disagreements between the local and the central authorities.

During 2000, the government released the legal statute for the regulation of all the productive activities, which requires the constitution of a new supervising authority for each activity, in every region. The local authorities initiated the necessary procedures and at the same time froze the renewal of the operational licences so the new authority could undertake this responsibility. Yet, due to legal overlaps and gaps as well as the increased bureaucracy and the frequent replacement of the responsible public employees, the constitution of this authority is still not finalised (October 2012). As a result of these problems, all pole-type farming establishments and half the long-line-type establishments are currently deprived of operational licences. The owners of these units are subject to prosecution, resulting in both private and social costs (Konstantinou et al., 2012).

The situation described has created deficiencies in the institutional control of the activity. The mussel-farmers are using excessive cultivation practices to maximize their profit, although there is strong evidence that these actions create a negative impact on the quality of the production in the whole area (NCMR, 2001; ATEITH, 2007; Konstantinou et al., 2012). Additionally, the mussel-farmers that are deprived of a valid operational licence, in order to "legalize" their production, are forced to sell it through mussel-farms that are still operating legally, thus being subjected to oligopsony. On certain occasions, due to personal conflicts, they are also excluded from this option and they trade through black market. The situation creates illegality, labour insecurity, stagnant profits, inequality and severe conflicts. The incomplete legal framework creates insurmountable obstacles for the public authorities as well, as they cannot claim sufficient control over the activity and thus are incapable of imposing any control over the aforementioned excessive cultivation, "legalization" of production and the black market.

### *9.4.2 Stakeholders' engagement*

The operational framework revealed a dysfunctional situation creating socio-economic pressure on the local community and socio-political pressure on the supervising authorities. Mussel-farmers and public representatives are separated into smaller inter-conflicting categories, supporting different interests. The trust relationships among stakeholders and between stakeholders and scientists were very poor, creating obstacles during the engagement process.

During the first round of interviews, all the local supervising authorities, except the central government authority, were interested in discussing the activity and the SAF implementation. Yet, the majority of the officials were unwilling to participate in any preliminary stakeholders' meetings as they could foresee conflicts and were reluctant to undertake responsibilities for the management of the area. During the second round of private interviews some of them refused to provide data, claiming it was private information.

The mussel-farmers' response to the engagement process was sufficiently altered during the process. The initial mistrust was based on the opinion that any scientific approach would result in "unnecessary" environmental protection against their own interests, something also observed by Zanou et al. (2005). Effort was invested in gaining trust, mainly during private meetings, where the goals of the SAF were explicitly explained and the problems of the activity were discussed from their perspective. Despite the development of good communication, most of them were reluctant to participate in the anonymous questionnaire survey, as the questions concerned economic parameters of the activity that they were unwilling to share, suspecting that it could be used for taxing and legitimacy purposes. The results of the questionnaire survey revealed that most participants answered with what should be the legally accepted answer.

During the time between the interviews and the first stakeholders' meeting, a change in the government resulted in replacement of several of the representatives of the public authorities, including the representative of the central government authority. A new, brief, engagement process took place, resulting in two outcomes: (a) the establishment of communication with the central authority, and (b) the obtainment of mussel production data from the local authorities that were previously unavailable.

The deliberations in both meetings revealed important aspects for the nature of the stakeholders' interrelationships and the evolvement of any integrated management procedure in the area. The most important aspects are listed below:

1. The complexity and inadequacy of the relative institutional framework along with the frequent replacement of the higher level officials are the main reasons behind the delays in the establishment of functional regulation, which would resolve most of the issues connected to governance.
2. The local authorities' representatives are not willing to take responsibility for the situation, so they transfer the problem to a higher level authority. The central authority claims that some of the local authorities are creating obstacles to the formation of new regulation by withholding information and the necessary documentation.
3. The mussel-farmers are segregated into four conflicting associations. Their conflicts, which are minor and usually personal, are fostered by actors that gain from the lack of collaboration, through oligopsony and black market.
4. The engagement of the majority of stakeholders during the SAF implementation was well accepted and raised hopes for a solution regarding the regulative issues.

The stakeholders' collaboration group is an innovative structure for the area, as participatory and collaborative processes are not well developed in Greece. Yet, as it is evident from the communication to date, this group is not ready to operate independently and requires external stimulation in order to meet. Additionally, the communication remains fragile and is kept at a personal level. If the participants are replaced for any reason, the communication bond will break and will require a new engagement process. It is necessary to pursue iteration of the procedure in order to persuade stakeholders for the value and power of public participation and thus to promote their incorporation in the local culture.

### *9.4.3 Incorporating institutional deficiencies in a management tool*

The alteration of the texture of the policy-issue, from environmentally orientated with socio-economic projections, to governance orientated with environmental and socio-economic projections, created a number of dilemmas regarding the development of the management tool.

The first dilemma was connected to oligopsony, the black market and in some cases the deliberate negligence of the responsible authorities. These aspects, although evident, can neither be proved nor quantified. At the same time, no stakeholder would undertake the responsibility to start a discussion around these issues and even if this was undertaken by the scientific team, the outcomes could result in further conflict, undermining the deliberation.

The second dilemma concerned the formulation of the current situation in the management tool. From a scientific perspective, the current situation is not a management alternative but the result of management absence. Yet a representation of the current state is necessary in order to provide a means of comparison to other management states. In what ways can the absence of management be "modelled" and how can this formulation be presented in a way that would stimulate stakeholders' interest without imputing liability to a specific group of people?

To overcome these dilemmas, the scientific team decided to provide a descriptive economic comparison between the status quo of insufficient management and the desirable one, of operating under a coherent regulative framework. The comparison was based on the extra costs created from the imposition of fines to mussel-farms deprived of operational licenses, the corresponding leaking profits deriving from the annual perquisite that the mussel-farmers would pay to obtain the operational licenses and the extra profit from a potential rise of the selling price (Konstantinou et al., 2012). This estimation neglects the "grey zones" under which the activity operates, but underlines the hazard between the two situations, providing a base for further dialogue. Most importantly, the economic effect of the institutional deficiencies pointed stakeholders' interest to the management tool's results. More questions were set on the discussion table providing directives for the expansion and future development of the tool.

The draft version of the management tool designed for Chalastra incorporates mainly environmental and economic aspects of the mussel-farming activity. It can be accessed at http://dataportals.pangaea.de/spicosa/SPICOSA_model_library.html. Although it is lacking important data and is still under development, it is a means through which a quantification of the effect of various management decisions is feasible. The value of the tool regarding the stakeholders' engagement is that: (a) it enabled a deeper understanding of the system's functioning, both in environmental and economic terms, and (b) it raised new management questions from the stakeholders, enabling future evolvement of the science–policy interface.

## 9.5 Discussion

The SAF implementation in Chalastra, regarding stakeholders' involvement and the incorporation of institutional deficiencies into a management tool, proved a challenging and continuous learning process. The effort invested in science and stakeholders' interface revealed the full nature of a policy-issue that was until now received as mainly environmental (NCMR, 2001; Zanou et al., 2005; ATEITH, 2007). At this point, it is clear that if the institutional deficiencies are not resolved in a way that will provide a new regulative framework and institutional control of the activity, any effort for sustainable management will be in vain (Noble, 2000; Konstantinou et al., 2012). Thus, it is of major importance that organized communication between stakeholders is fostered. In that way, all the actions taken will receive a more transparent and official character, insuring equity and equality (Noble, 2000; Tett et al., 2011a) and minimizing any "grey zones" that might be nourished by institutional chaos.

Furthermore, the implementation of the SAF in the area revealed the significant trust issues that Greek stakeholders have in what concerns science, a finding also highlighted in other studies (Glenn et al., 2011). Most stakeholders are reluctant to share information and cooperate either in what concerns the "misuse" of personal data or in the way that sensitive sustainability issues are being confronted. They believe that scientists have a "hidden agenda" concerning environmental protection against human welfare. The time and effort that has to be invested to build a well-founded, trusting cooperation is significant. Yet in this case, the relationships lack officialdom: it is the person that is being trusted, not the discipline that the person represents. This issue has multiple projections, the most important of them being the role that science must claim in supporting management decisions and the importance of investing in a more communicative form of science that will produce integrated results that are understandable and useful for the community (Tett et al., 2011a).

Under the current Greek politico-economic situation, the sustainable development and management of productive activities, especially those connected to the coastal zones, is a matter acquiring crucial national importance. Yet, the SAF implementation in the area of Chalastra is a small example of the implications that a management attempt can confront. At the same time, the results concerning the stakeholders' response are promising although fragile. The society, both policy-makers and stakeholders, will respond in organized participatory management attempts but, as these kinds of structures are not yet well developed, the scientific community has to undertake the responsibility of communicating the importance of this interface, promoting and underling the added value that can be gained through integrated, scientific management strategies (Mette, 2011).

## References


ATEITH (Alexander Technological Educational Institution of Thessaloniki) (2007). Water circulation in Organized Areas of Aquaculture Development and land-planning and environmental management interventions. Final Technical Report [in Greek]. Archimed II. Scientific coordinator A. Moriki. 344 pp.

Glenn, H., Tingley, D., Sánchez Maroño, S., Holm, D., Kell, L., Padda, G., Runar Edvardsson, I., Asmundsson, J., Conides, A., Kapiris, K., Bezabih, M., Wattage, P. and Kuikka, S. (2011). Trust in the fisheries scientific community. *Marine Policy* **36**(1): 54–72.

Hopkins, T. S., Bailly, D. and Støttrup, J. G. (2011). The Systems Approach Framework adapted to coastal zones. *Ecology and Society* **16**(4): 25.

Karageorgis, A. P., Skourtos, M. S., Kapsimalis, V., Kontogianni, A. D., Skoulikidis, N. Th., Pagou, K., Nikolaidis, N. P., Drakopoulou, P., Zanou, B., Karamanos, H., Levkov, Z. and Anagnostou, Ch. (2005). An integrated approach to watershed management within the DPSIR framework: Axios River catchment and Thermaikos Gulf. *Regional Environmental Change* **5**(2–3): 138–160.

Konstantinou, Z.I., Krestenitis, Y.N., Latinopoulos, D., Pagou, K., Gallinou-Mitsoudi, S. and Savvidis, Y. (2012). Aspects of the mussel-farming activity in Chalastra, Thermaikos Gulf, Greece: an effort to untie a management Gordian knot. *Ecology and Society* **17**(1): 1.

Kreuger, R. A. and Casey, M. A. (2000). *Focus Groups: A Practical Guide for Applied Research*. California, Sage Publications Inc.

Noble, F.B. (2000). Institutional criteria for co-management. *Marine Policy* **24**: 69–77.

McCreary, S., Gamman, J., Brooks, B., Whitman, L., Bryson, R., Fuller, B., McInerny, A. and Glazer, R. (2001). Applying a mediated negotiation framework to integrated coastal zone management. *Coastal Management* **29**(3): 183–216.

McFadden, L., Priest, S. and Green, C. (2010). Introducing institutional mapping: a guide for SPICOSA scientists. SPICOSA Project Report. Flood Hazard Research Center Middlesex University, London.

McFadden, L. and Priest, S. (2011). Guidance for identifying the Policy Issue. SPICOSA Project Report. Flood Hazard Research Center Middlesex University, London.

Mette, A., Borner, J., Sandberg, A., Vanderlinder, J.-P., Priest, S., McFadden, L., Lowe, C., D'Hernoncourt, J., Fernandes, T., Hirschfeld, J. and Raux, P. (2011). Guide to System Output, SPICOSA Project Report, Berlin, KMGNE. Available from http://www.coastal-saf.eu/output-step/refs.shtml (accessed 10 September 2012).

Mette, A. (2011). Bridging the gap between science and society. In Tett, P., A. Sandberg and A. Mette (Eds.) *Sustaining Coastal Zone Systems*. Dunedin Academic Press, Edinburgh, Scotland.

NCMR (National Center for Marine Research) (2001). Management study of the mussel production zones of the Thessaloniki and Thermaikos gulfs. Final Technical report [in Greek]. National Center of Marine Research on behalf of the Prefecture of Thessaloniki. Scientific coordinator Anagnostou, Ch.

Tett, P., Mette, A., Sandberg, A. and Bailly, D. (2011a). The systems approach. In Tett, P., A. Sandberg and A. Mette (Eds.) *Sustaining Coastal Zone Systems*. Dunedin Academic Press, Edinburgh, Scotland.

Tett, P., Mongruel, R., Levrel, H., Hopkins, T., Sandberg, A., Hadley, D., Fernandes, T., Hendrick, V., Mette, A., Vermaat, J., Gilbert, A., McFadden, L., Priest, S., Green, C. and d'Alconà, M. R. (2011b). Guide to System Design, v.3.09, SPICOSA Project Report, Scottish Association for Marine Science, Oban. Available from http://www.coastal-saf.eu/issue-identification/refs.shtml (accessed 10 September 2012).

Treby, E.J. and Clark, M.J. (2004). Refining a practical approach to participatory decision making: an example from Coastal Zone Management. *Coastal Management* **32**: 353–372.

UNEP (1999). Conceptual framework and planning guidelines for integrated coastal area and river basin management. United Nations Environment Programme.

Van Asselt, M.B.A. and Rijkens-Klomp, N. (2002). A look in the mirror: refection on participation in Integrated Assessment from a methodological perspective. *Global Environmental Change* **12**: 167–184.

Vanderlinder, J.-P., Stojanovic, T., Schmuëli, D., Bremer, S., Kostrzewa, C. and McFadden, L. (with others) (2011). *The SPICOSA Stakeholder-Policy Mapping User's Manual.* SPICOSA Project Report, Guyancourt, Paris, Université de Versailles-Saint-Quentin-en-Yvelines. Available from http://www.coastal-saf.eu/issue-identification/refs.shtml (accessed 10 September 2012).

Voinov, A. and Bousquet, F. (2010). Modelling with stakeholders. *Environmental Modelling & Software* **25**: 1268–1281.

Zanou, B., Anagnostou, Ch. and Papathanasiou, E. (2005). Seeking the factors to stimulate the users in coastal zone planning. Case study: Open discussions with mussel farmers in the Axios river (GR). *Mediterranean Marine Science* **6**(1): 107–115.

Chapter 10

# Knowledge Issues in ICZM and EBM Applied on Small Geographic Scales: Lessons from a Case Study in Risør, Norway

*Håkan T. Sandersen, Eirik Mikkelsen, Erlend Moksness and Jon Helge Vølstad*

**Abstract**

The demand for scientific knowledge in coastal zone governance is increasing, and multidisciplinary science is now a requirement for most policy initiatives. Coastal zone management is highly complex, as many of the governance objects, drivers and effects are transitory and dynamic. In addition, coastal governance is usually conducted within a complex and fragmented institutional framework. This drive towards more knowledge-intensive coastal management is discussed here in the context of the apparent change in approach from integrated coastal management towards ecosystem-based management. This chapter presents some lessons learned from a case study to implement integrated coastal zone management in the Risør municipality on the Norwegian south coast, conducted as part of the EU-sponsored SPICOSA research program. Focus is on knowledge requirements and geographical scale and these issues are discussed in relation to the change in approach to coastal management.

## 10.1 Introduction

Public policy-makers rely increasingly on science to make better decisions, and to strengthen the legitimacy of their decisions. But the relationship between decision-making and science is far from simple, and becomes even more complicated by increased demands for public participation that further increase the tensions between users and producers of knowledge. This is particularly so in highly visible policy arenas of natural resource management (Steel et al., 2004). Such problems are likely to intensify in the coastal zone, which

*Global Challenges in Integrated Coastal Zone Management*, First Edition. Edited by Erlend Moksness, Einar Dahl and Josianne Støttrup.

is characterized by complex interacting institutional (i.e. multilevel governance structures, open access and common property) and ecological conditions.

Integrated coastal zone management (ICZM) is one of the more complex governance tasks in public policy. This complexity increases even further when ICZM is expected to adopt also the principles of ecosystem-based management (EBM). Like ICZM, EBM is an integrated approach to management. Although both ICZM and EBM are committed to protect the environment, ICZM has its main focus on conflict resolution, while EBM has the strongest emphasis on keeping the ecosystem healthy, productive and resilient (McLeod and Leslie, 2009). Managers and researchers all over the world struggle with how to put the EBM principles into practice. As for ICZM, the experience with practical implementation of EBM is so far not very successful. This is due to factors such as poor funding, competing stakeholder interests, institutional barriers, inadequate science and lack of political will or lack of communication with, and involvement of, user-groups (Aswani et al, 2011; Espinosa-Romero et al., 2011).

This chapter is about experiences from the Risør study site[1] in the EU 6th Framework Program Integrated Project Science and Policy Integration for Coastal Systems Assessment (SPICOSA). The main objective of SPICOSA was to develop and test a Systems Approach Framework (SAF) to better include science into coastal zone management. The chapter presents challenges and dilemmas with our attempt at applying science in a local governance context. Even if this was only a partial and experimental implementation of EBM, it gave insights about EBM implementation in specific coastal settings, and how it interacts with an established system for ICZM.

Several lessons developed from this case relate to the lack of relevant data, challenges in the modeling of data, difficulties in balancing the disciplines and methodological approaches involved to keep it truly interdisciplinary, and the difficulty of demonstrating the relevance of the developed knowledge to local policy-makers. Finally, the ICZM–EBM debate in relation to geographical scope, knowledge requirements and urgency of the management issue is discussed.

## 10.2 ICZM and EBM – complementary or competing?

The dominating management framework in coastal zone management has since the early 1970s been ICZM (sometimes called ICM). More recently the EBM approach seems to increasingly take over or at least heavily influence the knowledge base for ICZM. ICM is a "process designed to overcome the fragmentation inherent in single-sector management approaches . . . in the splits in jurisdiction among different levels of government, and in the land-water interface" (Cicin-Sain and Knecht, 1998). It had its beginnings in the practice of conflict resolution and dispute mediation (Forst, 2009) and the focus is usually on cross-institutional collaboration and formal and informal governance processes (Christie, 2011). The concept emerged in the USA, but entered the global political scene during the Rio Earth Summit in 1992 where ICZM became one of the principal recommendations of Agenda 21. Sustainable use and bringing together human and ecological dimensions is

[1] All four authors were part of the Risør SPICOSA study site research team.

essential in most definitions of ICZM. Sorenson (1997:9) defines ICZM as: "The integrated planning and management of coastal resources and environments in a manner that is based on the physical, socio economic and political interconnections both within and among the dynamic coastal systems, which when aggregated together, define a coastal zone".

Ecosystem-based management (EBM) is "an integrated approach that considers the entire ecosystem, including humans. The goal of ecosystem-based management is to maintain an ecosystem in a healthy, productive and resilient condition, so that it can provide the services humans want and need" (McLeod and Leslie, 2009). Another widely quoted EBM definition is "... ecosystem management is integrating scientific knowledge of ecological relationships within a complex sociopolitical and values framework toward the general goal of protecting native ecosystem integrity over the long term" (Grumbine, 1994). The call for EBM has its background in the many observable and widespread declines in ecosystem functions and services, since the EBM approach has a stronger commitment to reversing these negative tendencies than ICZM.

According to Patrick Christie, the rapidly expanding influence of EBM seems to be caused by "a sense of urgency and frustration promoted by a sense of failure by ICM and fisheries management to slow or reverse the decline of critical habitats and associated marine organisms on a broad scale and the belief that ecological considerations are ignored during policy-making processes" (Christie, 2011:4). Whether EBM should be seen as an expansion and refinement of ICZM or as a paradigm shift is debated (Aswani et al., 2011).

The concepts of ICZM and the EBM are integrative frameworks that aim to overcome different kinds of fragmentation, and they are in many ways overlapping and similar. Forst (2009:302) even claims that the two approaches are increasingly converging.

While ICZM more often has a short-term pragmatic approach aiming at arbitrating conflicts, the EBM approach has the longer-term but definite goal of preserving and maintaining the functions and capacity to produce of ecosystems. However, ICZM also has ambitions about safeguarding the environment, but this aim is not as dominant as in EBM, which requires a defined vision of desired conditions for an ecosystem and for which ecosystem services should be given priority. It implies more ambitious and complex management tasks than species-specific or sector-specific management, and creates a new series of challenges and problems in addition to the ones already experienced in ICZM. For these reasons, EBM is frequently criticized as being too complicated, daunting, expensive and with prohibitive data requirements (Tallis et al., 2010; Espinosa-Romero et al., 2011).These challenges become particularly urgent when it is implemented in more vulnerable, small-scale coastal contexts (Aswani et al., 2011). ICZM embraces the institutional and stakeholder complexity of the coastal zone, to which EBM also adds its ecological complexity.

Both ICZM and EBM should be seen as processes, rather than certain states or endpoints. They can be incremental or comprehensive, slow or fast, and implemented in multiple ways (UNEP, 2011). However, the broad and rhetorical character of the concepts and their lack of clear definitions means there is little guidance on what should be the first steps in the implementation of ICZM and EBM.

The disciplinary starting points seem to have influenced heavily the frameworks' construction and goals (Degnbol et al 2006; Christie, 2011). The seminal works in ICZM were written mainly by representatives of applied social science, planners and supported by natural scientists experienced in coastal management (e.g. Cicin-Sain and Knecht 1998; Kay and Alder, 2005). EBM is more closely associated with applied ecology and, supported by economy and other social sciences. As the focus is on maintaining large-scale ecological processes and ecosystem integrity, EBM is also typically applied on larger geographical scales (e.g. McLeod and Leslie, 2009). Recognizing that ecological and human well-being is coupled, and that natural resources management is mainly about politics, it is clear that sustainability needs to be present in both arenas. Making ICZM more ecosystem-based is one way to improve the balance between the social and the natural scientific disciplines.

Both approaches are supported by influential epistemic communities that create policy-specific policy agendas. Interestingly they advocate some of the same solutions. Both have, for example, given considerable attention to and actively promoted marine protected areas (MPAs) as valuable management tools (Christie, 2011).

The ecological and biological data going into EBMs is predominantly based on scientific methods and a big effort from scientific personnel. In Norway, EBM has been attempted only for its three large ocean ecosystems; Barents Sea, Norwegian Sea and the North Sea (Olsen et al., 2011). It seems unlikely that a similar level of scientific effort will be made to obtain ecological monitoring data for local-scale coastal zone management in Norway in the near future. Norway has 263 coastal municipalities, each responsible for making its coastal zone plan. How can the necessary knowledge for ecosystem-based coastal zone management be provided? Perhaps stakeholders and users of coastal resources must be more directly involved in data production, with scientists assuring the quality of the data, rather than being their sole provider?

## 10.3 The SPICOSA project

The SAF developed in the SPICOSA project aimed at mobilizing the best available scientific knowledge to support deliberative and decision-making processes in coastal zone management (Hopkins et al., 2011). The SAF can be seen as an EBM-inspired approach that is based on natural science perspectives, but also incorporates economic, social, cultural and political elements. A central element of the SAF was to translate the policy issue into one or more quantifiable impacts measured by proxies/indicators. From these dependent variables, casual chains were established to the independent variables from the environmental and socio-economic components. The SAF deliberately excludes the dynamics of policy- and decision-making, by emphasizing the delineation between "the objective role of science and the democratic role of governance" (Hopkins et al., 2011). The interface between science and policy is meant to be taken care of through the inclusion and participation of the decision-makers in the project.

The SAF was developed and tested in 18 Study Site Applications (SSA) in 16 European countries, in a diverse set of habitats, human activities, policy issues and governance systems. The SAF included four steps: Design, Formulation, Appraisal and Output. The *Design step* aimed at defining policy issues, developing a conceptual model and gathering of basic information and data for the subsequent modeling and simulations. The *Formulation step* focused on the development of the system modeling, identification of problems, unacceptable approximations, issues that were not possible to quantify, etc. Experimental runs of the sub-models were made. The *Appraisal step* represented the merging of the sub-models into an integrated model, model calibration and validation, and preparing scenarios. The

*Output step* focused on presenting information and results of simulations to policy-makers, stakeholders and the general public in suitable forums.

## 10.4 The Risør case study

Risør is a municipality of some 7000 inhabitants located on the south Norwegian Skagerrak coast. The region is popular among summer tourists, and there are about 1550 second homes in the municipality. Risør town – the beautiful and picturesque center of the municipality – receives a very high number of boaters every summer. The main conflicts in the coastal zone are between locals and second home owners, and there are also problems related to the high number of boaters who make temporary stops along the coast in the peak season of June–August. Fisheries are no longer important, as the local catches and numbers of fishers have dropped steadily over the past 20 years. Of the species caught, shrimp *Pandalus borealis* is most important both in terms of quantity, value and economy (Directorate of Fisheries, 2011).

Following the 1985 Planning and Building Act, the Norwegian coastal zone management system is based on physical planning that extends into the sea one nautical mile beyond the baseline. The municipalities designate the different zones of their waters for certain types of use, exclusively or in combination. The Planning and Building Act has a number of provisions on procedures of the planning process, including hearings and other measures to ensure stakeholder involvement.

In collaboration with the mayor and head bureaucrats of Risør, the SPICOSA research team decided to evaluate a policy issue centerd on *fishing tourism* with the cod *Gadus morhua* stock as an ecosystem indicator. Although this was not a critical or strong political issue in the municipality, the policy-makers thought it would be useful to evaluate if the local economy could be improved through increased fishing tourism, without harming the marine ecosystem or causing other effects negative for the local inhabitants. Which policy instruments, or combination of policies, would be most effective for achieving this goal?

To support the policy-makers, the research team developed an integrated model[2] with ecosystem and socio-economic components that was used to assess the effects of different policy options and the trade-offs between them. The model was implemented in the software ExtendSim (www.extendsim.com)[3] and five scenarios were constructed to represent management options. In our case study, all steps except Formulation involved the participation of the municipal policy-makers. Running the model gave the following results for the different scenarios, compared to a baseline scenario:

- Scenario 1, where mortality of juvenile cod was reduced by eliminating predation by birds, seals and eel fishing, gave a significant increase in cod biomass, increased number of fishing tourists and increased local economic benefit from both fishing tourists and commercial fishers.

[2]See Moksness et al. (2011) for a detailed description of the model.

[3]To use ExtendSim in the modeling was a requirement set by the central managers of the SPICOSA-project.

- In Scenario 2, where all cod fishing was banned, a strong increase in cod biomass followed, tourist fishers disappeared, local economic benefits dropped, and the level of conflict between tourists and local inhabitants increased.
- Scenario 3, where only commercial cod fishing was banned, caused an increase in cod biomass, more tourist fishers, and an overall increase in local economic benefits.
- Scenario 4 included stock enhancements through the release of juvenile cod in the fjord system, and led to a strong increase in cod biomass and increased number of tourist days and local economic benefits.
- In Scenario 5 the quality of accommodation dedicated to tourist fishers were improved to the highest level in the Norwegian classification system and a fivefold increase in new second homes over the baseline scenario was allowed. It had positive effects on the number of tourist days and local economic benefit, but only small effects on other stakeholder groups and the cod population.

So far, the results from the SPICOSA project have not been used directly in local ICZM policy-making in Risør. The time is simply not right, and there is no strong or shared sense of urgency about this issue. A local vocational school plans to integrate the model in their work on coastal issues, and to refine and develop the model through their use.

## 10.5 Top-down requirements

Stakeholder involvement is important in any ICZM or EBM process (Espinosa-Romero et al., 2011). Without defining objectives that are somewhat consistent with local stakeholder views, the legitimacy and support for plans and measures will be weak. Stakeholder involvement is important for developing management initiatives that are credible, enforceable and that can be realistically implemented (Leslie and McLeod, 2007).

Risør was selected as a study site for SPICOSA primarily because of the long and rich marine biology data series available, including empirical data on annual recruitment indices for cod over 90 years (1918–2009) from the same fjord system (Moksness et al., 2011), and was not initiated by the municipality or other stakeholders. The initial public meeting with stakeholders to define a policy issue that could also meet the SPICOSA project's requirement of interdisciplinary (social science + economy + ecology) and policy-relevant nature was unsuccessful for two reasons. The number of stakeholders attending the meeting was very small, and the issues they suggested did not fit the SPICOSA project's requirements. Given that the case study area was already selected, a participatory bottom-up approach did not prove useful. The policy issue was therefore chosen after discussions with municipality representatives, including the Mayor. Considering that stakeholder participation is important in coastal management this was problematic, but since the SPICOSA implementation was only experimental, this limitation was not of concern to the community.

In an instrumental problem-solving perspective, knowledge is usually applied and channelled into the policy process after the main problems have been identified. However, policies often evolve in a manner where the choice of solutions is often made even before the problems have been identified. This means that the sequences in a "traditional" instrumental policy cycle are not always as expected. In our case a "solution" (i.e. the development of an inter-disciplinary modeling approach, and the promise of the knowledge this model would provide) was proposed for a problem that the stakeholders had not specified, and the municipality had not considered how and when to apply the 'solution'.

The initial selection of the study site, based on natural science data availability before the project started and the SPICOSA requirement for an interdisciplinary approach, made the other SPICOSA requirement of stakeholder and user-group participation difficult to achieve. These problems illustrate the importance of a flexible, open approach in the implementation of ICZM and EBM. There are many predefined elements, compulsory operations and procedures common in the various general frameworks for both EBM and ICZM. While ICZM following the Norwegian Planning and Building Act has many procedural requirements, EBM typically has requirements concerning the kind of knowledge that should be gathered about the ecosystem. Neither of these may suit the local context, with its mix of stakeholder interests, resources and arrangements.

## 10.6 The quality of the knowledge

A shortage of reliable monitoring data for modeling environmental, social and economic conditions and relationships at a local scale is normal. Most biological and socio-economic surveys are designed to obtain data for inferences at relatively large spatial scales. Due to cost considerations, sufficient biological data are rarely obtained for local areas, and socio-economic data for local studies are also often restricted due to confidentiality issues. Although the amount and quality of data from the case study on the south coast of Norway probably exceed what can be obtained for local studies in most other countries in Europe, missing data was a persistent problem for the project. Even though the case study site was selected to take advantage of the long and reliable biological data series, it proved difficult to find biological data to calibrate the ecological component of the model. It also proved problematic to delineate a spatial management unit in the rather open fjord system. Currents and migratory stocks have strong and variable influence on the ecological system within the fjords. Most of the fishing takes place outside the fjord system, but coastal cod stocks in the fjords are nevertheless more stationary than those in the open sea, with cod stocks in many fjords genetically distinct from each other, and from those in the open sea (Knudsen et al., 2003).The research team also experienced a serious lack of reliable economic data for the study area (Moksness et al., 2011). The lack of data called for a parsimonious modeling approach, as a more realistic representation of the complex dynamic systems with more variables could have increased the uncertainty and reduced the predictive power of the model.

The findings from the model runs are consistent with what anyone interested in fjord issues might have anticipated. It can be argued that this strengthens the findings and that the findings strengthen the common-sense expectations and thus make the results more robust. But one could also argue that this makes the study and the modeling redundant. We consider this to be unproblematic, as the model can still be used to guide the choice of the best policy option among a set of options. Modeling results with high uncertainty may also open for a de-legitimisation of policies based on these results (van Haastrecht and Toonen, 2011).

The SPICOSA-approach is most easily applied in settings where the data availability already is good, but then the practical need for SPICOSA-modeling is likely not very large. The same can be said for any data-demanding, ecosystem-based approach. The usefulness of a model is inversely proportional to the availability of good and reliable data necessary to make the model. Rather than seeing our model as providing accurate forecasts of various scenarios, it should be seen as a tool to evaluate more qualitatively the effects (positive or negative effects; order of magnitudes) of the relationships between the ecological, economic and social elements (Moksness et al., 2011).

How efficient is the SPICOSA-model in other empirical settings? It is necessary here to distinguish between the model (the theoretical assumptions, variables and relations within the model + the technical/computer part) and the empirical approximations derived from the model (the results of the modeling). The output from the model will in each local context be used to further improve and calibrate the relations within the model. However, the approach requires a substantial amount of local and highquality data to make good assessments, and we see this as the major problem with the SPICOSA-approach as well as the EBM-approach, since these requirements are not met in most small-scale settings[4].

The relations and knowledge provided by the model are general enough to be of some value also in other coastal areas, similar to that of Risør. In a situation where there are little local data available the model may provide some general and crude rule-of-thumb knowledge about the main relations and proportions that is not necessarily very accurate, but still more solidly based than sheer guesswork. During the work, the research team benefitted from data from the local fish landing facility. It is clear that assistance from local fishers and other user-groups could have substantially improved information on local conditions. If coastal zone management on a fairly small spatial scale is to be based on the type of ecosystem knowledge that EBM in Norway seems to aim for, data collection must use local resource users and stakeholders to a large degree. Otherwise, data collection will be too expensive. Scientists should play a role by assuring the quality of the data, and by interpreting the data in relation to management options. In Norway we have already seen examples of this. Local fishermen have helped the Institute of Marine Research to make maps of fish resources (spawning areas and feeding grounds) on a relatively fine spatial scale (see http://kart.fiskeridir.no).

## 10.7 The communication of the knowledge

According to Leslie and McLeod (2007) synthesizing and communicating scientific knowledge to policy-makers, managers and other stakeholders is an important part of the implementation of EBM. Poor integration of science in decision-making is frequently due to ineffective communication of the science to the relevant actors (Byron et al., 2011). Phenomena such as technical jargon, abbreviations and differences in training and socio-economic status often distort the communication and knowledge exchange between scientists and non-scientists. One objective of SPICOSA was to improve the communication interface between researchers and policy-makers and stakeholders. Communication in the Risør case could have been better. The selection of the policy issue by the research team and the municipality leaders in a rather closed process meant there was no obvious arena for stakeholder involvement and dialogue.

[4]Not all SPICOSA study areas are as small and rural as Risør.

That the findings from the simulations produced no unexpected or counter-intuitive results made the project less prone to press coverage and local interest. The findings merely supported the common-sense knowledge, which made them harder to communicate. The research team presented the project on several occasions, including open meetings during the local annual lobster festival, meetings with the municipal Executive Committee, and presentations in the local newspaper. To explain highly technical methodologies to a diverse audience with little or no knowledge about the field is challenging. To assess the value of the knowledge, it has to be used for critical analysis of trade-offs and estimations of the cost and benefits of the different scenarios to the individual stakeholder group.

Due to time restrictions, the model was presented to the municipality representatives in an ecological EBM context, without much reference either to the Norwegian coastal zone planning system or to the approved municipal coastal zone plan. Hence it was difficult for them to grasp to what extent the model was a radical new instrument requiring completely new institutional and legal tools, or if it was merely a decision-supporting tool not requiring any new practices. To introduce and present dynamic and complex ecosystem models to lay-people requires big efforts from both parties and is far more difficult to communicate than simple technical fixes (Degnbol et al., 2006).

## 10.8 Applicability of knowledge

The existing governance system and institutional legacies constrain management actions. Implementation of ICZM and EBM should therefore always be rooted in a thorough knowledge of the traditions and structures of the present governance system as it usually requires cooperation and coordination among government bodies with different jurisdictions and abilities to introduce and enforce relevant laws and regulations (Tallis et al., 2010).

Although the model has resulted neither in actual policy changes, nor in implementation of new measures, our case illustrates clearly how institutional boundaries and overlapping jurisdictions can influence the chances for implementation and use of the knowledge provided. The present Norwegian governance structure severely limits the management strategies available to coastal municipalities. Hence the developed scenarios have rather slender chances of being implemented.

Scenario 1 implies a ban on eel fishing, which is not very likely to be introduced, even though there is only a small handful of retired men who take part in this fishery, as the Norwegian fishery regime is generally very reluctant to introduce local regulations that excludes some fishers for the benefits of others. The scenario also implies an increased and organized hunt for seals and cormorants that must be financed and coordinated by the municipality. Both seals and cormorants eat considerable amounts of fish. In March 2010, the Mayor of Risør applied for permission from the County Governor to resume public seal hunting, but it was not granted. It is debated whether the seals in the area are stationary or migratory. If they are migratory, local programs to control the seal population will have limited effect. Hunting of cormorant is open in the hunting season, and only a general hunting license fee applies. The regional Fishermen's association applied for permission to extend the hunting season in 2007 and 2009, but this was not granted either. To coordinate and increase the cormorant hunting in Risør is not costly, but it still has to be organized and funded. To what extent the municipality is willing and capable of do this is still unclear.

We see that the municipality's lack of jurisdiction over local wildlife makes it difficult to implement the policy suggested in this scenario.

The municipalities lack the legal power to create marine protected areas in their coastal waters, and scenario 2 and 3 can only be implemented in cooperation with the fishery authorities. In general, the commercial fishers' right to fish is highly valued and protected by law. Today, only special circumstances will legitimize an infringement of their rights, and then most likely only to protect the fish resource itself or motivated by other nature conservation issues. Particularly scenario 3, where tourist fishers are allowed to fish, but not the local commercial fishers, is highly unlikely to be implemented. It will clearly challenge the local sense of fairness and traditional rights, and will also go against national policy in this field. A total ban on cod fishing (scenario 2) is more likely to be implemented, but may not be very efficient as by-catch problems will occur. The policies suggested in these scenarios are thus very unlikely to be implemented under the current regime. What might happen in the long term, if tourism gradually becomes more important economically and fishing less important, is less certain.

Scenario 4 implies that a local hatchery will run a local cod stock enhancement program. However, it is not obvious that the municipality council will approve this measure even if it is likely to have effect. Despite a number of research projects to investigate marine fish stock enhancement (Svåsand, 1998), it has not been implemented anywhere in Norway as a policy measure. This maybe as the costs will be substantial for a small Norwegian municipality, and the effects are seen as uncertain. Marine stock enhancement may thus not be a realistic policy measure, despite the relatively good effects predicted from our model.

Scenario 5, with a fivefold increase in the number of new second homes in the municipality, does not seem very realistic either. However, the municipal council has the legal authority to change this policy. Local land-owners work to increase the numbers of second homes in Risør, but so far, the municipal policy, with strong support in the municipal council, has been more restrictive due to nature protection and public access concerns, than that of the neighboring coastal municipalities which have allowed many new second homes. It is not very likely that fishing tourist operators will have the necessary capital to establish or upgrade all tourist fishing accommodation to a five-star level, based on the current profitability of this segment of the Norwegian tourism industry. The tourist fishing season is also quite short and can be strongly affected by events outside local control, such as economic crisis, changes in ferry services between Norway and Denmark, etc. The municipality can also only encourage the entrepreneurs to go for high-quality accommodation.

Summing up, the interventions proposed in the scenarios either exceed the legal instruments available and the institutional and financial capacity of the municipality, or contradict local cultural, political norms and priorities, or what seem to be the economic realities. This apparent mismatch does, however, not mean that the work has been futile. Exploring the effects of policies that today seem very unlikely can be important when considering changes in the *policy-making system,* and not only for *policy measures.* Trying to identify needs for institutional reform, or for new forms of collaboration between local municipalities, fisheries authorities and counties should be done regularly. Also, while norms in some areas may be strong today, they can sometimes change surprisingly fast.

## 10.9 The call for interdisciplinary and methodological brinkmanship

Building interdisciplinary scientific capacity is regarded as vital for EBM (UNEP, 2011), and Leslie and McLeod (2007:546) state that: "There is a tremendous need to build bridges among disciplines, particularly between the natural and social sciences. This is no small endeavour. "It is quite clear that ecological and social dynamics cannot be studied in the same way, even if from an EBM-perspective the two spheres are intrinsically coupled. To reach an integrated understanding of these two systems is certainly not easy.

By focusing on and pushing for quantitative modeling SPICOSA takes side in the quantitative-qualitative methodological debate. The inter-disciplinary aim of SPICOSA is much easier to fulfill between the quantitatively oriented parts of biology, economics, and the other social sciences, than between the quantitative and qualitative camps of, for example, political science. It is quite obvious that in order to solve complex coastal issues neither of these "two cultures"[5] can be ignored, and the balance between them should be decided by the nature of the management problem at hand. Qualitative research asks other questions and consequently gives other answers than quantitative approaches. They are not just different ways to answer the same questions. The largely quantitative model-focused SPICOSA design leaves little room for less tangible political and social issues. The struggle to find a good quantifiable indicator of user-conflicts in the Risør case may serve as an example. Conflicts often come with a deeply embedded, subjective and context-dependent character. Conflicts are therefore often hard to model in accurate and meaningful ways.

This preference for quantification in SPICOSA reflects the domination of a scientific worldview in the project. The project was interdisciplinary in nature and many disciplines where represented, but nevertheless, in a number of the study sites the imbalance between the disciplines was striking (McFadden et al., 2011). Partly this was due to the individual study site problem formulation, partly as a consequence of the main SPICOSA framework. This call for a quantitative approach and coherence and standardization to make the study sites comparable, facilitate information exchange, etc., limited the numbers and types of problems that could be tackled.

In some cases, natural science or policy analysis can improve coastal management on their own, without being integrated. Horizontal integration and coordination between private actors, stakeholders and municipalities or vertical coordination between governance actors on several levels may be more important than scientific integration. The strong focus of the SPICOSA project on scientific integration will for some cases and policy issues place the integration emphasis in the wrong place. The focus on the science-integrating tool narrows the scope of problems that can be solved. There is a risk that the availability of the SAF as a tool could limit the range of policy issues considered in actual coastal management, similar to how Degnbol et al. (2006) described available policy measures in fisheries management as limiting. Thus, one insight developed from the Norwegian case

[5]An allusion to the British scientist and novelist C.P Snow's classic work *The Two Cultures and the Scientific Revolution,* (Snow, 1960) which addresses the barriers between science and humanities which he saw as hindrances towards further scientific progress.

is that a generic ecosystem-based management approach also has to allow for more open, qualitative and process-oriented approaches, in particularly when complex and intangible policy issues are in focus. In small-scale coastal/fjord settings, the importance of including qualitative approaches is probably even greater than in large-scale settings.

These experiences resemble the observation in the ICZM–EBM debate that there is often an imbalance in the efforts allocated to monitoring the social and ecological processes (Christie, 2011). In EBM, the phenomena of concern are mainly located in the world of nature (Steel et al., 2004), and natural science tend to dominate the framing or construction of environmental problems. It is far too often assumed that what is needed for efficient policy crafting and making is better knowledge of the ecological processes and interconnections, while largely ignoring the incomplete understandings of policy making and ethical and social processes. The importance of interdisciplinary work is more clearly conveyed if we understand environmental problems as "imbalances in coupled social-ecological systems" (Christie, 2011:1).

EBM is typically carried out with natural scientists in prominent positions. They seem inclined to consider the nature as fixed and with objective limitations. They often give primacy to these limitations, and may fall into an authoritarian position towards politicians and other stakeholders, and try to push measures in a top-down manner with little consideration of the existing management systems (Aswani et al., 2011). It is, however, not sufficient to take into consideration the best available ecological science or the socio-economic needs of the community; deeper cultural and political issues also have to be addressed. Usually a pragmatic balance has to be struck between the various goals and interests. Natural scientists sometimes seem to think that we need natural science understanding for efficient policies, and that social science is important only for efficient policy making – i.e. for getting an efficient policy-making process. Efficient policies can only be achieved if they actually get implemented according to their intentions. Social conditions usually affect not only how policies are implemented, but also what the policies actually end up being. Both natural and social science is needed to achieve efficient policies.

When integrating science and policy, the question is whether the balancing act includes sufficient science and enough policy to meet the local political and cultural concerns and stakeholder needs and expectations. EBM should therefore, in line with Aswani et al., (2011), be built on, and integrated with, existing ICZM platforms and principles. Treating the two as incompatible paradigms may lead us further away from sustainable solutions. The link between existing coastal management regimes and EBM is rarely made clear (Aswani et al., 2011). If this link is not considered one risks both making new EBM programs inefficient and ruining the function of existing coastal management regimes.

## 10.10 Five concluding lessons

It is important to bear in mind that SPICOSA was not mainly an attempt to provide national, regional or local political authorities with scientific knowledge relevant for policy- and decision-making. The 18 study sites were mainly testing and development grounds for the System Approach Framework (SAF) application, but with likely positive side effects for the study sites and communities. As stated by Hopkins et al. (2011), the systems approach faced two main challenges: that of simulating complex systems and that of creating an effective science-policy interface.

The case of Risør is so far about scientific knowledge production and not about stakeholder participation and policy implementation. This experimental case has nevertheless given some insights of relevance also to unstudied contexts. It has made clear some challenges for the development and implementation of an integrated ecosystem-based model of different uses of the local cod stock in a south-Norwegian coastal municipality. With emphasis on knowledge production and dissemination, the issue of scale, and the ICZM-EBM-debate, we have developed five lessons.

*Lesson 1: Taking EBM and modeling to support it to a smaller geographical scale creates new challenges for data reliability and reduces the model's usefulness compared to EBM at larger scale.*

A lack of reliable data is common in coastal management, and even more so in EBM-inspired approaches. The availability of relevant, reliable and valid quantitative local data is usually poor. The long and costly data series typically required for good modeling are rarely available. Data that can be used to establish relationships in quantitative modeling usually come from large-scale settings where individual and local differences are evened out. When a model based on such data is used to predict behavior, the uncertainty in the predictions will be bigger, the smaller the geographical scale the model is applied at.

Better data and models may improve the quality of predictions in ecology and other nature sciences, but the same is not necessary true when modeling human behavior. Model-derived knowledge can lead people to change their behavior and attitudes and politicians to change their priorities and policies. Such strategic behavior, where the choice of actions depends on an analysis of how others will react to your actions, is probably more likely in a small-scale setting. The larger the setting, the less influence a single individual's behavior will have on others.

We disagree with Olsen et al. (2011:257) who contend that "ecological, governance and management complexity increases with increasing geographic scale, with associated increases in uncertainty". We think it often is the other way around, and that the complexity of local issues is vastly underestimated. There is, in addition, no reason why complexity and scale should co-vary in the same way for ecology, governance, institutions, user-group structure, and society.

Rather than seeing models as objective and accurate instruments of predictability, they should be treated as producers of rules-of-thumb useful for identifying the order of magnitude of relations, change, etc. A useful side of models is that where stakeholders have some useful data, the models may help them identify which other data should be prioritized for collection, should they choose to use limited resources to gather more data.

*Lesson 2: Heavy emphasis on quantitative modeling reduces the room for input about social and political contexts and processes.*

Aswani et al. (2011) writes that standardized approaches to EBM will fail unless they are contexted appropriately. This is because they will be unable to take full advantage of existing knowledge and institutional structures. EBM needs to be multi-disciplinary and multi-methodological, even though the composition of the various elements may vary from case to case. In complex, small-scale coastal settings the EBM efforts need to resonate with the local cognitive frames of references, to be accepted and integrated in the existing local systems (Aswani et al., 2011).

In the EBM-tradition much attention is paid to measurement and indicators of peoples' values, what trade-offs among ecosystem services people are willing to make, measurements of human well-being, etc. The SPICOSA modeling approach also mainly involved and required quantitative data. The emphasis on quantification allows the use of national or international data sets and approximations in different local contexts, but with inherent uncertainty as predictions of local conditions or developments are made based on this, as we have stated earlier.

This approach also tends to leave out valuable, but not quantifiable, variables and information. The drive towards inter-disciplinarity must include both quantitative and qualitative approaches, not only disciplines. To understand contemporary political issues and processes on different levels, qualitative research is required, and particularly so in local, small-scale settings where quantitative data are neither available nor very useful for describing and understanding the processes. To include also qualitative research on social and political systems may become immensely demanding. We face not only data limitations, uncertainties and challenges related to combining quantitative models, but also the problems of integrating quantitative and qualitative approaches and results.

The scientific approaches chosen and the data used and produced will in itself push the focus of the project in certain directions. Social science-based qualitative data are oriented towards core ICZM issues like interests, perceptions, stakeholders, power, and decision-making, whereas natural science may push in direction of biology, ecology, sustainability, resilience and other typical EBM issues. Nature management is about influencing individual and collective human behavior, and it is clear that to effectively achieve this, both types of knowledge are necessary. Analysis of the given context should be the point of departure for the optimal mix between social and natural science. Such an initial analysis should be made by individuals or groups understanding both camps.

ICZM defined as a tool for solving user-conflicts is strongly dependent on natural science when the given management challenge involves such issues. But generally ICZM is less dependent on natural science than EBM is dependent on the social science perspectives. This is mainly due to the generally more comprehensive ecosystem-related ambitions and more difficult management tasks of EBMs.

Another aspect of SPICOSA that made the approach less sensitive to local conditions was that the aims and requirements of SPICOSA restrained the scope and range of coastal problems and issues that could be addressed in the project. The policy-issue in the Risør case had little local attention and was not publicly understood as an urgent and important issue that required research and policy change. The way the policy-issue came about made it somewhat artificial, and this had consequences and repercussions throughout the project span. This illustrates that too rigid and detailed frameworks can be counterproductive by reducing the opportunities to be context-sensitive.

*Lesson 3: In local EBM collaboration and information exchange between stakeholders, policy-makers and scientists are relatively more important than on large-scale EBM.*

EBM often implicitly make a distinction between knowledge providers and knowledge users. Science provides knowledge and the policy-making system applies this knowledge in policy formation processes. However, knowledge production for EBM at small geographical scales will generally be expensive. Models other than pure science-based ones are required, usually with some forms of stakeholder involvement. There are, at least in local, small-scale settings, good reasons to see the relationship between scientists and stakeholders as a mutual one, where both parties contribute with different, but necessary knowledge inputs. In small settings also, the relationship between researchers and policy-makers can be seen as a kind of partnership, where both parties are dependent on each other and have common interest in what they are collaborating to achieve. Deliberations between the involved are also required to determine which ecosystem services should be prioritized and which ecosystem functions are needed for providing them (Leslie and McLeod, 2007).

The political logic for small coastal settings is often quite different from that for larger scale settings. People are generally more reliant on, and intimately linked to, these smaller coastal systems. The actors on all sides are also different in terms of power, organizational capacity and competence in larger and smaller contexts. In a local context, the differences between the stakeholders, managers and scientists will usually be larger. This means that local knowledge and involvement of local stakeholders and other interest groups are more important, as the reliance of their knowledge becomes more relevant than in large-scale EBM settings. Stakeholder involvement will bring better knowledge of local cultural and institutional practices and most likely also increase the legitimacy of the decisions made.

Stakeholders also have an important role in the reception and dissemination of the provided scientific knowledge. Even if the model developed for Risør is simple and parsimonious, it is still quite complicated. For lay people not used to this way of thinking, it takes much intellectual courage and energy to understand. Participation in the model-generating process would enhance stakeholder ability to understand the outcomes from running the model, and to assess their meaning and relevance. This will also be the case in science-intensive EBM-efforts.

The experiences from the Norwegian SPICOSA study site illustrate that relations and trust between the knowledge providers and the knowledge users, including stakeholders, has to be established to make the implementation efficient. If the scientific findings call for a change in management regime or management measures that will reallocate resources, income, etc., the reliability and legitimacy of the new knowledge are likely to be questioned. Involvement of stakeholders through the whole knowledge generating process and the policy development phase will then help to provide legitimacy.

*Lesson 4: Local government's limited legal power becomes even more problematic in EBM than in ICZM.*

EBM is not only about unifying science disciplines; it is also about unifying decision-making processes. Hardly any country has ecosystem-centered departments, where several or many ecosystem services are subject to holistic, unified decision-making processes. Truly ecosystem-based management seems to require new and appropriate governance frameworks. The challenge is to strike the right balance between change and continuity in the given context.

The Risør case shows that many of the measures suggested by the scenarios have to be implemented by the regional or national fisheries authorities. To introduce such measures, the municipality would have to change the modus operandi of several county, regional state and central institutions. However, if the new knowledge and implementation problems show that the present policy and institutions are inappropriate and has poorly understood consequences, it may be a driving force towards change in the management system itself.

The future of coastal, marine ecosystems is in the hands of policy-makers, resource managers, scientists and stakeholders. It depends on their ability to handle the collected information, integrate it and operationalize it for EBM or ICZM. Such management efforts should therefore take the local political, socioeconomic, and governance realities as their point of departure in its conceptualization and implementation (Aswani et al., 2011), and not introduce new management measure or regimes without considering the existing management frameworks (UNEP, 2011). Natural conditions are of course also an important departure point for management, and for management reform. The likelihood for success is probably larger when EBM can be built on and incorporated in existing ICZM practices. However, ecosystems often cross the boundaries and jurisdiction of several municipalities, counties and state agencies, and fish and many other marine resources move around. This makes the EBM aims harder to achieve than even ambitious ICZM programs.

*Lesson 5: ICZM and EBM should be viewed as different, but complementary approaches on a spectrum where policy issue, scale, urgency and knowledge requirements should determine the approach*

The challenges encountered in the course of this case study illustrate that the step from ICZM to EBM on a local coastal level increases the complexity and the uncertainty of the processes. ICZM has here been understood as a less complex and ambitious management framework than EBM. While ICZM is more often seen as a practical and pragmatic approach to solve user-conflicts, EBM is in addition also explicitly oriented towards ecosystem recovery and maintenance. In practice, the differences between the two approaches are often rather modest. ICZM usually has a very strong focus on the two-dimensional sea-space, and various zoning systems are the most common way to handle user-conflicts. However, EBM is also frequently considered to be inherently place-based and dependent on different kinds of zonings (Crowder and Norse, 2008; Olsen et al., 2011), and MPAs in particular are popular EBM tools worldwide.

Marine Spatial Planning (MSP)[6], MPAs and other spatial approaches can be seen as tools more closely associated with the ICZM approach, but with an additional or stronger focus on ecological criteria. MPAs may ease or solve many of the problems with EBM related to limited data availability, large uncertainties, and costly research. Using MPAs is an efficient response to the insight that for particular species[7], including humans, some places are of much greater importance than others, making it the main task of scientists to identify these places (Crowder and Norse, 2008). There is something simple and intuitive about spatial planning and zoning. In addition, MPAs simplify management tasks, are easy

[6]In EU referred to as Maritime Spatial Planning.

[7]Some species are also more important than others in maintaining the function and resilience of the ecosystems (Crowder and Norse, 2008).

to explain to constituencies (Degnbol et al., 2006), and open up political processes about where to zone, the borders, etc. Invisible and abstract ecosystems are only comprehensible through models, and numbers can be perceived as definite and authoritarian and do not easily lend themselves to political processes. You can negotiate with stakeholders, but not with Mother Nature. Nature management is inherently political and we need governance systems designed to work – both politically and environmentally. MPAs, however, are mainly efficient for relatively stationary species, and cannot solve problems related to pollution, diseases, climate change, shoreline developments etc.

EBM is relatively cumbersome and time-consuming and risks causing delay in addressing the problems at hand. Hence, when dealing with an issue is urgent, ICZM-dominated management processes may be preferable. Typically, an ICZM-approach is more appropriate compared to EBM, the smaller the scale that management occurs at, the more urgency there is in getting issues resolved, and the less relevant codified knowledge that is easily available.

## Acknowledgements

This research was generously supported by grants from European 6th Framework program and the Norwegian Research Council. Thanks to the large number of colleagues in the SPICOSA project that helped us solve interesting problems. We also would like to thank our informants in Risør for their help and contributions to the projects, as well as two anonymous referees whose suggestions greatly improved the text.

## References

Aswani, S., Christie, P., Nyawira, A., Muthiga, N., Mahon, R., Primavera, J.H., Cramer, L.A., Barbier, E.B., Granek, E.F., Kennedy, C., Wolanski, E. and Hacker, S. (2011) The way forward with ecosystem-based management in tropical contexts: reconciling with existing management systems. *Marine Policy* **36**(1), 1–10.

Byron, C., Bengtson, D., Costa-Pierce, B. and Calanni, J. (2011) Integrating science into management: ecological carrying capacity of bivalve shellfish aquaculture. *Marine Policy* **35**(3), 363–370.

Christie, P. (2011) Creating space for interdisciplinary marine and coastal research: five dilemmas and suggested resolutions. *Environmental Conservation* **38**, 1–15.

Cicin-Sain, B. and Knecht, R.W. (1998) *Integrated Coastal and Ocean Management: Concepts and Practices*. Island Press, Washington DC.

Crowder, L. and Norse, E. (2008) Essential ecological insights for marine ecosystem-based management and marine spatial planning. *Marine Policy* **32**(5), 772–778.

Degnbol, P., Gislason, H., Hanna, S., Jentoft, S., Nielsen, J.R., Sverdrup-Jensen, S. and Wilson, D.C. (2006) Painting the floor with a hammer: technical fixes in fisheries management. *Marine Policy* **30**(5), 534–543.

Directorate of Fisheries (2011) Fisheries Statistics, Directorate of Fisheries, Bergen, Norway. http://www.fiskeridir.no/statistikk/fiskeri/fangst-og-kvoter/fangst-paa-kommunenivaa. Retrieved 10 September 2012.

Espinosa-Romero, M.J., Chan, K.M.A., McDaniels, T. and Dalmer, D.M. (2011) Structuring Decision-making for Ecosystem-based Management. *Marine Policy* **35**(5), 575–583.

Forst, M.F. (2009) The convergence of integrated coastal zone management and the ecosystem approach. *Ocean & Coastal Management* **52**(6), 294–306.

Grumbine, R.E. (1994) What is ecosystem management? *Conservation Biology* **8**(1), 27–38.

Hopkins, T.S., Bailly, D. and Støttrup, J.G. (2011) A Systems Approach Framework for coastal zones. *Ecology and Society* **16**(4), 25.

Kay, R. and Alder, J. (2005) *Coastal Planning and Management*, 2nd edn. Taylor & Francis, London.

Knudsen, H., Jorde, P.E., Andre, C. and Stenseth, N.C. (2003) Fine-scaled geographic population structuring in a highly mobile marine species: the Atlantic cod. *Molecular Ecology* **12**(2), 385–394.

Leslie, H.M. and McLeod, K.L. (2007) Confronting the challenges of implementing marine ecosystem-based management. *Frontiers in Ecology* **5**(10), 540–548.

McFadden, L., Priest, S., Sandberg, A., Bailly, D., D'Hernoncourt, J., Reis, J. and Mette, A. (2011) Hindsight is wonderful: simple lessons for increasing science and policy integration. Paper presented at the ICZM 2011 – International Symposium on Integrated Coastal Zone Management 3–7 July 2011 in Arendal, Norway.

McLeod, K.L. and Leslie, H.M. (2009) Why ecosystem-based management? In: McLeod, K. and H. Leslie (eds.) *Ecosystem-based Management for the Oceans*. Island Press, Washington DC, pp. 3–12.

Moksness, E., Gjøsæter, J., Lagaillarde, G., Mikkelsen, E., Olsen, E.M., Sandersen, H.T. and Vølstad, J.H. (2011) Effects of fishing tourism in a coastal municipality – a case study from Risør, Norway. *Ecology & Society* **16**(3), 11.

Olsen, E., Kleiven, A.R., Skjoldal, H.R. and von Quillfelt, C.H. (2011) Place-based management at different spatial scales. *Journal of Coastal Conservation* **15**(2), 257–269.

Snow, C.P. (1960). *The Two Cultures and the Scientific Revolution*, 2nd edn. University Press, Cambridge.

Sorenson, J. (1997) National and international efforts at integrated coastal management: definitions, achievement, and lessons. *Coastal Management* **25**(1), 3–41.

Steel, B., List, P., Lach, D. and Shindler, B. (2004) The role of scientists in the environmental policy process: a case study from the American west. *Environmental Science & Policy* **7**(1), 1–13.

Svåsand, T. (1998) Cod enhancement Studies in Norway – Background and results with emphasis on releases in the period 1983–1990. *Bulletin of Marine Science* **62**(2): 313–324.

Tallis, H., Levin, P.S., Ruckelshaus, M., Lester, S.E., McLeod, K.L., Fluharty D.L. and Halpern, B.S. (2010) The many faces of ecosystem-based management: Making the process work today in real places. *Marine Policy* **34**, 340–348.

UNEP (2011) Taking steps toward marine and coastal ecosystem-based management – an introductory guide. Authors: Agardy, T., Davis, J., Sherwood, K. and Vestergaard, O. UNEP Regional Seas Report and Studies No. 189. UNDP, Nairobi.

van Haastrecht, E.K. and Toonen, H.M. (2011) Science–policy interactions in MPA site selection in the Dutch part of the North Sea. *Environmental Management* **47**(4), 656–670.

Chapter 11

# Governance Systems for Marine Protected Areas in Ecuador

*Vincent Gravez, Robert Bensted-Smith, Pippa Heylings and Taylor Gregoire-Wright*

**Abstract**

This chapter analyzes the emergence of a variety of governance systems for Marine Protected Areas (MPAs) in Ecuador. Recent political change and international commitments have brought about a sharp increase in the number of MPAs and an increased attention to the effectiveness of participatory governance systems for Ecuador's MPAs. The 2008 Constitution affirms the right of citizens to participate in crafting government policy and in the management of protected areas. Some legal mechanisms exist for participation in MPA governance, but participatory management remains the exception. In many cases the lack of preferential access rights and the inability to directly influence resource use rules have discouraged stakeholder participation. In the few MPAs that have adopted participatory governance systems, participation has led to debate and action on long-neglected problems, with positive social and ecological results in some cases, most notably mangrove concessions. Our examination of the current state of MPA governance in Ecuador reveals that legislation does not separate management categories from governance categories and that the government has shown ambivalence, promoting participation and decentralization but constraining real empowerment. Nevertheless, there is an openness to experiment with new systems for participation and new approaches to resource access rights, and the chapter offers recommendations in this regard.

## 11.1 Introduction

Motivated by its commitments under the Convention on Biological Diversity, analyses highlighting the gaps in marine protection (Terán et al., 2006) and increasing concern about the problems facing coastal communities, Ecuador has, since 2005, created seven new mainland marine protected areas (MPAs). Their total area of 238 350 ha (Figure 11.1) is less than 1% of Ecuador's 200-mile zone and is dwarfed by the 13 million ha of the Galapagos Marine Reserve; nevertheless, it represents a significant decision to start

*Global Challenges in Integrated Coastal Zone Management*, First Edition. Edited by Erlend Moksness, Einar Dahl and Josianne Støttrup.

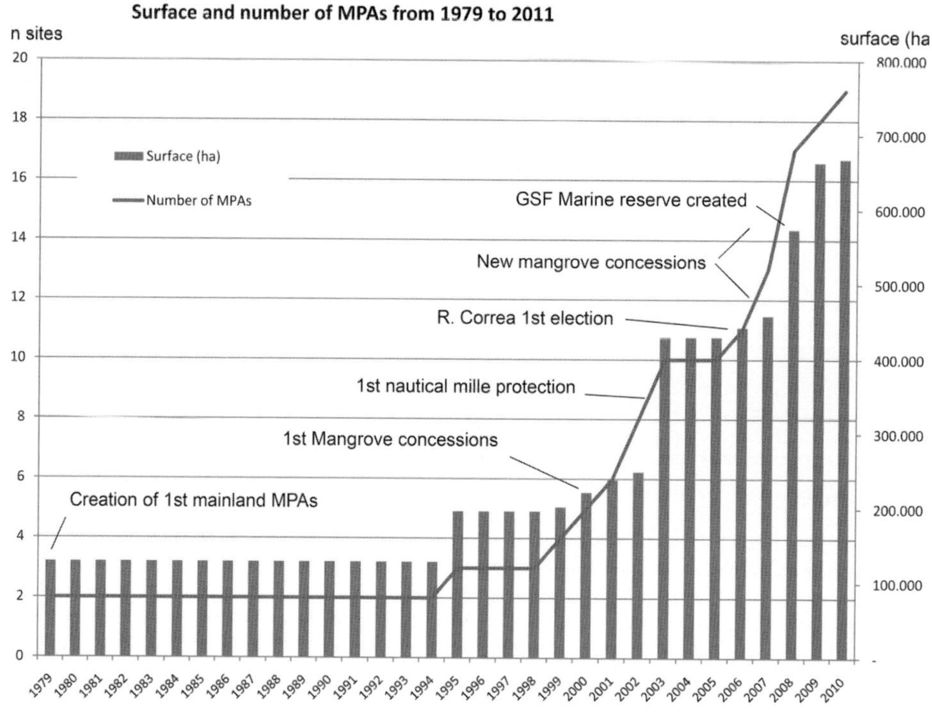


**Figure 11.1** Number and surface of the mainland's MPAs created by the Central Government Ecuador from 1979 to 2010 (Ministry of Environment and Ministry of Agriculture and Fisheries). Data except Galapagos Marine Reserve (created in 1998).

tackling the decline of marine biodiversity and associated livelihoods. By 2013, Ecuador's current National Plan will have increased the mainland MPA area by a further 252 100 ha, that is almost doubling the area under protection in 2009 (SENPLADES, 2009). This sharp increase in MPA creation demands that Ecuador also strengthen MPA governance, which in many cases has been ineffective. In this context, Ecuador needs to re-examine the roles of local stakeholders and communities in MPA governance, both because the constitutional and legal framework for participation has changed, as explained later, and because of the growing evidence that well-designed participation mechanisms are an important factor in MPA effectiveness. In a study of coral reef MPAs in Indonesia, Philippines and Papua New Guinea, Green et al. (2011) conclude that "Common indicators of success such as increased fish catch and habitat quality variables were consistently associated with several independent variables: sustainable financing for management, clarity of MPA network rules, enforcement by community level enforcers, local skills development, and involvement in management by local elected politicians, functional management board, multi-stakeholder planning mechanisms and participatory biophysical assessments". Christie & White (2007) cite several studies to support the statement that "empirical research demonstrates that when collective action and conflict resolution mechanisms break down, MPA effectiveness rapidly deteriorates". In a study of 127 marine reserves, Pollnac et al. (2010) concluded that high levels of compliance with reserve rules were related to complex social interactions rather than simply to enforcement of reserve rules. Thus, while participation can have high transaction costs and does not in itself ensure MPA effectiveness, there is evidence that it can be an important factor in MPA effectiveness and can make MPA governance more resilient in the face of complexity or conflict.

Ecuador's 2008 Constitution establishes the right of citizens to participate in crafting policy at all levels of government and requires that the State "foster [citizen] participation" in the administration and management of protected areas. The 2010 Citizen Participation Law defined the mechanisms for participation in policy formulation[1]. These two instruments, along with the 2010 Code for Territorial Organization, Autonomy and Decentralization[2], form a new constitutional and legal framework intended to make the Ecuadorian State more decentralized and democratic in its distribution of power.

Nonetheless, Ecuador's effort to decentralize state power and increase citizen participation in government has been insufficient to bring about participatory management of MPAs. It could be that the legal framework itself is over-prescriptive about participation mechanisms. It could be that efforts to transition from authoritative to participatory management are being hindered by conflicts relating to the increasing pressure on resources in areas of high poverty (40% of poverty and 11% of extreme poverty in the coastal area; SENPLADES, 2009). Issues of institutional capacity, coordination and confidence in the effectiveness of participatory management may also be a factor. With this investigation we assess approaches to governance in established MPAs and in various recent marine initiatives, highlight key factors affecting progress, and consider the implications for those seeking to establish effective participatory governance systems in MPAs.

## 11.2 Governance systems and participation in Ecuador's MPAs

Ecuador's National System of Protected Areas (NSPA) is composed of four subsystems, Natural Heritage Areas (NHA), Protected Areas of Local Governments, Community Areas, and Private Areas. The Ministry of the Environment (MoE) is responsible for crafting regulations and coordinating the management of NSPA. As can be seen from the categories of the subsystems, Ecuador has not separated conservation categories from governance categories as independent dimensions of any protected area, in the manner promoted by the revised IUCN guidelines on categories and governance of protected areas (Dudley, 2008). For example, the areas of highest biodiversity are awarded the Natural Heritage Area status and are automatically governed by the State, thereby disregarding any opportunities for

[1]Asamblea Nacional del Ecuador, "Ley Orgánica De Participación Ciudadana", Suplemento del Registro Oficial 175, 20-IV-2010, 26, 2010.

[2]Asamblea Nacional, "Código Orgánico De Organización Territorial, Autonomía y Descentralización" (COOTAD), Publicado en el Suplemento del Registro Oficial N° 303 de martes 19 de octubre del 2010, 2010.

local governance mechanisms or for more complex cross-scale institutions (Rydin, 2006). To date, all the marine and coastal protected areas fall within the NHA subsystem, with the exception of a large number of small mangrove areas that can be classed within the Community Areas subsystem.

## *11.2.1 Natural Heritage Areas (NHA)*

The MPAs of the NHA subsystem are State controlled and governed, and are managed by the MoE through the Undersecretary of Marine and Coastal Management. At the operational level, each protected area is administered by the relevant Provincial Office of the MoE. There are two exceptions to this, in which the Ministry of Environment co-manages the area with another State actor, the Ministry of Defense: Arenillas Ecological Reserve and Santa Elena Wildlife Reserve.

Although decision-making in the mainland MPAs is generally top-down, the legal framework does provide for a consultative level of local stakeholder participation through Management Committees (*Comités de Gestión*), as defined by Ecuador's Environmental Law[3]. A Management Committee can provide input to the MoE about the community's management preferences and can support the implementation of local management activities. They can be made up of local government, communities, and legally recognized organizations, such as fishing cooperatives.

At the beginning of 2010, only four of the fourteen mainland MPAs had legally recognized Management Committees[4], while four other Management Committees were in the process of being created[5]. Six had no Management Committees and, therefore, were lacking in any recognizable framework for participatory governance (Jones et al., 2011) (Figure 11.2).

Management Committees can be useful allies to the MPA managers, but they face serious challenges: a lack of proper administration skills, a lack of clarity regarding the mechanisms for influencing decision-making and the limits of participation, a lack of communication channels, internal administration that is usually not proactive, and a lack of technical and operational training (FFLA, 2011).

Though established under the same legal framework as existing mainland MPAs, the Galera-San Francisco Marine Reserve (GSFMR) is pioneering a different approach to participatory governance. The creation of the Reserve in 2008 was a genuinely grass-roots initiative, supported by a local non-governmental organization, the Nazca Institute for Marine Research. That local protagonism has been largely maintained through the lengthy management planning process, which itself has broken new ground in Ecuador. The GSFMR Management Plan, scheduled for approval in 2011, gives local communities preferential access to marine resources and establishes a zoning scheme with substantial no-take zones and community-managed zones, both firsts for mainland Ecuador. If the

---

[3]Agreement No. 051 of the MoE.Libro IV, *Texto Unificado de Legislación Secundaria del Ministerio del Ambiente*.

[4]Machalilla National Park, Manglare El Salado Wildlife Production Reserve, Manglares Churute Ecological Reserve, Arenillas Ecological Reserve (Memoria de taller de coordinación para la construcción de la Red de Áreas protegidas Marino-Costeras, FFLA-SGMC, 23 de abril 2010, Manta).

[5]Galera - San Francisco Marine Reserve, Isla Corazón y Fragatas Wildlife Refuge and Puntilla Santa Elena Bioaquatic Reserve Zone.

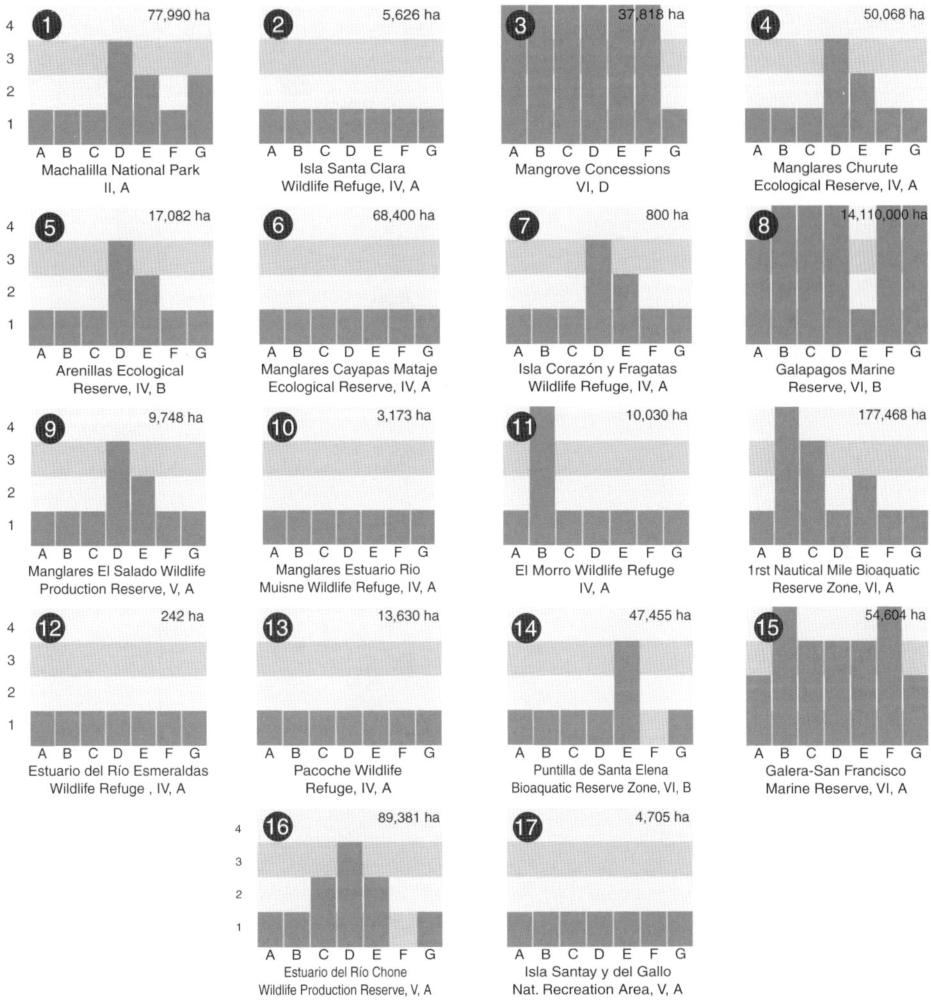


| | A<br>Participation Level (Borrini-Feyerabend model, 2007) | B<br>Implementation of the participation level | C<br>Implementation of preferential acces to resources | D<br>Institutionalization of participation | E<br>Capacity of stakeholders to influence decisions about rules | F<br>Biological monitoring for/by local stakeholder | G<br>Institutionalization of interinstitutional cooperation |
|---|---|---|---|---|---|---|---|
| 4 | f) Private / customary rights recognized & stakeholders assist in management | Yes | Preferential access rights effectively enforced | Participative management system | Can make final, binding decisions | Yes | Functioning interinstitutional agreement |
| 3 | d) Stakeholders involved in decision-making negotiations | | Preferential access rights exist but are poorly enforced | Consultative participation | Can make decisions, but lack the final say | | |
| 2 | c) Consensus sought | | Preferential access rights under discussion | | Consultative | | Agreement planned but not implemented |
| 1 | b) Stakeholders informed/consulted | No | No preferential access | No formalized participation | None | No | None |
| 0 | a) Effectively no participation | | | | | | |

**Figure 11.2** Location map. For site names, see Figure 11.3.

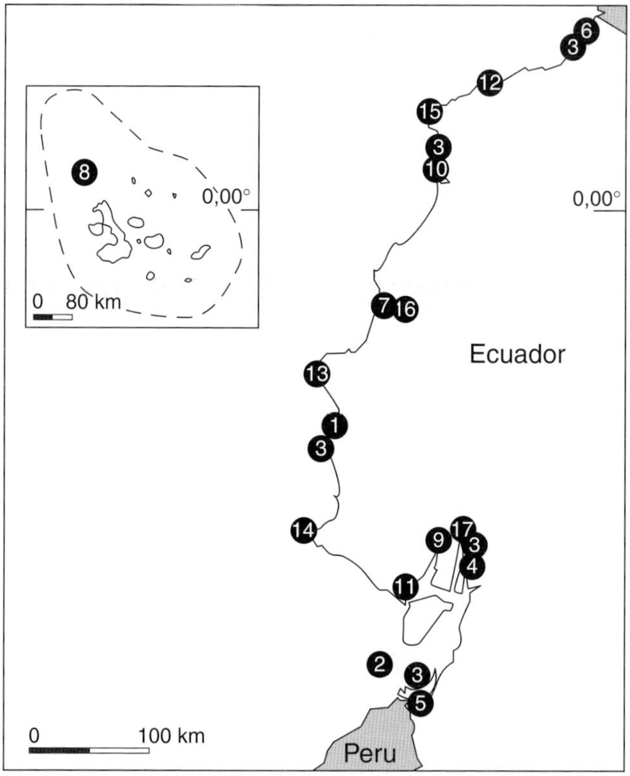


**Figure 11.3** Comparison of some characteristics of the governance system for each MPA in Ecuador. Data are represented on a 5 level scale, depending on the status of the themes of participatory management, empowerment and inter-institutional cooperation. Names of the MPAs, IUCN management categories and governance type (I-VI; A-D: Dudley, 2008) are below each graph. Sources: MoE (Undersecretary of Marine and Coastal Management), FFLA (2011), Workshop for construction of MPA network (MoE-FFLA, 2010, unpublished), Coello et al. (2008), authors' experience in Machalilla National Park, El Morro MPA, Galera San Francisco, and Jambeli Mangrove concessions.

*de facto* local protagonism leads to formalized empowerment and improved MPA management efficacy, then GSFMR can become an important model for other MPAs of the National Heritage Areas system. Even before formal approval of the Plan, there is anecdotal evidence (Soledad Luna, Nazca Institute, pers. comm.) that the participatory governance initiative has brought marked social benefits in terms of local communities' organization, leadership skills, technical knowledge and self-esteem. Some ecological benefits are also emerging, with local proposals for No Take Zones and an incentive-based scheme to restore lobster populations, which is already yielding encouraging results.

One exception within the NHAs is the Galapagos Marine Reserve (GMR). It was created in 1998 through the *Special Law of the Galapagos*[6], which provides for an innovative,

[6]Ley Orgánica de Régimen Especial para la Conservación y Desarrollo Sustentable de la Provincia de Galápagos (LOREG; Registro Oficial 278 del 18 de marzo del 1998).

two-tiered participatory governance system, with a local, consensus-based board positioned under the Inter-institutional Management Authority (IMA; Figure 11.2). At the local level, participation is organized through a Participatory Management Board (PMB), composed of the Galapagos National Park authority, the fishing sector, the tourism sector, conservation and education interests, and tour guides. Above the PMB sits the IMA, made up of government ministries and members of the PMB, which allows binding decisions to be made in the case of a failure to arrive at a consensus at the local level (Heylings & Bravo, 2007). Berkes (2006) highlights the importance of complex systems of governance that include cross-scale institutions for MPA, and in this regard the GMR is an example of a governance system that brings actors together horizontally across sectors and vertically between local and national actors and authorities.

In its 10 years of experience, this participatory governance model has been successful in excluding industrial fishing from the GMR, in managing some but not all areas of conflict between stakeholders and administration of the park, institutionalizing dialogue, increasing acceptability of No Take Zones to fishermen (by negotiating formal consensus on their location), obtaining a high degree of compliance with consensus-based quotas for lobster, agreements to close depleted fisheries, controlling fishing fleet capacity, building strong inter-institutional cooperation and building capacity and a culture of consensus and of participation in the Galapagos. However, despite the cross-scale nature of the governance mechanism and its social resilience over the past decade, the system has been unable to cope with external driving forces, in particular from the international markets for high-value products, such as sea cucumber and shark fins (INGALA, 2007; Defeo et al., 2009).

### *11.2.2 Mangrove concessions*

The MoE can also authorize concessions of mangrove ecosystems – legally recognized as public goods[7] – to traditional user communities[8]. These Agreements of Sustainable Use and Custody of Mangroves are 10-year contracts between the MoE and organized groups of users, that allow a community or cooperative to take custody of the mangrove ecosystems in order to harvest crabs and molluscs and prevent deforestation (timber harvesting is prohibited). At the time of writing, 40 of these mangrove concessions exist, totaling 37 818 ha.

There is no standard mode of participation and each concession-holder is responsible for organizing its own system of governance and self-regulation, which is generally a product of its internal institutions and legal framework (e.g. the fishing cooperative's rules and procedures; Figure 11.2). In the 10 years since the first concession was granted, many of the early problems and threats have been overcome and the majority of concessions are now proving successful in curbing deforestation, sustaining increased yields of crab and molluscs, improving livelihoods, empowering the concession holders and reducing conflicts with shrimp farms (Coello et al., 2008). The mangrove concessions are notable because they are the only kind of MPA in Ecuador where local stakeholders have exclusive

[7]Ley Forestal y de Conservación de Áreas Naturales y Vida Silvestre. Ley No. 74. RO/ 64 de 24 de Agosto de 1981 integrada en Texto Unificado de la Legislación Ambiental Secundaria publicado en la Edición Especial No. 2 del 31 de marzo de 2003, mediante Decreto Ejecutivo 3516.

[8]Art. 19, Libro de Recursos Costeros, Texto Unificado de la Legislación Ambiental Secundaria (Ibid) and Acuerdo Ministerial 129 del 11 de agosto del 2010.

usufruct rights and a clear legal basis for their management role. Due to their success and supportive government policy, mangrove concessions are expanding rapidly; half of the current concessions have been created between 2006 and 2011, though in total area they only include 6% of all mainland MPAs.

### 11.2.3 *Fishing reserve*

MoE is not the only government institution authorized to declare MPAs. The Ministry of Agriculture and Fisheries (MAF) declared in 2003, and revised in 2007, a Reserve Zone for the Reproduction of Aquatic Species that stretches from the shoreline to one nautical mile out to sea[9]. This established a 177 468 ha multiple-use protected area with specific regulations for the entire coastline, including the regulation of artisanal fishing gear and methods and the exclusion of industrial fishing.

In 2009, the MAF declared a second Reserve Zone, to manage and conserve the bioaquatic resources of the Chone River estuary (89 381 ha[10]). A management plan, prepared with full community participation, is awaiting approval.

Both inside and outside the Reserve Zones, decisions about size limits, closed seasons and other regulations are made by the Undersecretary for Fishery Resources, on the basis of advice from the National Fisheries Institute and with influence by the fishing sector, including the politically powerful national artisanal fisheries organization[11] and the industrial fishing lobby. However, a basic problem is that the Undersecretary for Fishery Resources lacks the resources and inter-institutional cooperation necessary to enforce these regulations along Ecuador's extensive coastline, so they exist only on paper. The category of Reserve Zone for the Reproduction of Aquatic Species does not imply any participatory governance system (Figure 11.2). However, the drafting of new fisheries legislation, which is currently under way, could be an opportunity to authorize new approaches to governance of Reserve Zones that would give coastal communities a role in deciding the local regulations and in ensuring compliance, with the incentive of preferential access to the fisheries resources in their Reserve and mechanisms established to deal with local conflicts. These new approaches draw on the successful examples cited in the Introduction, would have to be tested and may not be sufficient in themselves to overcome the whole range of problems.

## 11.3 Ecuador MPA governance systems analysed according to the IUCN governance matrix

The IUCN governance classification for MPAs highlights the fact that governance dimensions should be independent of management categories that define levels of protection (Dudley, 2008). For any category of protected area, four governance types are

[9]Acuerdo ministerial 134 del 24 de julio del 2007, Reforma al Acuerdo Ministerial No. 03 316, Registro Oficial N° 151 del 20 de agosto del 2007.

[10]Acuerdo Ministerial No. 009, del 30 de enero del 2009 para el manejo de los Recursos bioacuáticos en el estuario del río Chone.

[11]The FENACOPEC, National Federation of Artisanal Fisheries Organizations.

posited as possibilities depending on the local enabling conditions for each governance system to achieve the biodiversity conservation and livelihood improvement objectives: (A) Government Managed Protected Areas; (B) Co-Managed Protected Areas; (C) Private Conserved Areas; and (D) Community Conserved Areas. This classification system allows us to observe the diversity of governance types being practiced in Ecuador (see Figure 11.2).

Of the seventeen MPAs, that are not mangrove concessions, fourteen are Type A, directly administered by central government through provincial offices of the MoE or the MAF, without any delegation to other institutional actors. Three areas classify as Type B, Co-Managed Areas: the Galapagos Marine Reserve with its special regime described above, and two areas under joint management between the MoE and the Armed Forces. Under Type D governance are the numerous but small mangrove concessions, for which management is delegated from central government to local organizations.

Among the range of governance types presented, various options exist for the levels of participation provided for by the prevailing authority. These options create what Borrini-Feyerabend (2007) calls a "continuum of governance options" (Figure 11.2). It is notable that of the 14 Type A areas, 13 have participation characterized as "Stakeholders informed/consulted" whilst only one other is characterized as "Consensus sought" together with benefit sharing and preferential access incentives. This exceptional case is the Galera San Francisco Reserve.

## 11.4 Discussion

Following a history of ineffective marine governance in Ecuador, with little opportunity for diversity in governance mechanisms, the past few years have seen a push by the government to create more MPAs and the enactment of legislation about participation. We discuss here three issues in relation to these changes: (i) the respective roles of government and stakeholders in governance, (ii) governmental organization and coordination, and (iii) incentives for participation and self-regulation.

### *11.4.1 Roles of government and stakeholders*

In light of the MoE's lack of resources and the well-documented tendency for common pool resource stakeholders to undermine rules imposed by an outside authority (Ostrom, 2005), a strategy of imposing and enforcing regulations is almost certainly cost prohibitive. As stated in the introduction, experience from elsewhere suggests that the perceived legitimacy of rules can be enhanced and compliance increased, if stakeholders participate in setting the rules. Yet the information presented here reveals the government's ambivalence regarding participation in decision-making. The government has shown far greater commitment and innovation, in relation to MPA establishment and management, than any predecessor and promotes participation in MPA management, yet it has been hesitant about if and how to empower stakeholders. The Constitution gives the central government exclusive jurisdiction over national protected areas but simultaneously provides for local participation in their governance[12], which is encouraged by the Citizen Participation Law. The government

[12]Constitution of Ecuador, Art. 405.

is enthusiastically expanding mangrove concessions, which have a high degree of local empowerment, *but* has done little to develop new legal instruments that expand community participation in decision-making in new, national MPAs. Even the purely consultative Management Committees are operating in only four mainland MPAs. The government has responded positively to the community-led initiative at GSFMR and acted on various community requests, but in GMR is proposing legal reforms that would weaken stakeholder participation in decision-making by changing the Participatory Management Board into a purely consultative body and eliminating the Inter-institutional Management Authority that was providing the cross-scale nature of its decision-making structure. Underlying this ambivalence may be concerns that participation could prevent the taking of tough conservation decisions in the national interest, impede efficient decision-making, obstruct large-scale economic development, or simply be "inappropriate" for national heritage. Similar concerns may have driven the mixing of conservation and governance categories in the Constitutional definition of the NSPA. One could assess the extent to which each concern is valid, but in any case the design of participatory governance systems needs to address them and demonstrate that they can function better than current approaches. In particular:

- The decision-making process must respect the MoE's responsibility to set the policies and boundaries for local decision-making and to have the ultimate say. This must be balanced by mechanisms to promote consensus-seeking with stakeholders and to ensure that the MoE cannot over-ride local decisions arbitrarily but only after due process and with documented justification.
- The mangrove concessions and the new, locally driven MPAs, especially GSFMR, present opportunities to build the confidence of the MoE in local communities as valuable, trustworthy partners, able to participate efficiently in taking and implementing decisions, including tough ones.
- Legal empowerment can be complemented through strengthening local stakeholder capacities in organization, access to information, use of information, communication, alliances, etc. The GMR and GSFMR are both good examples of this.

## *11.4.2 Governmental organization and coordination*

The above comments about refining the participatory governance system to balance a variety of factors are consistent with recent literature on governance of marine resources. Berkes (2006) and Rydin (2006) highlight the need for governance analysis to move beyond the early Ostrom (1990) analysis of local community-based, self-governing and self-regulating strategies to an understanding of complex, interdependent systems that recognize the interplay (Young, 2006) between multi-sectoral and "cross-scale" institutions.

The data in Figure 11.2 shows that only GMR scores highly on institutionalized coordination between institutions; in fact, the key body for that purpose, the GMR Inter-institutional Management Authority is likely to be disbanded soon, according to the proposed reform of the Galapagos Special Law submitted for debate to the National Assembly in December 2011. Elsewhere, inter-institutional coordination is inconsistent and there is a need for Ministries, Navy and local authorities to show greater commitment to collaboration and coordination for effective MPA management.

A significant factor in the development of cross-scale institutional relations is the process of decentralization in Ecuador. Local governments are assuming new responsibilities in such areas as development planning, biodiversity conservation, and environmental management (including waste and pollution)[13]. The law requires that these new responsibilities be fulfilled through participatory processes. In the case of protected areas, the Constitution gives the central government exclusive jurisdiction but allows for collaboration with other levels of government, leaving open the possibility of MPA governance systems that are multi-sectoral, cross-scale and participatory. Thus, decentralization is both a challenge and an opportunity to empower local governments and communities. Much will depend on the pioneering mainland MPAs, notably GSFMR, to demonstrate the feasibility and effectiveness of such an approach to governance.

### *11.4.3 Incentives for participation and self-regulation*

The recent UNEP MPA case study analysis (Jones et al., 2011) highlights the importance of incorporating multiple incentives within the design of governance mechanisms rather than limiting analysis to a rigid set of particular regimes and a debate over the relative values of top-down or bottom-up strategies. These authors define incentives as "*institutions that are instrumentally designed to encourage people to choose to behave in a manner that provides for certain strategic policy outcomes, particularly biodiversity conservation objectives, to be fulfilled*". The results of this current analysis are consistent with the UNEP results in terms of the importance of mixing legal, economic and participatory incentives in order to achieve social and ecological objectives; what the UNEP study calls the interplay of State steer, market steer and people steer. Our study highlights the importance of preferential access to marine resources as a critical incentive for participation and self-regulation.

Ecuadorians have been well educated about their right to participate, but the free-rider problem (Ostrom, 1990) dissuades many from actually participating. That is, when participating involves crafting regulations that constrain the ability of one's own community to appropriate natural resources upon which it depends, one must be confident that the community will be the beneficiaries of their own self-restraint and that people from outside the community (the free-riders) will not be allowed access to those resources without compensating the community. Otherwise, there is no incentive to craft the conservation regulations in the first place. Though the principle may be self-evident, there are ecological and social complexities, resulting issues of scale, not to mention a long history of open access with little or no enforcement outside GMR. The results in Figure 11.2 show that, while open access is still the norm, there are significant pioneering initiatives. The mangrove concessions are the outstanding example, where exclusive access to crab and mollusc resources has in most cases motivated participation in decision-making, self-regulation and protection in the field. In GMR, the artisanal fishing sector as a whole have exclusive access to fisheries. However, the sector is numerous, heterogeneous and dispersed between islands, so, while the exclusivity led initially to widespread compliance with negotiated regulations, there was insufficient enforcement to prevent free-riding within the sector.

[13]Art. 264 and 267 of the Constitution, and Art.54, 64, 430, 431 and 432 of the "Código Orgánico De Organización Territorial, Autonomía y Descentralización", ibid.

This soon undermined compliance by other individuals and conflicts arose. This highlights the need for exclusivity of access to be at the appropriate scale and supported by enforcement when needed. Preferential access is envisaged, but not yet applied, in the first nautical mile zone and in the Chone Fisheries Reserve. If its proposed management plan is approved, GSFMR would be the first mainland MPA of the NHA to design and implement preferential access, highlighting once again the role of GSFMR as a pioneer for the whole coast.

## 11.5 Conclusions

Responding to a new Constitution and legislation, as well as international commitments, and recognizing that marine governance has hitherto been largely ineffective, Ecuador is developing new approaches to MPA governance. However, the majority of MPAs are still governed centrally by the State and their effectiveness is limited by being isolated within the Ministry of Environment, with poor inter-institutional collaboration, despite the complex, multi-level and multi-sectoral challenges facing MPAs. Effective MPA management in Ecuador will also require greater government commitment to inter-institutional collaboration and coordination. Decentralization presents both a challenge and an opportunity for cross-scale coordination and local empowerment. These conclusions are consistent with those of Berkes (2006) and Rydin (2006) in terms of the multi-sectoral and cross-scale nature of governance.

Our study identifies several areas of innovation but also reveals the government's ambivalence about taking definitive legal steps to empower local communities to participate in MPA decision-making. This stems in part from constraints in the Constitution itself and in part from concerns about whether participatory governance can indeed deliver the effective conservation of national assets. There is apparent openness, however, to enabling GSFMR – and potentially other sites – to experiment with new approaches involving greater empowerment, provided that they respect the MoE's ultimate decision-making responsibility.

The mere existence of a participatory body and mechanisms does not guarantee stakeholder empowerment and participation; incentives for participation and self-regulation are also essential. Ecuador is taking steps towards the crucial incentive for local communities of preferential access to the marine resources, which they depend on and are helping to protect.

The conclusion of this study is consistent with the findings of the recent UNEP MPA case study analysis (Jones et al., 2011) concerning the interplay of legal, economic and participatory incentives, characterized as State steer, market steer and people steer.

In all the above, the areas pioneering new approaches have a vital role, not only for their innovation but also in terms of building the confidence of government institutions in the commitment and capacity of local communities to conserve their MPAs. The mangrove concessions, in particular, have paved the way for increased community responsibility and self-regulation in return for exclusive access to resources. Amongst the MPAs of the NSPA, GSFMR, which has always been a locally driven initiative, stands out as the testing ground for innovation in participatory governance, preferential access and use of local knowledge. It is essential that both government and supporting research and conservation organizations seize this opportunity to learn from these experiments in participatory governance, by monitoring both their implementation and their social and ecological impacts. This will provide evidence on which both government and coastal communities can decide if and how to travel further down the road of empowerment and incentives.

## Acknowledgements

The project for Innovative Governance Models for Marine Protected Area Management in Ecuador is helping to establish participatory governance systems at various sites along the coast of Ecuador. It is using the experiences gained at these sites to inform the development of a national network of marine protected areas. The project is supported by the Darwin Initiative, FFI's Halcyon Fund, Walton II project (through Conservation International-Ecuador) and the IUCN Netherlands Ecosystem Grant Program.

## References

Berkes, F. (2006) From community-based resource management to complex systems. *Ecology and Society* **11**(1), 45: http://www.ecologyandsociety.org/vol11/iss1/art45/ (accessed 10 September 2012).

Borrini-Feyerabend, G. (2007) The "IUCN protected area matrix"—A tool towards effective protected area systems. Report: IUCN World Commission on Protected Areas Task Force: IUCN Protected Area Categories. 12 pp.

Coello, S., Vinueza, D. and Alemán, R. (2008) Evaluación del desempeño de los acuerdos de uso sustentable y custodia de manglar de la zona costera del Ecuador. Informe: Ministerio del Ambiente del Ecuador, CI, UICN y PRODERENA. 52 pp.

Christie, P. and White, A.T. (2007) Best practices for improved governance of coral reef marine protected areas. *Coral Reefs* **26**, 1047–1056.

Defeo, O, Castilla, J.C. & Castrejón, M. (2009) Pesquerías artesanales de invertebrados en América Latina: paradigmas emergentes de manejo y gobernanza. In: C. Lodeiros, J. Alió, L. Freites, N. González, A. Guerra and M. Rey (eds). *II Foro Iberoamericano de los Recursos Marinos y la Acuicultura*, pp 89-117. Cumaná, Estado Sucre, Venezuela.

Dudley, N. (Ed.) (2008) *Directrices Para la Aplicación de las Categorías de Gestión de Áreas Protegidas*. Gland, Switzerland: UICN. x + 96pp.

Fundación Futuro Latinoamericano (2011) Gobernanza en las Áreas Protegidas Marinas y Costeras: el caso del Ecuador, FFLA publ., Quito, 40 pp.

Green, S.J., White, A.T., Christie, P., Kilarski, S., Meneses, A.B.T., Samonte-Tan, G., Bunce Karrer, L., Fox, H., Campbell, S. and Claussen, J.D. (2011) Emerging marine protected area networks in the coral triangle: lessons and way forward. *Conservation and Society* **9**(3), 173–188.

Heylings, P, and Bravo, M. (2007) Evaluating governance: A process for understanding how comanagement is functioning, and why, in the Galapagos Marine Reserve. *Ocean & Coastal Management* **50**, 174–208.

INGALA (2007) Informe Galápagos 2006–2007. Parque Nacional Galápagos, Instituto Galápagos, Fundación Charles Darwin edit., 160 pp.

Jones, P.J.S., Qiu, W., and De Santo, E.M. (2011) Governing Marine Protected Areas—Getting the Balance Right. Technical Report, United Nations Environment Programme.

Ostrom, E. (1990) *Governing the Commons. The Evolution of Institutions for Collective Action*. Cambridge University Press, 280 pp.

Ostrom, E. (2005) *Understanding Institutional Diversity*. Princeton University Press. 335 pp.

Pollnac, R., Christie, P., Cinner, J.E., Dalton, T., Daw, T.M., Forrester, G.E., Graham, N.A.J. and McClanahan, T.R. (2010) Marine reserves as linked social–ecological systems. *Proceedings of the National Academy of Sciences* **107**, 18262 -18265.

Rydin, Y. (2006) Institutions and networks: the search for conceptual research tools. In Rydin, Y. and Falleth, E. (Eds) *Networks and Institutions in Natural Resource Management*. Cheltenham, UK: Edward Elgar Publishing, pp. 15–33.

SENPLADES (2009) Plan Nacional Para El Buen Vivir 2009–2013: Construyendo Un Estado Plurinacional e Intercultural, 520 p., Secretaría Nacional de Planificación y Desarrollo – SENPLADES, Quito, Ecuador.

Terán, M.C., Clark, K., Suárez, C., Campos, F., Denkinger, J., Ruiz, D. and Jiménez, P. (2006) Gap Analysis and Identification of High-priority Areas for the Conservation of the Coastal and Marine Biodiversity of Continental Ecuador. Executive summary. Ministerio del Ambiente. Quito, Ecuador, 32 pp.

Young, O. (2006) Vertical interplay among scale-dependent environmental and resource regimes. *Ecology and Society* **11**(1), 27. www.ecologyandsociety.org/vol11/iss1/art27/ (accessed 10 September 2012).

Chapter 12

# The Battle for Space – the Position of Norwegian Aquaculture in Integrated Coastal Zone Planning

*Bjørn Hersoug*

**Abstract**

Even if Norway has a long coastline and a small population, there is a battle for space with a steady increase in the number of stakeholders. Besides the traditional activities of fishing and sea transport, the oil and gas sector, energy (windmills at sea), tourist sector and nature conservation are all important stakeholders in the coastal zone. The most important in the near-shore area is aquaculture, which has increased production from zero to one million tonnes within 40 years. In spite of this success, aquaculture farming is a contentious activity, creating considerable conflict. This chapter demonstrates the actual area occupation by aquaculture, and shows how the industry has to compete with other stakeholders within the framework of a complicated planning system.

## 12.1 Introduction

Even if Norway has a coastline of 103 000 km (including inlets and islands), an Exclusive Economic Zone (EEZ) of nearly one million km$^2$ and only 5 million people, the battle for space is becoming more critical by the day. There is a steady increase in the number of stakeholders demanding more space, and occasionally, exclusive use. Apart from obvious activities like fishing and sea transport, there is a rapidly increasing oil and gas sector all along the Norwegian coast. Aquaculture has made increasing demands on sea areas since the early 1970s, while the energy sector (windmills at sea and tidal water mills) is a new actor. The tourist industry has been an important stakeholder for many years, especially regarding tourist fishing, while recreational activities have been increasing all along the coast, for both Norwegian and foreign tourists. At the same time, conservation interests are increasing, partly inspired by international agreements signed by Norway (Ese, 2009), but

*Global Challenges in Integrated Coastal Zone Management*, First Edition. Edited by Erlend Moksness, Einar Dahl and Josianne Støttrup.

also by a stronger desire to protect valuable areas, not only on the coast but also at sea, through marine protected areas (DN, 2010).

What is new with these activities is that they have all increased their ambitions, occasionally they require exclusive use, they are often protected by special laws, and certain sectors (like oil and gas exploration and aquaculture) require rapid answers. Finally, it is important to emphasize that not all sea space is equally valuable. In the aquaculture sector, the actors are interested in the so-called "*super localities*", that is, the best localities in terms of depth, currents, and exposure to waves, besides good access from land for improved logistics. This chapter will concentrate on the aquaculture sector and its demands for more space in the coastal zone – by many considered the most important limiting factor for future development. Why is aquaculture such a contested activity? What is the actual use of sea space of this sector? What are the conflicts with other stakeholders, and how are these conflicts dealt with?

This chapter aims to give a short introduction to some of the research issues dealt with by the large iCoast research project[1], set up to improve the processes of integrated coastal zone management, with a special emphasis on the use of *sea areas*. While conflicts over use of the near-shore marine areas seem to be a common phenomenon all over the world (Tett et al., 2011), the management systems developing to handle these conflicts will largely depend on the nature of the interests involved and the specific needs they have. For obvious reasons the management of the Great Barrier Reef in Australia will be quite different from managing the Norwegian coastal commons, although some *planning techniques,* like area zoning, will be similar. In Norway, the rapid expansion of the aquaculture industry in the early 1980s prompted the set-up of a system of integrated coastal zone management with specific focus on the use of sea areas (Bennett, 2000). Before we go into the aquaculture industry, we need to have a quick glance at the Norwegian planning system, which constitutes the framing of all development activities in the coastal zone.

## 12.2 A complicated planning system

Before 1970, there was limited need for planning of sea areas in the coastal zone. The two main activities, fishing and sea transport, had co-existed for hundreds of years, and even with dramatic changes in the size of vessels and fishing technologies, both industries were able to adapt. Sea lanes were fixed, and fishers knew that putting lines or nets in the main transport corridors would always imply risks, while the transport sector would respect fishing boats actively engaged in harvesting. Allocation of sea space took place during the large seasonal fisheries, as in the important cod fisheries in Lofoten, but not on a permanent basis (Jentoft and Kristoffersen, 1989; Holm et al., 2000). Sea space was allocated to different gear groups for the duration of the fisheries, not in order to regulate the resources, but to maintain order and escape traffic chaos. Fishers with different gears and

[1]The project is initiated by the Norwegian Research Council and comprises 25 researchers belonging to 11 research institutions, mainly located in Trondheim and Tromsø. The project is multi-disciplinary and due to be finalized in 2012.

with vessels of different sizes should all have the possibility of obtaining catches, without being disturbed by unregulated competition.

Then aquaculture entered the seascape. In Norway, the production of trout in dams on shore has a long history, but it was with the introduction of *salmon in net pens at sea* that the industry really expanded, increasing production from zero to one million tonnes within 40 years. Aquaculture was soon regulated in the same manner as the traditional fisheries, by licenses. The idea was to use the new industry as a means to create new employment in rural, coastal areas, thus allowing only one farm per local owner, with production strictly regulated by the volume of the sea pens. The provisional Aquaculture Act was introduced in 1977, after years of planning for the new industry, while the permanent act was passed in 1981, later to be amended on several occasions (Hersoug, 2005).

Other stakeholders are also regulated by sector laws, such as the Petroleum Act of 1996, which set up the requirements for oil and gas installations at sea and on shore. Central to both is the demand for comprehensive environmental impact analysis, including potential effects on the natural environment as well as social and economic effects. The fisheries are also regulated by a sector law, the recent Marine Resources Act of 2008, where the main element is that marine living resources shall all be regulated based on the idea of *sustainable use*. This implies, among other things, that all commercially important fish stocks are regulated by global quotas (Total Allowable Catch), in addition to a large number of regulations pertaining to fishing gears, fishing areas and time. The new offshore energy sector (windmills and tidal water mills) is regulated by a new law of 2010, giving the state, that is, the Ministry of Oil and Energy, the discretion to decide where to allocate such farms and under what conditions. Sea transport is also regulated by a separate sector act, giving the Coastal Directorate (*Kystverket*) the right to decide the use of sea areas connected to all major transport lanes, and furthermore, the right to review (and if necessary, to raise objections against) a number of activities that potentially can affect sea transport. Finally, nature conservation was for many years based on an act designed for conserving important land habitats. With the new Biodiversity Act of 2010, it has become considerably easier to suggest also areas at sea, a fact reflected in a new proposal of conserving 36 marine areas, all along the coast.

However, all these sector laws must somehow be coordinated, and in Norway the idea is that coordination takes place through the Planning and Building Act, originally introduced back in 1965 as the Building Act, but recently renewed and amended in 2008. Here it is explicitly stated that the counties (*fylke*) and the municipalities (*kommune*) shall be responsible for area planning as well as all planning that can affect the inhabitants socially and economically. Originally this was part of a large national system, designed to provide economic growth and the rational use of natural resources, with planning on three levels (national, county and municipality), gathering information bottom-up and allocating resources top-down. But realities soon turned out to be different from the ideals. Integrated plans on national level soon faded, while national *sector* plans became the norm, the county level never gained the prominence that was intended in 1976, when democratic elections were introduced for the county assemblies, and on municipality level planning soon developed in very different directions. Strong municipalities were able to use physical planning strategically, in order to create new infrastructure and new employment, while weak and remote municipalities were lagging behind, being dependent on external consultants to prepare and finalize their plans. In 1985 the Building Act of 1965 was considerably extended, and in 1989 the municipalities obtained the right to plan also their sea areas, that is, out to the base lines – the straight lines connecting the outlying land areas and islands – an area approximately 90 000 km$^2$ or the size of Portugal. Planning of sea areas was, however, only *optional* and not mandatory. Later, in 2009, the municipalities were allowed to extend their sea areas even further; to the baselines plus one nautical mile (1852 m). This change was to harmonize the Norwegian planning system with the EU regulations regarding marine water bodies (the Water Directive). At the same time, this extension also benefited the aquaculture industry, now moving from the interior waters and further out on the coast.

On top of this complicated system of sector planning and physical area planning, Norway implemented a completely new system of Integrated Management Plans for large marine ecosystems, starting in 2005. At present, there are plans for the Barents Sea and the Norwegian Sea, while the plan for the North Sea is in preparation. These are plans integrating the effects of petroleum development, fisheries and sea transport, in a context of global warming and including the effects of long distance contamination of air and sea (Knol, 2010).

Shortly summarized, in Norway there are area plans, resource plans, ecosystem plans, sector plans – all with great ambitions, but involving considerable problems of coordinating the different stakeholder interests. Local concerns may easily be overrun by national priorities and plans, while national goals may be undermined by local foot dragging and protests. Nevertheless, Norway has managed to develop a world leading position in petroleum, fisheries and aquaculture, while marine tourism is increasing and most Norwegians still enjoy easy access to marine waters for their leisure activities. This *could* be an indication of coastal conflicts being exaggerated, but let us turn to aquaculture, the industry most actively claiming more (and better) space.

## 12.3 Aquaculture: the new entrant in the coastal zone

When aquaculture started in Norway in the late 1960s, no one could foresee that within less than 40 years the new industry would be larger than the traditional fisheries measured in economic terms. The start was very modest, characterized by trials and failures, and run by practical people with experiences from fishing. Political authorities saw the new industry as a welcome opportunity to create employment and settlement in remote coastal areas, all feeling the effects of rationalization in fisheries and agriculture. Hence, for the first 20 years the regulations implied that owners could have only one farm, and access was restricted through a licensing regime, familiar from the fisheries. Only when the mandated sales union (*FOS*) went bankrupt in 1991, was the ownership regulation lifted and the companies allowed to freely trade their production. By that time the standard volume limitations had been increased several times, from 3000 m$^3$ to 5000 m$^3$ to 12 000 m$^3$ for a standard license. Despite setbacks, mainly caused by diseases or market failures, the industry increased steadily, reaching nearly one million tonnes in 2010. At that time most of the original entrepreneurs had either sold out or gone bankrupt, and salmon farming had become an industrial activity, dominated by large companies.[2] The industry is largely based

[2]In 2009 the five largest salmon farming companies were responsible for 56% of the total Norwegian production (Lassen et al., 2011).

on scientifically developed knowledge, although public funding for research is still central. Regulations in the industry had also changed in the meantime, first substituting volume (of net pens) with feed quotas, ending up with biomass regulations, where a standard license could contain maximum 780 tonnes of fish or 900 tonnes in the extreme north. Each license was allowed to have a maximum of four production sites (or six if two farms were cooperating), thus opening for a system based on rotation, in order to reduce the environmental impacts of fish farming.

The administrative set-up is complicated: the Aquaculture Act comes under the Ministry of Fisheries and Coastal Affairs which also allocates licenses, while diseases and quality control fall under the Norwegian Agency for Food Safety, arranged under the Ministry of Agriculture and Food. The Norwegian Climate and Pollution Agency is responsible for granting discharge permits, while the actual allocation of space takes place through the municipality plans, arranged under the Ministry of Environment with the county administration (*fylkeskommunen*) responsible for coordinating all new applications. Without going into the details, the point is that *sea space* becomes an important production factor, and sea space is made limited through the strict licensing system. As could be expected, sea space in terms of localities also acquired value, even if licenses up to 2002 were free of charge.[3]

The main change in the regulatory regime over time has been that previous legislation was primarily concerned with *who should be the owners,* while present policy is geared towards *how the farms are operated.* This reflects both the development of the industry (from small-scale to industrial) and the changing priorities of the Norwegian government (from focus on rural development to focus on sustainability and ecosystem-based planning and management).

## 12.4 From first to third generation of aquaculture farming

When salmon farming in net pens at sea started in the late 1960s, the entrepreneurs preferred to have the farms close to shore, in order to have easy access. That meant that most farms were located in protected fjords, often in shallow waters and within short range from the owners' house. With larger net pens, expanding from 5 m depth up to 50 m with a diameter of up to 160 m, farmers moved to deeper waters in what has been termed the second generation of salmon farms. With stricter regulations pertaining to diseases, new feeding technology (via tubes from a floating barge) and the need for more coordinated logistics, we are now witnessing the third generation of farms, moving out to deeper waters, with stronger currents and less protection. Besides technological developments, the amalgamation of farms, due to increased concentration, has meant that the larger companies prefer to produce more fish on the best locations, thus obtaining better logistics in terms of bringing in workers and provisions. This development has implied a steady reduction of production sites, in spite

[3]In reality it is the license, that is the permission to produce, that acquires value. With transfers, the license is sold as part of a package, including the right to use a certain area (the location), the technical installations (net pens and feed barges) as well as the fish. Hence, it is difficult to assess the value of the site separately.

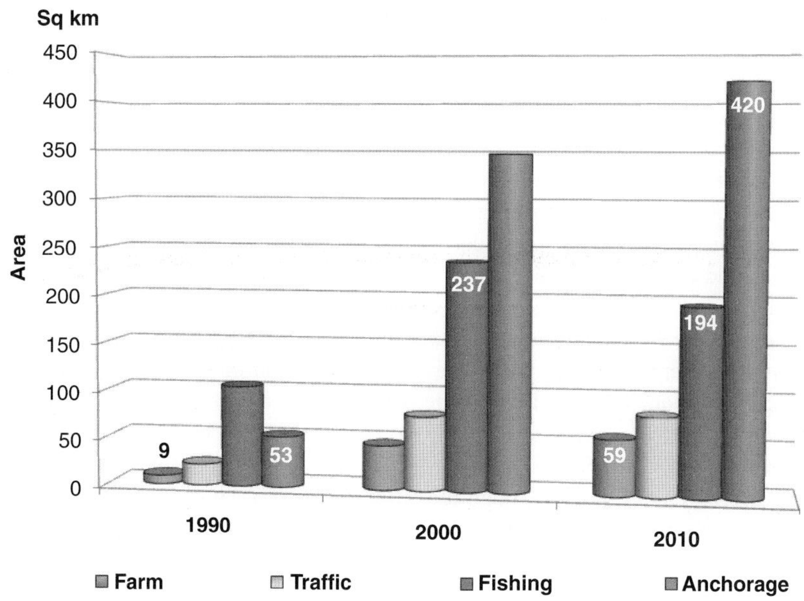


**Figure 12.1** Total area occupation of Norwegian salmon production.

of a rapidly increasing total production. While the number of production sites in 1999 was close to 2000, the number in 2011 reduced to 1000 (Arealutvalget, 2011). This has meant a massive reordering of space, which has been managed by the municipalities through coastal zone planning.

## 12.5 How much space is occupied by the aquaculture industry?

While the reduction in number of localities has been used as an indication of reduced conflicts with other stakeholders, the true picture is a little more complicated. Area occupation means different things to different people, depending on how the calculations are made. While the actual farm is protected by a security zone of 20 m where traffic is prohibited, the zone where fishing is prohibited is larger (100 m), while the anchorage zone is the largest. As can be seen from Figure 12.1, the anchorage zone was about 420 km$^2$ in 2010, or 0.5% of the inshore area. While still a very limited part of the interior waters, the area is considerably larger than the 59 km$^2$ occupied by the actual farms.

While the farm site and the traffic prohibition zone effectively limits all other interests, traffic such as paddling or sailing may also take place inside the fishing prohibition zone, and fishing may take place within the anchorage zone. Even within conservation areas (*naturvernområder*), aquaculture may be allowed, if this does not affect the values to be conserved.

## 12.6 How to get access to more space?

In order to secure more space for the expanding aquaculture industry, the Ministry of Fisheries and Coastal Affairs established an expert group by the end of 2009 to suggest strategies for further development. After only one year of work the group presented its findings – a comprehensive report covering altogether 25 proposals (Arealutvalget, 2011). During 2011, the report was subject to a wide public hearing, before production of a white paper, ending up with possible amendments of laws and regulations in 2012.

The starting point is that aquaculture needs more space, *if the industry is going to expand in a sustainable manner.* The Ministry of Fisheries and Coastal Affairs (2009) has drawn up the strategy for a sustainable aquaculture industry, where the efficient use of sea space is one out of five prioritized areas. Certain marine areas in the south (where the industry started) seem to be overloaded, while the potential for further expansion is found to be in the north, particularly in the three northernmost counties. The main tool suggested by the group is to establish a number of *production areas* with barriers created between them in order to prevent, or at least limit, the spread of pathogens and sea lice. The production areas should then be divided into *put-out zones,* where farmers are required to follow the "all in–all out"-principle, that is, putting the smolt in the pens at the same time and harvesting the fish at the same time. Then the zones will be laid fallow for a certain period. This principle has already been practiced in some regions on a voluntary basis, but here the recommendation is that the principle should be *mandatory.* Needless to say, such a proposal is bound to create resistance, not least among the smaller farmers, who have limited possibilities for shifting their production between different production areas and put-out zones, while the larger companies are positive to this reordering of the industry. Other proposals include the possibility of forced removals of farms from overloaded areas, easier access for shore-based production, and use of indicators in management, such as loss of fish through the entire production cycle. The idea then is to use this indicator to decide in which areas the farmers should be allowed to expand and in which they should be restricted. Finally, the group suggests that no further expansion should be allowed until the production has been reorganized into production areas and put-out zones, a process that could take at least two years *if* the politicians decide to go for the proposals. So far it is safe to say that the industry is rather sceptical to the most dramatic proposals. According to the Norwegian Seafood Federation (FHL, 2011), the generic industry association, the proposed production regime implies "a restart of the entire industry".

## 12.7 "No pay, no cure"– on the use of area or production fees

With the original allocations in the 1970s, the licenses were free of charge. Risks were high and the bonus for the municipalities, if the farmers succeeded, was employment, company tax, income tax as well as multiplier effects in other industries. This system was maintained all the way until 2002, when it was evident that successful farmers obtained windfall gains, when selling farms, that is, the production permit, net pens, feeding barges, fish and not least localities. For the best localities the price had in 2010 reached NOK 55 million (7 million Euro) per license. At that time the state had introduced a fixed price per license of NOK 8 million (1 million Euro), to be used by each new round of allocations. However, the municipalities did not receive a cent, neither as down payment nor as an annual fee for the use of their sea areas. On the other hand, it was evident that the municipalities had all the downsides, that is, the monetary costs connected with planning the areas, the political costs involved in negotiating with other stakeholders, and quite often also the pollution involved, not least when farms went bankrupt, leaving net pens, ropes and anchors at sea.

In 2008, an initiative by important aquaculture municipalities of introducing an annual area or production fee failed after strong resistance from the farmers' own organization; the Norwegian Seafood Federation. Hence, these municipalities felt the need to continue lobbying for a larger payback, and they created a new organization of aquaculture municipalities. So far they have had limited success, and the newly introduced system of taxing the technical value of the farms (excluding the value of the fish) has so far produced only marginal incomes for the municipalities involved. But the writing on the wall is pretty clear; without more payment to the municipalities, the willingness to allocate more space will be limited. Some municipalities have even refused to allocate any space to aquaculture in their new area plans. This situation has long been registered by a number of the larger aquaculture companies, which have started paying voluntary contributions to the municipalities where they are involved. However, the picture is diversified; some municipalities with large farms, slaughtering facilities and processing factories are well served, both in terms of employment and in terms of tax income. Other municipalities have only grow-out farms, frequently with absentee owners and with employees commuting in from other places, thus creating very few spinoff effects. These municipalities are now the ones most sceptical about allocating more space for aquaculture.

## 12.8 No shortcut to progress

As already demonstrated, aquaculture is not a large consumer of space at the national level. Even when applying the most extensive zone, the total area is in the order of 420 km$^2$, or 0.5% of the total area inside the base lines. Most farms are located where other stakeholders are involved, and quite often the activities cannot be combined. Fishing and farming are, for example, not possible in the same area. When the aquaculture industry, supported by the Ministry of Fisheries and Coastal Affairs, put the area question on the agenda, it was clearly seen that further development of the industry would imply the use of larger areas along the coast. However, the expert group (*Arealutvalget*), the ministry and the farmers forgot that somebody had to take up the bill – or more precisely – they forgot that although fish farming is an important industry deserving national political attention, the right of municipalities to decide on area use is firmly entrenched in the Planning and Building Act. In principle there are three ways out of the dilemma:

1. Top-down planning, using national guidelines in order to safeguard the continued expansion of the aquaculture industry
2. Planning on county level, making binding plans, involving all the coastal municipalities in the county
3. Bottom-up planning, initiated by the municipalities themselves, as demonstrated in the case of Sør-Trøndelag and others, where planning of coastal areas involving several municipalities is done on a voluntary basis.

The expert group (*Arealutvalget*) goes a long way towards the first alternative, although it remains to be seen what will be the final political outcome. At present it seems like political suicide to *force* municipalities to receive aquaculture farming against their own decision not to accept such activity. The second alternative is now possible, according to the new Planning and Building Act, but it has so far not been implemented anywhere. Only indicative plans have been made, thus leaving the final decision to the municipalities. The third alternative has been practised in several places, most actively in the case of Sør-Trøndelag county, where 11 coastal municipalities have created a project of working out a joint area plan for the coastal (sea) areas (*"Kysten er klar"*). So far none of the initiatives have resulted in a finished and updated plan.

Three factors are clearly influencing the current tradition of working out coastal zone plans on the municipality level. The first applies to the new guidelines of planning based on ecosystem considerations. This is strongly entrenched in all modern laws regarding the use of coastal areas and resources. What is defined as the ecosystem may vary, but it is evident that most ecosystems do not follow municipality borders. In the case of aquaculture, we often find that a particular fjord may be shared by five or more municipalities, each planning as if their marine areas do not belong to a larger ecosystem, where effects in one area easily may spill over into the areas of other municipalities.

The second applies to the new regulations pertaining to fish diseases and sea lice. According to the regulatory authorities, these regulations are aiming at larger production areas, with solid corridors between them, serving as a safety mechanism against the rapid spread of diseases and sea lice. An optimal system will clearly depend on the working out of a "current catalogue", showing how pathogens may spread from one farm to the other, but in the meantime the regulators use minimum distances between farms, slaughteries and processing factories. With larger production areas, many municipalities will end up with their sea areas defined as buffer areas, where farming will not be allowed. In this case we may experience the opposite effect of the one described above; municipalities that have (or would like to have) aquaculture farming, may not be allowed to allocate areas for such activity.

The third applies to expertise. For years, remote municipalities have struggled to find (and keep) qualified planners on land use. Even more complicated is finding experienced planners to handle complicated conflicts in the coastal zone with due regard to ecosystem considerations. For this reason alone, coastal zone planning is lagging behind.

Together these three factors all point in the direction of planning for larger marine areas, involving several municipalities belonging to the same ecosystem. In order to implement such planning, new and more innovative approaches have to be found. At present, the stakeholders have a "Catch 22-situation"; the industry, supported by the central authorities, requires more and better areas, while the local level is waiting for a better deal. In the meantime new areas will not be made available. However, at present, sea lice and escapees from the farms are the most prominent deterrent features of Norwegian aquaculture. Until these challenges have found a reasonable solution, no further expansion of the aquaculture sector will be politically possible.

So far the major instrument used in integrated coastal zone management has been *zoning* (Cicin-Sain and Knecht, 1998). Areas for fishing, aquaculture, recreation, oil and gas development, conservation, etc., have been allocated along the coast, either for exclusive or for joint use. Only interests able to document their claims and aspirations through maps will be accepted, as clearly demonstrated with the traditional fisheries (Kidd et al.,

2011). However, area zoning is principally a two-dimensional planning device, while an ecosystem-based approach implies at least three dimensions, involving interactions in the water column as well as on the sea floor. Several authors claim that Marine Spatial Planning can be developed further in the direction of ecosystem-based planning (see e.g. Douvere, 2008; Douvere and Ehler, 2009), but so far the goals have been unclear and goal attainment difficult to measure. The same applies to the Norwegian case, where conflicting use so far has been "solved" by zoning, while the real challenges ahead are connected with the impacts on the larger ecosystem. It remains to be seen whether these challenges can be solved through the current planning system, or if new and more innovative approaches have to be found. In any case, it is clear that the aquaculture industry is only one of many legitimate users in the coastal zone.

## References

Arealutvalget (2011) Effektiv og bærekraftig arealbruk i havbruksnæringen – areal til begjær. Rapport fra et ekspertutvalg oppnevnt av Fiskeri- og kystdepartementet. Oslo: Fiskeri- og kystdepartementet.

Bennett, R. (2000) Coastal planning on the Atlantic fringe, north Norway: the power game. *Ocean & Coastal Management* **43**, 879–904.

Cicin-Sain, B. and Knecht, R. W. (1998) *Integrated Coastal and Ocean Management. Concepts and Practices*. Island Press, Washington, DC.

DN (Direktoratet for naturforvaltning) (2010) Nasjonal marin verneplan. http://www.dirnat.no/content/500040226/Nasjonal-marin-verneplan [accessed 10 September 2012].

Douvere, F. (2008) The importance of marine spatial planning in advancing ecosystem-based sea use management. *Marine Policy* **32**(5), 762–771.

Douvere, F. and Ehler, C. N. (2009) New perspectives on sea use management: Initial findings from European experience with marine spatial planning. *Journal of Environmental Management* **90**(1), 77–88.

Ese, A.-C. (2009) Skjær i sjøen. Utvikling av marin verneplan. Masteroppgave i Statsvitenskap. Institutt for statsvitenskap, Universitetet i Tromsø.

FHL (2011) Høringssvar til "Effektiv og bærekraftig arealbruk i havbruksnæringen". http://www.fhl.no/getfile.php/BILDER/Div_illustrasjoner/H%C3%B8ringssvar%20Gullestad-utvalget.pdf [accessed 10 September 2012].

Hersoug, B. (2005) *Closing the Commons—Norwegian Fisheries from Open Access to Private Property*. Eburon Academic Publishers, Delft.

Holm, P., Hersoug, B. and Rånes, S. A. (2000) Revisiting Lofoten: Co-Managing Fish Stocks or Fishing Space? *Human Organization* **59**, 353–364.

Jentoft, S. and Kristoffersen, T. (1989) Fishermen's Co-management: The case of the Lofoten Fishery. *Human Organization* **48**, 355–365.

Kidd, S., Plater, A. and Frid, C. (2011) *The Ecosystem Approach to Marine Planning and Management*. Earthscan, London and Washington, DC.

Knol, M. (2010) Marine ecosystem governance in the making. Planning petroleum activity in the Barents Sea – Lofoten area. PhD dissertation. University of Tromsø.

Ministry of Fisheries and Coastal Affairs (2009) Strategi for en miljømessig bærekraftig havbruksnæring. Oslo.

Tett, P., Sandberg, A. and Mette, A. (Eds.) (2011) *Sustaining Coastal Zone Systems*. Dunedin Academic Press, Edinburgh.

Section 4

# Linking Science and Management

# Chapter 13

# Integrated Coastal Zone Management – Preview and Evaluation of its Application on the Coast of Cyprus

*Nicholas Kathijotes and Athina Papatheodoulou*

**Abstract**

Increasing populations in coastal areas generate significant pressure on the supporting ecosystems. The need for a more integrated, long-term approach instigated certain initiatives on building the concept of integrated coastal zone management (ICZM), which is a powerful tool for the control of coastal areas. The European Union (EU) Legal Framework is emphasized as applicable, together with an evaluation of the ecological degradation of the Mediterranean coasts. The city of Limassol in Cyprus is evaluated, stating the ecological risks and the possible mitigation measures, resulting from the European Union directives and policies. Types of pollution are discussed, including urban wastewaters, aquaculture, surface runoff, industries and agriculture. The concept of 'Urban Metabolism' is also introduced as a tool used to quantify and assess urban sustainability of coastal cities. Suggestions, measures and good practices are given to direction the creation of model ecological cities, based on an integrated coastal zone management practice.

## 13.1 Introduction

Cyprus is the third largest island in the Mediterranean with a total of 772 km of shoreline, only 296 km (38%) of which is within the area over which the Government exercises effective control. The largest part of the country's territory can be considered as coastal zone and this has a dominant influence on many aspects of Cyprus (Ministry of Agriculture Natural Resources and Environment, 2005). Cyprus, as other islands in the Mediterranean basin, is characterized by great physical complexity, important biological diversity (mainly due to the island's isolation) and has always been subjected to great human pressure. Therefore

*Global Challenges in Integrated Coastal Zone Management*, First Edition. Edited by Erlend Moksness, Einar Dahl and Josianne Støttrup.

by presenting the island's status on the implementation of ICZM (European Commission, 2002) will give a snapshot of the case in other Mediterranean countries. For example, Malta – which similar to Cyprus – is facing challenges regarding the implementation of ICZM recommendations, not only due to its size and population density, but also due to conflicting demands for coastal uses, which require suitable management to achieve sustainable coastal management, recognized as a need for coastal areas in Chapter 17 of Agenda 21 (United Nations Conference on Environment and Development, 1992; Malta Environmental and Planning Authority, 2011).

The term 'integrated coastal zone management' is used throughout the paper, to reflect its conformity and dominance in European legislation, even though over time the concept of coastal management was given different names (Shipman and Stojanovic, 2007). Note that there is no single legal or functional definition of the 'coastal zone' in Cypriot legislation (Ministry of Agriculture Natural Resources and Environment, 2006). In the following paragraphs the condition of coastal areas in Cyprus is presented, as well as management practices employed in the context of ICZM.

Coastal areas generate by far the largest source of household income, and most of the urban development. Tourism is probably the most important economic activity of the island, whose coastal zone is the primary destination for tourists. The tourist industry contributes > 10% to the country's gross domestic product and 29% in the occupational sector (Statistical Service, 2011).

Most Cyprus coastal waters are used for bathing and recreational activities. The results of monitoring programs verify that coastal waters comply with the guideline values set by the Bathing Water Directive 2006/7/EC. Cyprus has been recognized as 'star performer', with 100% of its bathing water sites in compliance to the Directive for 2010 (European Environment Agency, 2011).

Under the provisions of the European Water Framework Directive, five coastal water bodies have been characterized as 'heavily modified' due to navigation infrastructure (ports and marinas). As part of the Government's development policy, more marinas have been proposed for construction. Furthermore, human activities have greatly modified the seafronts of all four major coastal cities evidenced by quays, extensive earth filling, reinforcement walls, modifications for flood protection, etc. Figure 13.1 shows the heavily modified coastal bodies of Cyprus resulting from the presence of navigation infrastructure, the areas of environmental concern and the pollution hotspots (WL Delft Hydraulics et al., 2004).

## 13.2 Analysis of the major stakes and challenges for the coastal zone

During the 'Coastal Protection Study' a definition of the 'coastal strip' being synonymous to 'coastal zone' was suggested. This describes an area of 2 km extending inland from the coastline and is used as the most popular definition of the coastal zone. This area covers 23% of the island's total area, within which about 50% of the total population lives and works. It also accommodates about 95% of all tourist industry.

The population and economic growth, as well as the sprawl of building development at the suburban edges of urban areas (dominant features of urbanization), have been rapid and

**Figure 13.1** Heavily modified coastal bodies due to the presence of navigation infrastructure, areas of environmental concern and pollution hotspots in Cyprus (Coastal bodies of the island, in which the Republic does not exercise effective control, are not reported).

impressive, mainly during 1981–2001. This development has generated several problems and obstacles concerning the effectiveness of the management of coastal resources and the future sustainability of the development process (Ministry of Agriculture Natural Resources and Environment, 2006). Some examples are discussed below:

(a) Land use conflicts: The policies available are not capable of maintaining the required balance between coastal environment and development at those fragile coastal zones. Land use conflicts often lead to the loss of agricultural land, coastal mobility problems, shore erosion, overcrowding of beaches, infrastructure deficiencies, high building densities in many urban coastal areas and insufficient protection of areas of natural beauty. This is the case not only for Cyprus but also for the southern coast of Spain and Malta, where concrete development overrode other essential coastal land uses, creating various conflicts.
(b) Social disparities and loss of rural heritage: Coastal urbanization and coastal centred economic growth has overshadowed the original rural heritage and the quality of coastal environment in many parts of the island.
(c) Land ownership and local economic interest: A significant feature of most jurisdiction related to coastal zone management, is thought to be public ownership (Gibson, 2003), which is widely distributed among the population, often giving rise to legitimate expectations for private building development in coastal areas.
(d) Limited environmental awareness: Public environmental awareness is still at very early stage and this has important consequences for coastal management. Such an example is the construction of two premises in the western peninsula of the island in a turtle nesting protected area, in the summer of 2011.
(e) Poor harmonization: The benefits from development are not related to the environmental cost on society, making the protection of the environment a secondary issue.
(f) Lack of common vision: Parallel policies affecting the development and management of coastal resources (land use, fisheries, water, agriculture, tourism, ports, roads, etc.)

are designed and implemented by separate competent authorities not always under a coherent overall broadly approved framework. Therefore, problems in the administrative framework of coastal planning and management have been observed.

## 13.3 Coastal pollution evaluation

### *13.3.1 Urban wastewater and storm water*

Cyprus is served by more than 220 wastewater plants, the majority of which are the privately owned plants of hotels and tourist villages. A common practice is to reuse treated effluent in agriculture for irrigation and aquifer recharge. Discharge to the sea is not generally applied and is practiced only in the cases of Limassol and Larnaca in exceptional cases (i.e. reduced demand during winter months, water tank maintenance, pipe network infrastructure restrictions). At present, 66% of urban areas are in compliance to the European Directive 91/271/EC requirements whereas only 8% of rural areas are (Ministry of Agriculture Natural Resources and Environment, 2010). It is anticipated that by the end of 2012, 30 new wastewater treatment plants (WWTPs) will be serving the areas with populations greater than 2000 inhabitants. All new plants have provisions for tertiary treatment, not only signifying the importance of effluent treatment as an alternative water resource of the island, but also the environmental concern of disposing of treated effluent in the marine environment and groundwater resources.

In Limassol, the largest coastal city of the island, there is one of the biggest WWTPs on the island. This plant has been in operation since 1995, initially treating 20 000 $m^3$ $day^{-1}$ and reached 40 000 $m^3$ $d^{-1}$ after its expansion in 2008. A new WWTP in Limassol is under study. The construction and operation of the Limassol's sewerage network has played an essential role in the control of sewage pollution, and eventually led to the award of 15 blue flags to its beaches.

Storm water runoffs may also exhibit a significant pressure on coastal water quality due to the extended impervious areas in urban sites, industrial areas, highways and airports. The major pollutants conveyed are of various types (oils, some heavy metals, nutrients, etc.).

### *13.3.2 Industries*

Even though the industrial sector of the island is not 'heavy', the industrial plants in operation do pose a pressure on coastal ecosystems, mainly in the form of industrial wastes and runoffs. Industries disposing wastes constitute point sources to the adjacent coastal water bodies, whereas each industry comprises a diffuse pollution source, as well as source of fallouts from its air emissions. Of the 900 large, medium and small industries of the island, 41 have been considered as significant pollution sources for surface waters (WL Delft Hydraulics et al., 2004). Significant point sources that pose pressure to the coastal environment are the beverage industries (wineries and breweries) in Limassol. Even though their connection to the central sewerage network in 2004 has relieved the marine ecosystem of the heavy burden of organic wastes, their cooling waters (with a high in biochemical oxygen demand – $BOD_5$) are still being discharged to the sea. Due to the ongoing construction of the new Limassol Marina in the general area of the beverage industries, it is foreseen that the industries will soon be translocated and the local point pollution will be minimized.

In terms of Industrial effluents the bay of Vasilikos, about 20 km east of Limassol, where factories with chemical effluents operate (cement company, power station, abandoned fertilizers chemical industry), create hot spots with inert materials from industrial activities possibly blanket the bay's sediments and have a negative effect on the benthic community of the area. The thermal effluents from the three power plants serving the energy needs of the island (Moni, Vassilikos and Dekelia) are still reaching the sea. As these industries are necessary and vital to the economy, new technologies and measures could be taken to minimize their effect on the environment, such as control of their effluents.

The Government's water policy for independency from climatic conditions for the island's aquatic resources, led to the construction of desalination plants. The first plant operated in 1997. Currently five desalination plants are in operation, with a total capacity of 122 000 $m^3\ d^{-1}$. All desalination plants discharge considerable quantities of brine to the sea, resulting in the formation of a stratified system due to density differences. Studies have also shown compositional changes to the macro-benthic communities in the vicinity of discharge points (Karavokiris and Associates Engineers Consultants and Kaimaki, 2010). The termination of the operations of the Cyprus Petroleum Refinery in Larnaca, in 2004 – which was considered as the main pressure for the waters of the area – ceased the discharge of its treated process waters into the sea (WL Delft Hydraulics et al., 2004). It then was suggested that the site be converted to a fuel terminal station, a practice that can still have severe impacts on coastal waters due to the washout of the station's pipes. There are delays to the process because of local reaction. Huge amounts of natural gas have been recently discovered offshore, within the Cyprus Exclusive Economic Zone. Cyprus, as a new state in the world's energy map, will process this resource soon, with planned plants on the coastal zone. In this case, great care and very careful planning will be required to avoid any damage to this fragile zone.

### *13.3.3 Aquaculture*

There are seven offshore aquaculture units in Cyprus, located 1–3 km offshore. The mean annual production of the units ranks between 100 and 500 tonnes $year^{-1}$, with the exception of a tuna ranching farm which produces about 1000 tonnes $year^{-1}$. Studies have shown that nitrate concentrations near the installations are relatively high to the concentrations of adjacent marine waters. As far as this pressure point is concerned, the district of Limassol experiences the greater pressure (WL Delft Hydraulics et al., 2004).

### *13.3.4 Agricultural runoffs*

Eutrophication is very common in sheltered marine water bodies such as harbors and semi-enclosed bays along the Mediterranean coast, mainly in the vicinity of coastal towns. Nutrient or mineral balances provide insight into links between agricultural nutrient use, changes in environmental quality, and the sustainable use of soil nutrient resources. A persistent surplus indicates potential environmental problems; a persistent deficit indicates potential agricultural sustainability problems. With respect to environmental impacts, however, the main determinant is the absolute size of the nutrient surplus/deficit linked to local farm nutrient management practices and agro-ecological conditions, such as soil type and weather patterns (rainfall, vegetation period, etc.). In Liopetri and Ayia Napa bays, nitrogen leaching from the intensely cultivated agricultural areas, due to the over-fertilization (150 tonnes of nitrogen per year) and aquaculture, are loading sea waters with nutrients that eventually create ecological problems to the marine environment (European Environment Agency, 1999).

### *13.3.5 Mines*

The mines on the island (one in operation, nine abandoned), can be considered as significant point pollution due to the volume of mining wastes. These can create pressures on the water environment, mainly as a result of their effluents and drains, related to high acidity, heavy metals and high chemical oxygen demand (COD). An example is the Limni mine in Polis Chrysochou area, where the mounds of toxic solid have undergone a high degree of chemical weathering leading to the dispersion of metals such as Fe, Cu, Zn, As, and Mn to the surrounding area and reaching the designated 'Natura 2000' coast. As a result great quantities of 'pyrite sand' and its oxidation products occur, posing a permanent source of toxic metals to the marine environment (Varnavas, 2010).

## 13.4 Erosion

About 30% of the total coastline of Cyprus suffers from erosion (Loizidou, 2008). Although erosion is rooted in natural causes, such as marine sediment transport, it can be amplified by human activities (e.g. sand quarrying, water impounding works and tourist development). The construction of the Limassol harbor resulted in beach erosion and the rectification measures (breakwaters perpendicular to the coastline), also resulted in serious degradation of water quality (Loizidou and Iacovou, 1995).

## 13.5 Coastal development in Cyprus – land uses

During the past two decades, coastal development was characterized by the conversion of formerly agricultural and natural zones to tourist development zones after each revision of the land use planning zones every four years. The situation along the coastline after the revision of the land use planning zones in 1997–98 was as follows: tourist zones cover 105 km (i.e., 37% of the length of the coastline); open areas/protected natural or archaeological areas cover 125 km (i.e., 43%); agricultural zones cover 36 km (i.e., 12%); residential zones cover 17 km (i.e., 6%); and industrial zones cover 9 km (i.e., 3%). As a result of the policies and the targets of the Cyprus Tourism Organization, the number of beds in the coastal areas of Cyprus has increased sevenfold over the past 20 years, that is from 12524 beds in 1980 to 88302 in 2001. At the same time, the numbers of beds in inland areas increased by only 11%, that is from 3902 in 1980 to 4358 in 2001(Kathijotes, 2010).

## 13.6 The 'urban metabolism' issue

Urban metabolism is a model to facilitate the description and analysis of the flows of the materials and energy within cities, such as undertaken in a material flow analysis of a city. Urban and coastal-city policy makers should be encouraged to understand the urban metabolism of their cities. It is practical for them to know if they are using water, energy, materials, and nutrients efficiently, and how this efficiency compares to that of other cities. They must consider to what extent their nearest resources are close to exhaustion and, if necessary, plan appropriate strategies to slow exploitation. Metabolism data have been established for only a few cities worldwide and there are interpretation issues due to a lack of common conventions; there is much more work to be done. Resource accounting and management are typically undertaken at national levels, but such practices may arguably be too broad and lack understanding of the urban driving processes. Beyond concerns over the sheer magnitudes of resource flows into cities, there are more subtle imbalances, feedbacks and spatial attributes which threaten the sustainable development of cities. Cities are sucking in resources from global ecosystems at increasing rates.

Urban metabolism is of major significance for the ongoing metabolic transition. A better understanding is needed for the patterns and drivers of urban metabolism to realize the substantial reduction potentials. Tools and methods are available but they need to be advanced and standardized.

## 13.7 Conclusions

Within the Mediterranean region, which is quite industrialized, there are *a priori* necessary prevention mechanisms, correction technologies and the appropriate legal framework. However, there is a lack of political willingness from countries to enforce proper environmental management. The problem of an erratic and fragmented approach to coastal development still exists. The complexity in the decision-making process, the large number of authorities involved, the lack of effective coordination and the huge financial interests involved in the coastal area have been addressed as the main reasons for this situation. As an island, the vulnerability of the coastal strip is extremely high in Cyprus, since all land-based activities have a direct impact on this narrow strip. There is a great need then, for adoption of integrated management of coastal areas. The ICZM involves a series of operations for planning and daily management of resources and to support activities that require synchronization. The practical need then is to establish governance mechanisms for ensuring ICZM action at all levels.

The ICZM training should be carried out through a sequence of stages namely, awareness of the subject, dialogue and cooperation among stakeholders and coordination of activities. A well-balanced coastal management with higher involvement of the private sector and academicians, supplemented by urban metabolism analysis is a demand of high priority for Cyprus in view of its newly discovered offshore wealth, at least if the target is sustainable development.

## References

European Commission (2002) Recommendation of the European Parliament and of the Council of 30 May 2002 concerning the implementation of Integrated Coastal Zone Management in Europe 2002/413/EC, Official Journal of the European Communities 6.6.2002, L14824.

European Environment Agency (1999) State and pressures of the marine and coastal Mediterranean environment. Copenhagen, *Environmental issues series No.* **5**, 52–56.

European Environment Agency (2011) European bathing water quality in 2010. Copenhagen, Report No.1.

Gibson, J. (2003) Integrated Coastal Zone Management Law in the European Union, *Coastal Management* **31**, 127–136.

Karavokiris and Associates Engineers Consultants and Kaimaki, S. (2010) Preliminary program of measures – Report 5. Nicosia, Water Development Department, Republic of Cyprus.

Kathijotes, N. (2010) Evaluation and control of marine pollution due to land-based activities – A preview of the Cyprus case. *Coastal Marine Science* **34**, 195–197.

Loizidou, X. (2008) Coastal erosion in Cyprus. *Monumenta* 2.

Loizidou, X. and Iacovou, N. (1995) The Cyprus experience in coastal zone monitoring as a basis for shoreline management and erosion control, in E. Ozhan (ed.) 1995 Proceedings of the 2nd International Conference on the mediterranean coastal environment, 24–27 October 1995, Tarragona, Spain.

Malta Environmental and Planning Authority (2011) Report on the implementation of the Recommendation of the European Parliament and of the Council concerning the implementation of integrated coastal zone management in Europe (2002/413/EC), Marsa, MEPA.

Ministry of Agriculture Natural Resources and Environment (2005) Water Framework Directive, Republic of Cyprus EU -Summary Report Articles 5 & 6. Nicosia, Republic of Cyprus.

Ministry of Agriculture Natural Resources and Environment (2006) A strategic approach to the management of the Cyprus coastal zones. Nicosia, European Commission.

Ministry of Agriculture Natural Resources and Environment (2010) Report on Urban wastewater treatment during 2007–2008. Article 16- Directive 91/271/EEC. Nicosia, Republic of Cyprus.

Shipman, B. and Stojanovic, T. (2007) Facts, fictions and failures of integrated coastal zone management in Europe. *Coastal Management* **35**, 375–398.

Statistical Service (2011) Tourism Statistics, January–December 2010. Series II. Nicosia, Republic of Cyprus.

United Nations Conference on Environment and Development (1992) Earth Summit. Agenda 21, Chapter 17, Protection of the Oceans, all kinds of seas, including enclosed and semi-enclosed seas and coastal areas and the protection, rational use and the development of their living resources. New York, United Nations, pp. 139–155.

Varnavas, S. (2010) Medical geochemistry – A key in the precautionary measures against the development of cancer and other diseases. *Bulletin of the Geological Society of Greece. Proceedings of the 12th international Congress* **XLIII** 1, 234–246.

WL, Delft Hydraulics, ENVECO, SA, D. Argyropoulos and Associates (2004) Analysis of Pressures – Report on the identification of pressures carried out in accordance of section 1.4 of Annex II of the Water Framework Directive, Volume 3. Nicosia, Water Development Department.

Chapter 14

# Framing a 'Post-Normal' Science–Policy Interface for Integrated Coastal Zone Management

*Scott Bremer*

**Abstract**

From its inception, integrated coastal zone management (ICZM) has sought to mobilise the best available knowledge in support of collective decision-making, and to this end promoted integration across the 'science–policy interface.' As ICZM has become described in terms of coastal governance, some authors have sought to democratise the science–policy interface to give effect to new epistemological norms of dialogue, inclusiveness, integration and knowledge quality. This has opened a discussion on how to give effect to these norms, with 'post-normal science' emerging as one potential approach for framing the science–policy interface. In this chapter the potential of post-normal science for contributing to 'high-quality' ICZM is explored. After introducing the key characteristics of post-normal science, ICZM practice is considered and eight lessons learned from case studies employing a post-normal science–policy interface at the international, national and local scale are discussed.

## 14.1 Introduction

Integrated coastal zone management (ICZM) has long sought to mobilise the best available knowledge in support of collective decision-making, across a 'science–policy interface.' With ICZM increasingly espousing models of governance, the interface is also changing; evolving from a uni-directional transformation of science into policy, to an inclusive setting that admits all knowledge systems, and seeks to arrive at 'high quality,' integrated knowledge through stakeholder dialogue (Boesch, 1999; Norgaard, Kallis, & Kiparsky, 2009). This has opened a discussion on how to give effect to these 'dialogic' norms, with 'post-normal science' emerging as one potential approach; *seeking to bring together all stakeholders as an 'extended peer community,' in a process of collective knowledge-production and*

*Global Challenges in Integrated Coastal Zone Management*, First Edition. Edited by Erlend Moksness, Einar Dahl and Josianne Støttrup.

*peer-review, steered by agreed principles of knowledge quality*. But to what degree can such an approach make meaningful contributions to ICZM in practice?

This chapter empirically explores the potential of post-normal science (PNS) as an alternative approach to framing the science–policy interface for ICZM. It begins in section 14.2 as a discussion of the shift in ICZM from management to governance, and the implications for the science–policy interface, before the potential benefits of a 'post-normal' science–policy interface for ICZM are introduced. Section14.3 introduces empirical study of explored post-normal initiatives in ICZM practice at the international, national and local scale, describing three case studies and the research method. Section 14.4 unpacks eight key lessons learned, and finishes by looking at what these lessons reveal for the promise of the PNS approach for ICZM.

## 14.2 The science–policy interface as an institution for coastal governance

### *14.2.1 A shift from coastal management to coastal governance*

In 2002, the World Summit on Sustainable Development in Johannesburg reviewed the experiences within ICZM worldwide, and recognised two significant developments: (i) that the coast has become framed as a complex, humans-in-nature ecosystem; necessitating (ii) that society's management response be framed in terms of 'governance.' To this end, contemporary developments in the field have tended toward participatory and collaborative models of 'governance-beyond-government' (Bille, 2007, 2008; Christie, 2005; Chua, 1993; McFadden, 2007; McKenna & Cooper, 2006; Stojanovic & Barker, 2008; Tobey & Volk, 2002). Such models depart from notions of a state-centred and unilateral approach led by a 'coastal manager;' conceiving instead of society's collective response to coastal issues as the joint responsibility of (i) the state, (ii) the private sector, (iii) civil society, and (iv) the scientific community, acting simultaneously and independently according to their own institutions. The model of 'interactive governance' for example, which has found expression in the ICZM literature, defines governance as; "... the whole of public as well as private interactions taken to solve societal problems and create societal opportunities. It includes the formulation and application of principles guiding those interactions and care for institutions that enable them" (Kooiman & Bavinck, 2005). The challenge for coastal governance then is to create the means for coordinating and harmonising these otherwise fragmented responses, and encouraging cooperation according to the principles espoused by ICZM (Cicin-Sain & Knecht, 1998). Stojanovic, Ballinger and Lalwani (2004) provide a useful account of these principles, listed in Table 14.1.

### *14.2.2 The changing shape of the science–policy interface*

Adopting models of coastal governance has implications for the way knowledge is mobilised in support of collective decision-making, often expressed in terms of a 'science–policy interface'. Under the management model the coast is potentially completely knowable, and science is charged with informing policy with objective and valid knowledge to allow decision-making with 'perfect insight' (Funtowicz & Strand, 2006). Governance scholars,

**Table 14.1** Principles of ICZM (Stojanovic et al., 2004).

1. Contingent: ICM initiatives tailored to local diversity
2. Comprehensive: at a scale determined by ecosystem boundaries
3. Long-term: at timescales pertinent to the ecological context
4. Participatory: epitomised by co-management
5. Cooperative: inter-sectoral (horizontal) and inter-hierarchal (vertical)
6. Precautionary: proceeding in a risk-adverse manner
7. Strategic: issues are prioritised for action
8. Incremental: ICM as an iterative and interactive process
9. Adaptable: flexible decision-making ready for 'surprise

however, recognise that the significant uncertainty characterising most coastal issues mean that the scientific community cannot provide certainty on the functioning and future of the coast. At the same time, they challenge the legitimacy of the scientific community's claimed monopoly over knowledge production; recognising that knowledge for coastal governance is not concentrated within any one group of stakeholders or institution, but diffused throughout an entire community (Stojanovic et al., 2004; Tobey & Volk, 2002; Vallega, 1997). All groups, from indigenous peoples, to local communities, to the scientific community construct their own legitimate knowledge of the coast; knowledge-building is described as a collective social activity (Zermoglio et al., 2005), according to what Vallega (1997) terms an epistemological shift in ICZM, 'from a disjunctive to a conjunctive approach'.

A number of authors have written on the characteristics of the science–policy interface needed for coastal governance (see e.g. Boesch, 1999; Cicin-Sain and Knecht, 1998; Costanza, 1999; Costanza et al., 1999; deReynier, Levin and Shoji, 2010; Fritz, 2010; Knol, 2010; Norgaard et al., 2009; Tobey and Volk, 2002). Across this multitude of authors there is some overlap around four broad themes. Firstly, the science–policy interface must be participatory and inclusive of all stakeholders and their perspectives, including their values alongside their knowledge. Secondly, it must be deliberative, facilitating high-quality dialogue between stakeholders for the mobilisation of knowledge. Thirdly, the interface must be integrated, according to attitudes of reciprocity that bring together the at-times inconsistent perspectives, and encourage their co-existence relative to each other. Fourthly, the interface needs to have an explicit treatment of knowledge 'quality'. This is quality both in terms of the process followed to mobilise knowledge, and the quality of the knowledge 'products' themselves, with quality often discussed in terms of salience, credibility and legitimacy (Cash et al., 2003; US National Research Council, 2007). There are a number of dialogic approaches put forward for framing the science–policy interface according to these four themes, with the 'post-normal science' perspective finding mention in the ICZM literature (Costanza, 1999; Kendra, 1997), and explored further here.

### *14.2.3 Introducing a 'post-normal' science–policy interface for ICZM*

Post-normal science was proposed by Funtowicz and Ravetz (1990, 1993) in the early 1990s, in recognition of the uncertainty, plurality and politics surrounding much collective decision-making, and challenging the relevance of reductionist and disciplined 'normal

**Table 14.2** Characteristics of the 'post-normal science' approach to framing the science–policy interface (Bremer, 2011).

1. Focussed on the science-policy interface in support of decision-making for issues characterised by uncertainty, plurality and high stakes
2. Complex social-ecological systems perspective
3. 'Epistemological' uncertainty is explicitly recognised as inevitable and irreducible, necessitating a precautionary approach
4. Dialogic setting wherein knowledge is mobilised and negotiated as 'evidence' in support of normative arguments.
5. Participatory setting admitting a plurality of equally valid perspectives
6. Relativize, and where possible reconcile, conflicting perspectives according to principles of reciprocity and co-existence
7. Conflicting perspectives are evaluated for their *'quality'* in terms of supporting decision-making for an issue. Quality is the organising principle of science rather than 'truth'
8. Stakeholders as members of an 'extended peer community,' which collectively evaluate knowledge according to collectively derived criteria of quality; 'extended peer review'
9. Reflexive
10. Social learning orientation in pursuit of collective understanding rather than truth
11. Adaptive cyclic process, with outcomes judged on the quality of the process
12. Strategic, long-term perspective

science' described by Thomas Kuhn (1962), for supporting such decisions. Beginning from assumptions of uncertainty and a plurality of valid knowledge perspectives, PNS introduces 'quality' as the organising principle of science; leaving the door open to new scientific norms of evidence and dialogue among stakeholders espousing multiple knowledge systems (Funtowicz & Ravetz, 1993, 1994; Ravetz & Funtowicz, 1999). In pursuing 'high-quality' knowledge for a given issue, stakeholders working at the science–policy interface represent an 'extended peer community;' charged with both effectively communicating their own perspective, and providing a critique of the other knowledge presented. The extended peer community represents a widening of the circle of 'critics' of knowledge for decision-making, and lends legitimacy to those stakeholders previously not recognised in this role (De Marchi & Ravetz, 1999; Frame & Brown, 2008). They work together as 'co-investigators' (Frame & Brown, 2008; Funtowicz & Ravetz, 1993); analysing knowledge as 'evidence' called upon to support different perspectives, with the burden of proof placed on the participant calling the evidence (Frame & Brown, 2008; Funtowicz & Ravetz, 1994; Ravetz, 2006). Knowledge for decision-making is evaluated relative to collectively defined criteria of 'quality' for a specific issue, which could extend to its credibility, salience and legitimacy, for example. Table 14.2 summarises these features of the PNS perspective.

What promise does a PNS perspective hold for framing the science–policy interface for ICZM? Those writing more broadly on participation at the science–policy interface (see e.g. Fiorino, 1990; Pahl-Wostl, 2002, 2005; Reid et al., (2006) invoke four key imperatives, which are highly relevant to this discussion. Most significantly, a PNS approach is likely to make *substantive* contributions to the knowledge available to support coastal governance. By challenging the primacy of scientific reports, PNS broadens the 'knowledge base' drawn on to support decisions, while ensuring that it is in an appropriate form and of an appropriate quality for the coastal issues at hand. Secondly, a PNS approach *normatively* addresses the plurality of perspectives within coastal communities. By maximising participation according to democratic principles, all values within the community are given consideration, the decision-making process is legitimised and decision-makers are held accountable. Thirdly PNS has an *instrumental* imperative, insofar as the science–policy interface includes all diverse stakeholder perspectives, this increases communication and understanding; building trust, reducing conflict, improving chances for consensus, and putting decisions into action. Finally, a PNS approach aids *social learning,* encouraging participants to learn about coastal issues, other actors' values, and the decision-making process. Accepting the complexity of the coast, where change is inevitable though unpredictable, 'learning-based' adaptation is essential to coastal governance. However, it is one thing to discuss the theoretical contributions of PNS, though another to look at the contributions to ICZM *in practice.*

## 14.3 Exploring a post-normal science–policy interface in practice

This chapter now turns to look at how a PNS-approach has found practical expression for ICZM. It summarises the results of doctoral research undertaken from 2008 to 2011, which explored how a post-normal science–policy interface can contribute to quality ICZM (Bremer, 2011). This research studied instances of coastal management where a PNS approach (or something approximating that approach) was used, and evaluated its contributions to coastal governance across various measures.

### *14.3.1 Cross-scale empirical studies*

Key to this research was its attempt to assemble a rich portfolio of different experiences from which to explore the influence of a post-normal science–policy interface. In this way, the case studies are split between New Zealand and Europe as two very different coastal management contexts, and according to a multi-scale imperative; spanning the international, national and local scales. The multi-scale nature of these studies was essential because the demands placed on ICZM initiatives differ across scales, with this particularly true of the science–policy interface, which has been shown to assume different forms at different scales; from science-centric at the international scale with examples like the Millennium Ecosystem Assessment, to a greater reliance on local knowledge at the local scale (Reid et al., 2006).

#### *14.3.1.1 International-scale case study: the SPICOSA Project*

At the international scale, this research focused on the European Union's integrated Science and Policy Integration for Coastal System Assessment (SPICOSA) Project. 'The overall objective of SPICOSA [was] to develop a self-evolving, holistic research approach for integrated assessment of coastal systems, such that the best available scientific knowledge can be mobilised to support deliberative and decision-making processes aimed at improving the sustainability of coastal systems by implementing Integrated Coastal Zone Management policies' (see Chapters 1 and 18 of this book). SPICOSA sought to develop tools for facilitating a science–policy interface that was arguably 'post-normal' in nature and implement these tools across 18 European study sites, many of which spanned international boundaries and extended beyond the EU. Key to this was encouraging dialogue between policy-makers, scientists and other stakeholders within study sites to collectively model coastal systems and use these models for deliberating management options.

The SPICOSA Project was studied via (i) a desk-top study of its evaluation reviews, and (ii) face-to-face structured interviews with a single SPICOSA Project representative (scientists and researchers from local research institutes) on condition of anonymity, from each of four study sites; (i) Cork Harbour; (ii) the Rhine-Meuse-Scheldt Delta; (iii) the Guadiana Estuary; and (iv) the Barcelona waterfront. Interviews were recorded and a transcript sent to respondents for validation, after which responses were condensed into a spread-sheet for analysis.

#### *14.3.1.2 National-scale case study: the New Zealand national experience*

At the national scale, this research sought to explore the influence of the science–policy interface on the ICZM regime in New Zealand, by attempting to arrive at some statement on the aggregated common national experience; as determined by both the national framework and the sum of regional practice. Rather than focus on any particular initiative, it aimed to provide a more general insight into the nationwide state of the science–policy interface, and its influence on the quality of coastal management institutions. This national review revealed a traditional dominance of science-centric coastal management, which is increasingly confronted across the regions with emerging practices of participation and dialogue at the science–policy interface.

The research method followed both (i) a desktop study of government reviews and published evaluations, and (ii) semi-structured interviews with a single coastal manager at all of New Zealand's 16 regional authorities on condition of anonymity. Interviews were recorded and notes taken from these recordings, which were sent to interviewees for confirmation, before being written into a draft report that was circulated to all interviewees for comment.

#### *14.3.1.3 Local-scale case studies: Whangamata, Waikaraka and Gisborne*

At the local scale in New Zealand, this research followed three particular coastal management initiatives in Whangamata, Waikaraka and Gisborne, which employed a science–policy interface approximating the post-normal approach. In Whangamata, local government convened a 'Stakeholder Forum' to mobilise knowledge on the spread of mangroves in the harbour, and arrive at consensus on their management. Waikaraka similarly faced an issue with the spread of mangroves, though this time the local community convened their own 'Estuary Managers Group,' drawing participation from local government and universities to arrive at a well-informed management plan for the estuary. Gisborne City Council faced a long legal battle over its wastewater outfall to Poverty Bay, and chose instead to convene the 'Wasterwater Adjournment Review Group' to design a new outfall. All three case studies represented an attempt to bring together a diverse group of stakeholders in extended deliberation, within a setting that at least ostensibly included all knowledge systems alongside the values and rhetoric of coastal stakeholders. Such settings were supported by various tools, ranging from formal presentations according to an agreed protocol, to collective mapping exercises, to knowledge partnerships. Common to all three cases was an emphasis on grounding decisions in the best quality knowledge possible, as a means for diluting the highly politicised value-divisions within the respective coastal communities. Besides providing insight into the contribution of this post-normal interface to the quality of ICZM locally, these case studies also provided a comparison to the general state of the science–policy interface in New Zealand nationally.

For each case study, the researcher visited the locality for one week to (i) study written material on the initiatives; (ii) undertake semi-structured interviews with 5–8 diverse stakeholders; and (iii) observe the local context and institutional settings, through attending meetings of the groups, and walking the coastline with stakeholders. The recorded interviews were transcribed, and the transcriptions were emailed to respondents to give them an opportunity to make any changes.

### *14.3.2 Method for exploring contributions to ICZM*

The framework that guided the exploration of the case studies, and the contributions of a PNS approach to ICZM, was essentially two-tiered. The first tier *described and analysed* the science–policy interface through the lens of the PNS perspective, relative to the characteristics listed in Table 14.2. This was to determine whether each case study represented a strong or weak instance of a PNS approach, and thus allow some comment on the applicability of PNS to coastal management. The second tier *explored* the influence of the science–policy interface for promoting successful coastal management, according to its contributions to the quality of coastal management institutions and the quality of stakeholder interactions for collective decision-making. *Institutional quality* was measured relative to the degree that coastal management institutions gave effect to key ICZM principles, which were drawn from Stojanovic, Ballinger and Lalwani, as shown in Table 14.1.

*Interactional quality* was measured relative to the 'social outcomes' that resulted from stakeholder interaction, as an alternative to measuring the quality of a decision, or consensus, itself (Innes, 1999). Such outcomes can be expressed in terms of stocks of 'capital' that stakeholders build up through constructive interaction and decision-making, and can draw on for subsequent decision-making processes. For this research, the quality of stakeholder interaction was measured in terms of its contribution to stocks of *financial, social and human capital.* Social capital was measured relative to increased participation and density of interactions between stakeholders, and the norms of trust and reciprocity that guide these interactions such that they promote collective action. Human capital was measured relative to the increased diversity of expertise and experience mobilised, learning among stakeholders and the emergence of leadership. Finally, financial capital served as a quantitative check of stakeholders' commitment to an ICZM initiative; 'putting their money where their mouth is'.

Using this conceptual framework, a semi-directed interview framework was created, comprising 15 broad questions (and prompts) spread evenly across three sections corresponding to: (i) the analysis of the science–policy interface; (ii) the exploration of institutional quality; and (iii) the exploration of interactional quality. An identical interview framework was used across all three case studies.

While not a statistically significant sample, the diversity of contexts and scales represented by the studies did allow for unique insights on how a PNS approach can find expression under different conditions; allowing for some interesting contrasts and comparisons. In fact, across this diversity of experiences there were some common themes. To this end, the paper unpacks eight key lessons learned from the empirical studies, and finishes by looking at what these lessons reveal for the promise of the PNS approach for ICZM.

## 14.4 Lessons from the empirical studies

From across the three empirical studies there emerged eight common lessons on the relevance and applicability of a 'post-normal' science–policy interface, which are summarised here.

*Lesson 1: Coastal stakeholders sought to democratise the science–policy interface out of dissatisfaction with science-based management.*

One common theme across all three studies was a dissatisfactory experience with science-based coastal management alone, with this a key motivation to try democratising the science–policy interface, including through a PNS approach. This dissatisfaction stemmed from (i) a lack of science with which to support decisions; (ii) poor integration of science with decisions; and (iii) distrust of science from other stakeholders.

*Lesson 2: The PNS approach has not found explicit practical application in ICZM, though there are many ways to give effect to its principles.*

This research could not find a single example of ICZM initiatives that have explicitly labelled themselves as an exercise in 'post-normal science;' indicating that the approach has not found widespread practical expression. However, it was not uncommon to find in the ICZM literature and in practice, initiatives that effectively give effect to the principles of PNS, even while not being 'labelled' as such. This research revealed that there is no *one* way to give effect to a PNS approach, though necessarily most initiatives begin from a dialogic forum, within which they attempted to give effect to inclusive and reciprocal dialogue between knowledge perspectives. However, the empirical studies in this research revealed that achieving this in practice is non-trivial; necessitating 'deliberation support systems/tools' for nurturing a 'better quality' of dialogue at the science–policy interface. These ranged from formal modelling tools, to maps, metaphors, and exercises for 'ground-testing' knowledge constructed at the interface.

*Lesson 3: A PNS approach can contribute to better quality ICZM institutions.*

The general experience taken from these particular contexts is that a participatory or post-normal science–policy interface can pose a better 'quality' ICZM institution compared to a purely science-centric institution, across a number of parameters. The research found a common increase across all three studies in the quality of ICZM institutions relative to (i) the *comprehensiveness* of their perspective on coastal issues, with an understanding more deeply based in ecosystems; (ii) their *contingency* to the context and issues; and (iii) their *strategic* treatment of coastal issues, particularly to the strategic collection of knowledge. This translated to an institution better equipped to mobilise high-quality knowledge in support with decisions. On the other hand the degree to which post-normal institution encouraged greater stakeholder *participation* and *cooperation* was largely dependent on the context, and they were rarely more *precautionary* or *adaptive.*

*Lesson 4: A PNS approach can contribute to better quality stakeholder interactions for collective decision-making.*

Both the SPICOSA Project and the New Zealand local-scale case studies demonstrated an increase in the quality of stakeholder interaction relative to most capital measures, with the degree of improvement contingent on the context. Primarily this was translated by increases in the human capital on which stakeholders could draw in support of decision-making, according to the significant 'social learning' among stakeholders. There were also demonstrable increases in the social capital stocks stakeholders could draw on according to norms of trust and reciprocity, and where this reached a critical threshold, it was translated into increased stocks of financial capital. The degree to which social and financial stocks increased was highly contingent on the socio-political conditions in a context, including the history of participation and stakeholder interaction.

*Lesson 5: Science is powerful and there is a danger that all knowledge perspectives become subsumed within a scientific framework.*

Across all three studies emerged one common obstacle to the successful implementation of a post-normal science–policy interface; the power and dominance of 'normal' science. Though a science–policy interface may admit stakeholders drawing on a rich diversity of knowledge systems, these diverse perspectives often become co-opted into the service of building a scientific picture of the coast. The danger is that in bringing together diverse knowledge systems in an institutional setting traditionally dominated by scientists, these other forms of knowledge become subsumed into a scientific framework.

*Lesson 6: Power is ever-present within any dialogic institutional setting, and influences the way knowledge is mobilised.*

Beyond the powerful influence of science, the successful implementation of a post-normal science–policy interface is largely dependent on the way power shapes institutions and the interactions between stakeholders. As for many processes drawing on a Habermasian ideal of 'real' reciprocal dialogue, this is only successful insofar as stakeholders do not seek to exert their power, either consciously or unconsciously, over others or over the process. Following the likes of Foucault (Flyvbjerg, 1998), who asserts that 'power is always present,' this research too found that power had a formative influence on the post-normal science–policy interface. This power manifested itself in both a negative way by sabotaging the process, and in a positive way through motivating leadership.

*Lesson 7: Context matters; a PNS approach is best used to improve the quality of pre-established dialogue and collective decision-making.*

The empirical research demonstrated how the specific contexts offered by the case studies had a significant influence on the degree to which a post-normal science–policy interface contributed to better quality institutions and interactions. Notably, a post-normal approach was less a means of encouraging new stakeholder participation and interaction as it was a means for improving the quality of those pre-established interactions. In this way, where a context already had a long history of participation, a strong network of stakeholder interactions, and existing dialogic institutions, a post-normal approach found success. On the other hand, where a PNS approach was exercised as a short-term intervention into a very fragmented context, with a poor history of stakeholder participation and interaction, it posed a very insular exercise with more limited success.

*Lesson 8: Scale matters: a PNS approach works best at a scale close and salient to the issue.*

By exploring three studies at the international, national and local scale, this research was able to offer some insights into the influence of scale on the shape of a post-normal science–policy interface. Scale was found to be important insofar as dialogue within a 'post-normal' setting was better enabled at a scale closely corresponding with a given issue – usually at the local scale. In this way, it was important to start from a well-defined issue, and match the science–policy interface dialogue to the appropriate scale of the issue; defined spatially, temporally, and in terms of the tier of decision-making and administration. Post-normal initiatives studied were perceived to be more legitimate where they engaged a diversity of stakeholders, which an affected community deemed to have credibility and a salient perspective on the issue. At the same time, the temporal scale over which a post-normal science–policy interface ran determined its success, with long-term initiatives much more successful than a short-term intervention.

## 14.5 Conclusion

By framing ICZM in terms of coastal governance we are forced to confront the implications of this perspective for the way we mobilise knowledge at the science–policy interface, with many authors promoting principles of inclusiveness, dialogue, integration and knowledge quality. This chapter has summarised empirical PhD research which explored the contributions of the particular approach offered by 'post-normal science' for coastal governance.

The three empirical studies demonstrated the significant promise offered by participatory and dialogic forms of science–policy interface, and particularly a PNS perspective. The cases studied all saw the creation of institutional settings that (to differing degrees) gave effect to principles of ICZM, and improved the quality of stakeholder interaction for collective decision-making, with particular regard for the way knowledge is mobilised and linked with decisions. In this way, initiatives resembling a PNS approach were able to realise a substantive increase in the availability and quality of knowledge on which to base decisions; offering a more comprehensive, contingent and long-term picture of issues meaningful to coastal communities, and integrating it closely with the political decision-making process.

At the same time these initiatives saw a more concrete expression of participatory democracy, by empowering previously marginalised stakeholders and encouraging new norms of inclusive knowledge production. At least within the local-scale case studies, this gave ICZM more legitimacy within the local community, and prepared the way for decisions to be put into action. Moreover, these institutions were seen to give rise to significant 'social learning' amongst stakeholders toward a common understanding of an issue, and provided a precursor to stakeholders compromising or even finding consensus.

However, it is important to note that a 'post-normal' science–policy interface is not a blanket solution. Rather it presents an approach to improve the quality of existing stakeholder dialogue to mobilise the best quality knowledge available for a specific category of complex issue, which is beyond the bounds of 'normal' science to solve. Moreover, any attempts to give effect to a PNS approach needs to take into account the eight lessons presented in this paper with particular regard to the influence of 'power' on stakeholder interactions, most often expressed through the dominance of the scientific community as knowledge provider.

## References


Bille, R. (2007). A dual-level framework for evaluating integrated coastal management beyond labels. *Ocean & Coastal Management*, **50**(10), 706–807.

Bille, R. (2008). Integrated coastal zone management: four entrenched illusions. *Surveys and Perspectives Integrating Environment and Society*, **1**(2), 75–86.

Boesch, D.F. (1999). The role of science in ocean governance. *Ecological Economics*, **31**(2), 189–198.

Bremer, S. (2011). *Exploring a 'post-normal' science–policy interface for Integrated Coastal Management*. Massey University, Palmerston North.

Cash, D.W., Clark, W.C., Alcock, F., Dickson, N.M., Eckley, N., Guston, D.H., et al. (2003). Knowledge systems for sustainable development. *Proceedings of the National Academy of Sciences*, **100**(14), 8086–8091.

Christie, P. (2005). Is integrated coastal management sustainable? *Ocean & Coastal Management*, **48**(3–6), 208–232.

Chua, T.-E. (1993). Essential elements of integrated coastal zone management. *Ocean & Coastal Management*, **21**(1–3), 81–108.

Cicin-Sain, B., & Knecht, R.W. (1998). *Integrated coastal and ocean management: concepts and practices* Washington, DC: Island Press.

Costanza, R. (1999). The ecological, economic, and social importance of the oceans. *Ecological Economics*, **31**(2), 199–213.

Costanza, R., Andrade, F., Antunes, P., van den Belt, M., Boesch, D., Boersma, D., et al. (1999). Ecological economics and sustainable governance of the oceans. *Ecological Economics*, **31**(2), 171–187.

De Marchi, B., & Ravetz, J. (1999). Risk management and governance: a post-normal science approach. *Futures*, **31**(7), 743–757.

deReynier, Y.L., Levin, P.S., & Shoji, N.L. (2010). Bringing stakeholders, scientists, and managers together through an integrated ecosystem assessment process. *Marine Policy*, **34**, 534–540.

Fiorino, D.J. (1990). Citizen participation and environmental risk: A survey of institutional mechanisms. *Science, Technology and Human Values*, **15**(2), 226–243.

Flyvbjerg, B. (1998). *Rationality and Power: Democracy in Practice*. Chicago: The University of Chicago Press.

Frame, B., & Brown, J. (2008). Developing post-normal technologies for sustainability. *Ecological Economics*, **65**(2), 225–241.

Fritz, J.-S. (2010). Towards a 'new form of governance' in science–policy relations in the European Maritime Policy. *Marine Policy*, **34**(1), 1–6.

Funtowicz, S.O., & Ravetz, J.R. (1990). *Uncertainty and Quality in Science for Policy* (Vol. **15**). Dordrecht: Kluwer Academic Publishers.

Funtowicz, S.O., & Ravetz, J.R. (1993). Science for the post-normal age. *Futures*, **25**(7), 739–755.

Funtowicz, S.O., & Ravetz, J.R. (1994). The worth of a songbird: ecological economics as a post-normal science. *Ecological Economics*, **10**(3), 197–207.

Funtowicz, S.O., & Strand, R. (2006). Models of science and policy. In T. Traavik & L.C. Lim (Eds.), *Biosafety First: Holistic Approaches to Risk and Uncertainty in Genetic Engineering and Genetically Modified Organisms* (pp. 263–278). Trondheim: Tapir Academic Press.

Innes, J. (1999). Evaluating Consensus Building. In L. Susskind, S. McKearnan & J. Thomas-Larmer (Eds.), *The Consensus Building Handbook: A Comprehensive Guide to Reaching Agreement.* Thousand Oaks: SAGE Publications.

Kendra, J.M. (1997). Seaport development versus environmental preservation: the case of Sears Island, Maine, USA. *Marine Policy*, **21**(5), 409–424.

Knol, M. (2010). Scientific advice in integrated ocean management: The process towards the Barents Sea plan. *Marine Policy*, **34**(2), 252–260.

Kooiman, J., & Bavinck, M. (2005). The Governance Perspective. In J. Kooiman, M. Bavinck, S. Jentoft & R. Pullin (Eds.), *Fish for Life: Interactive Governance for Fisheries* (pp. 11–24). Amsterdam: Amsterdam University Press.

Kuhn, T. (1962). *The structure of scientific revolutions* (3rd ed.). Chicago: University of Chicago Press.

McFadden, L. (2007). Governing Coastal Spaces: The Case of Disappearing Science in Integrated Coastal Zone Management. *Coastal Management*, **35**(4), 429–443.

McKenna, J., & Cooper, A. (2006). Sacred cows in coastal management: the need for a 'cheap and transitory' model. *Area*, **38**(4), 421–431.

Norgaard, R.B., Kallis, G., & Kiparsky, M. (2009). Collectively engaging complex socio-ecological systems: re-envisioning science, governance, and the California Delta. *Environmental Science & Policy*, **12**(6), 644–652.

Pahl-Wostl, C. (2002). Participative and stakeholder-based policy design, evaluation and modeling processes. *Integrated Assessment*, **3**(1), 3–14.

Pahl-Wostl, C. (2005). Actor based analysis and modeling approaches. *Integrated Assessment Journal*, **5**(1), 97–118.

Ravetz, J.R. (2006). Post-normal science and the complexity of transitions towards sustainability. *Ecological Complexity*, **3**(4), 275–284.

Ravetz, J.R., & Funtowicz, S.O. (1999). Post-normal science—an insight now maturing. *Futures*, **31**(1), 641–646.

Reid, W.V., Berkes, F., Wilbanks, T.J., & Capistrano, D. (2006). Introduction. In W.V. Reid, F. Berkes, T.J. Wilbanks & D. Capistrano (Eds.), *Bridging Scales and Knowledge Systems: Concepts and Applications in Ecosystem Assessment* (pp. 1–17). Washington, DC: Island Publishing.

Stojanovic, T.A., Ballinger, R.C., & Lalwani, C.S. (2004). Successful integrated coastal management: measuring it with research and contributing to wise practice. *Ocean & Coastal Management*, **47**(5–6), 273–298.

Stojanovic, T.A., & Barker, N. (2008). Improving governance through local Coastal Partnerships in the UK. *The Geographical Journal*, **174**(4), 344–360.

Tobey, J., & Volk, R. (2002). Learning Frontiers in the Practice of Integrated Coastal Management. *Coastal Management*, **30**(4), 285–298.

US National Research Council. (2007). *Analysis of Global Change Assessments: Lessons Learned*. Washington, DC: National Academies Press.

Vallega, A. (1997). *Fundamentals of Integrated Coastal Management*. Dordrecht: Kluwer Academic Publishers.

Zermoglio, M.F., van Jaarsveld, A., Reid, W.V., Romm, J., Biggs, O., Yue, T.X., et al. (2005). The multi-scale approach. In F. Berkes, M. Giampietro, T. Wilbanks & X. Jianchu (Eds.), *Millenium Ecosystem Assessment: Multiscale Assessments—Findings of the Sub-Global Assessments Working Group* (Vol. **4**, pp. 61–83). Washington, DC: Island Press.

Chapter 15

# Co-Learning in Marine Protected Areas for Integrated Coastal Zone Management

*Mohammad A.L. Siddique, S. Zafrin, S. Myers, T. Smith, R. Babcock and R.W. (Bill) Carter*

**Abstract**

The nature of marine systems and the multiple stakeholders directly (and indirectly) exploiting marine ecosystem services demands a stewardship and co-management approach to conserving marine resources. At the heart of co-management and stewardship is co-learning. It is in management's interest to determine the optimal approach to conserving biodiversity in marine protected areas. To achieve public confidence and trust, it is essential that the best management approaches are applied and that stakeholders are involved in the decision-making process. Little research has focused on shared or co-learning for improving management outcomes. This study aims to understand the mechanisms of sustainability learning systems and processes to improve existing management strategies. The study is based on a desk-top analysis of literature and planning instruments. The chapter identifies approaches to identifying pathways and barriers to developing a co-learning approach to management for achieving integrated coastal zone management through creating resilience among stakeholders and management processes.

## 15.1 Introduction

Marine protected areas are a key management strategy to achieve conservation of marine biodiversity and ecosystem services, which are increasingly under pressure, worldwide (IUCN-WCPA, 2008). Given the nature of marine systems and the multiple stakeholders exploiting marine ecosystem services, stewardship and co-management emerge as essential management approaches (Myers et al., 2012). Co-management promotes access to, and exchange of, material and non-material resources, such as money, technology, scientific and

*Global Challenges in Integrated Coastal Zone Management*, First Edition. Edited by Erlend Moksness, Einar Dahl and Josianne Støttrup.

local knowledge, as well as increased management legitimacy (Sandstrom and Rova, 2010). At the heart of co-management and stewardship is co-learning, where every intervention, including exploitative use, represents a learning opportunity to be shared towards improving mutual understanding.

Science, particularly in the form of research and monitoring, has the potential to provide a solid, defensible foundation to support park management through linking scientific information to management processes. However, integrating the science and management interface into a system-wide evaluation process has been challenging (Growcock et al., 2009). Integrating knowledge (scientific or otherwise) into management is a complex process that involves facilitation, translation and communication, and relies on stakeholder partnerships and the translation of knowledge into accessible and relevant media for communication between stakeholders at the appropriate times (Orbach and Karrer, 2010) and scales. This is acknowledged by the vision in IV/5 and VII/5 of the Nagoya Oceans Statement (Nagoya Oceans Statement, 2010), which stresses the need to apply the best available science, including the technological knowledge of local and indigenous communities. However, processes for accessing stakeholder knowledge, sharing and integrating knowledge towards co-learning and co-management remain problematic. The decision-making processes often remains compartmentalized, with individual governments or agencies frequently implementing policies in isolation from other government agencies (DERM, 2010: 51) and indeed, stakeholders. Sharing of information supports the standardised reporting of management issues and effectiveness, while improved knowledge of management strengths and weaknesses within the park system allows for the development of targeted work programs (Growcock et al., 2009).

To maximize the effectiveness of management of protected areas, managers and policymakers need information on the strengths and weaknesses of their management processes and on the threats and stresses that resources and management face (Hockings, 2003). Protected areas are established to conserve special values, so understanding these values and their significance at global, national or local scales is vital for both management planning and evaluation (Hockings et al., 2000). Managers need information on how to secure these values, what threats they face and about external influences, including the activity and impact of stakeholders – including local communities. While scientific data and understanding form a vital component of the information, other aspects such as traditional ecological knowledge and stakeholder/community's experience is also required for effective management. Because information concerning these aspects may essentially be different from science information, resource management decisions cannot be based entirely on the logical positivist approach of the modern scientific ethic (Holmes et al., 1992). Resource management requires information at a variety of levels, from philosophical to methodological, descriptive to explanatory, comprehensive to brief, true science to traditional. While scientific data are required for managers to make decisions, an interface or effective translation mechanisms are essential for conversion of data to information for all stakeholders to be able to effectively participate in the decision-making process. Co-learning can promote such an interface as a bridging pathway between managers and other key stakeholders to provide a resilient system for marine protected area management. This chapter synthesizes the published material addressing the learning process between managers and other stakeholders in a marine protected area (MPA) management system to develop a co-learning framework that would facilitate shared learning between stakeholders for mutual benefit.

## 15.2 Community involvement in marine protected areas

Collaborative natural resource management relevant to MPAs is a sub-set of broader planning and management. Collaborative planning 'accords with certain features of contemporary society; including the putative shift to new modes of governance that acknowledge the need to involve multiple stakeholders' (Brand and Gaffikin, 2007: 283). Specific stakeholder involvement in MPA development and management is a topic of growing interest, taking into account that social factors are the primary determinants of the success of MPAs (Mascia, 2003; NOAA, 2004). While there is growing awareness of the need for involvement, there is a lack of understanding of the implications of learning as an essential element in the process of stakeholder involvement. Significant stakeholder participation occurs in MPAs when contributions to the process are transparent and clear evidence exists that the contributions have helped shape a meaningful decision in planning and management. Such participation can be fostered by enhancing stakeholders' participation in the generation and application of knowledge, providing opportunities to exchange learning, and strengthening the ability to meet concerns and deal with changes throughout the process. Exchange of information can empower stakeholders to become involved in, and make an

**Table 15.1** MPA management priorities: changing from old to co-learning based (after Pimbert and Pretty, 1995. United Nations Research Institute For Social Development, Geneva).

| Characteristics | From old system | To the new learning system |
|---|---|---|
| Who sets priorities? | Management professionals | Local people with professionals |
| Management science and method | Scientific method is reductionist and positivist, with a strong natural science bias | Scientific method is holistic and post positivist, emphasizing social science issues |
| Strategy and context of intervention | Pre-specified research plan or project design; top down approach. Information and results are extracted from controlled situations; context is independent and controlled; blueprint oriented. | Open-ended learning process; understanding and focus emerge through interaction; context of inquiry and intervention is fundamental; process oriented. |
| Assumptions about reality | Assumption of singular, tangible reality | Assumption of multiple realities that are socially constructed |
| Relationship between all actors in the process | Professionals control and motivate client from a distance; they tend not to trust people (e.g. fishermen, indigenous and rural people) who are simply the object of inquiry or intervention. | Professionals enable and empower in close dialogue; they attempt to build trust through joint analyses and negotiation; understanding arises through this engagement, resulting in inevitable interactions between investigator and the "object" of inquiry or intervention. |
| Mode of working | Single disciplinary-working alone | Multidisciplinary working in groups. |
| Management technology or services | Rejected technology or service assumed to be fault of local people or local conditions; technology first. | Rejected technology or service is a failed technology; people first. |

impact on defining management priorities (Table 15.1) and the planning process in MPAs. An empowered community can address both the need for economic development and the conservation of natural resources (Pomeroy, 1995). In the end, resource conflicts may be diminished, access rights distributed more effectively and equitably, management initiatives better implemented, and resources better managed when stakeholders are more involved in management initiatives (Pomeroy, 1995).

## 15.3 Conceptual framework

The chapter initially conceptualizes the existing management system based on observation and reflection on experiences in MPA (Figure 15.1). MPA authorities, stakeholders and existing management instruments are considered the three major elements influencing co-learning. The framework is constructed against the broad question, 'Does co-learning form part of MPA management?' This leads to consideration of what knowledge/learning is being used in planning, implementing, monitoring and evaluating the management instruments and informing on-site actions. It also raises issues of what information/learning is shared, updated with experiences of implementation and whether information is continually shared. Based on desktop analysis, it is assumed that current management practices are mainly directive or enforcement based. Information/learning sharing is weak or absent in many cases (dashed lines in Figure 15.1), although resource users are rich in local, traditional and experiential knowledge and managers have better access to scientific knowledge. Stakeholder involvement in the planning process is almost encapsulated within submissions, rather than active participation. The framework also assumes that authorities are often less interested in addressing stakeholder needs and learning interests and likewise have

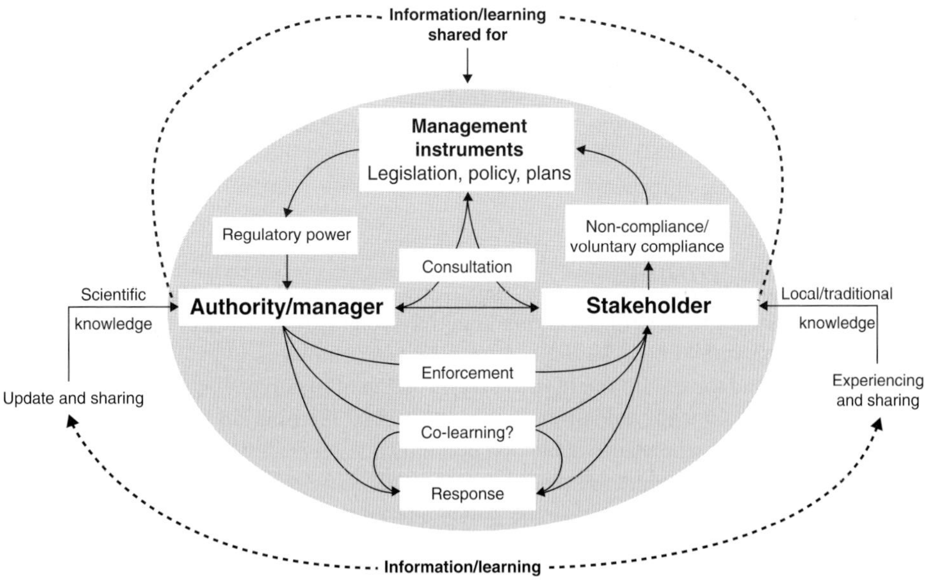


**Figure 15.1** Conceptual framework for existing co-learning in marine protected area management. Dashed lines denote weak/broken links.

'inappropriate' bias towards economic benefits, rather than protection of long-term ecosystem services.

## 15.4 Adaptive learning, management and governance issues in MPA management

Little work is found in the MPA literature identifying the ecological and social and economic linkages between MPAs and external areas in broader coastal zone management (Belfiore et al., 2004). Due to the dynamic nature of coastal ecosystems and the uncertainty associated with large-scale restoration, adaptive management principles need to be embedded throughout the organizational structure, planning process, implementation and monitoring programme to achieve local, national and global mandates associated with MPA management. Sound science is necessary for commons governance, but is insufficient (Dietz et al., 2003) because technologies are embedded in a network of social routines. Increasing emphasis has been devoted to processes of decision-making mainly based on formal decision theory in resources management, whereas processes of social learning, associated with those decisions, have largely been neglected (Pahl-Wostl and Hare, 2004). Despite advances in understanding over the past century, much remains a mystery about the linkages among species, habitats and oceanographic factors (MEAM, 2010); yet, policy-makers and managers make decisions with incomplete data, imperfect models and scientific disagreement. Under these circumstances, the environment under which organisations perform become increasingly unpredictable. As a result, organisations are forced to make significant transformations to adapt and survive, with the processes of learning and reflexivity among organisations being particularly critical (Singh, 2007; Fard et al., 2009; Storbjork, 2010). Mutual learning may also lead to multiple small-scale adaptations by individuals that can build resilience into their own systems (Bardsley and Sweeney, 2010), which may not exist in conventional natural management systems. So, key elements of adaptive learning for coastal management include social learning, sustainability learning, organisational learning and a bias towards reviewing and changing policy and management practice (Smith et al., 2009). The notion of co-learning may be a form of social learning with adaptive learning at its heart.

### *15.4.1 Social learning*

Social theories of learning define learning as active social participation in the practices of a community (Lave and Wenger, 1991; Wenger, 1998) and emphasise the dynamic interaction between people and the environment in the construction of meaning and identity (Muro and Jeffrey, 2008). The concept of learning, especially social learning, in the literature is quite diverse (Figure 15.2), where accessible updated information is a missing link. Information, knowledge and experiences are not shared and updated, or updated information is not always shared. This may also be reflected in the policy and management instruments applied and are often the source of conflict.

Despite substantial use of social learning in natural resources management and policy, there remains little consensus on its meaning or theoretical basis (Reed et al., 2010). A

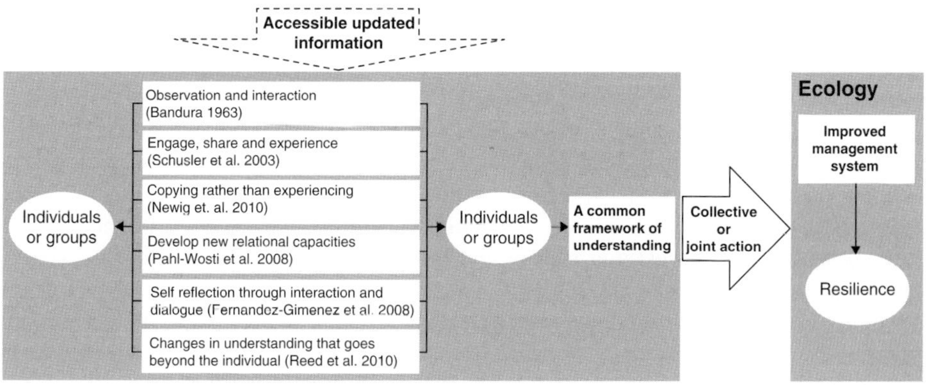


**Figure 15.2** Social learning process in social-ecological system of natural resources management. Dashed arrow indicates a missing link.

central hypothesis for the concept of social learning is that social involvement and the management of content are strongly interdependent (Lorenzoni et al., 2000) because social learning is a process of iterative reflection that occurs when people share their experiences, ideas and environments with others (Keen et al., 2005). Processes of social learning should thus involve shared problem perception in a group of actors, trust as the base for critical self-reflection, recognition of mutual dependencies and interactions in the actor network, assumptions about the dynamics and cause–effect relationships in the system, subjective valuation schemes and collective decision and learning processes (Pahl-Wostl, 2002).

In such a process, collective action and the resolution of conflicts require that people recognise their interdependence and differences, and learn to deal with them constructively. Management authorities with other key stakeholders need to learn about and increase their awareness of their biophysical environment and the complexity of social–ecological–economic interactions. An adaptive systems-based management process, based on social learning among all stakeholders, offers the potential for significantly increasing the sustainability of management decisions emerging from a decision-making process (McFadden et al., 2009). It is thus important to create informal space for developing relationships that are meaningful to individuals and groups, enabling continuous two-way flow of knowledge and learning. This constitutes a social learning process. For social learning to be considered a process, it must:

- demonstrate that a change in understanding has taken place in the individuals involved;
- demonstrate that this change goes beyond the individual and becomes situated within wider social units or communities of practice; and
- occur through social interactions and processes between actors within a social network (Reed et al., 2010).

Lack of interactions and thinking beyond their own boundary might be a gap, which can be minimized through co-learning. Co-learning may occur in a collaborative management system where the authority and community having joint responsibility for making decisions. The notion of a collaborative management system is where stakeholders are actively participating in all stages of management.

### 15.4.2 Organisational learning

To improve the efficiency of organizational learning, existing organizations must be rationalized to reduce bureaucratic redundancies and re-oriented toward the types of information provision that facilitate adaptive management (Thompson, 2006). Processes of adaptation involve changes to organisational 'routines', where routines represent much of an organisation's on-going activities, including a variety of tools and influences, such as rules, strategies, technologies, procedures, conventions, cultures and beliefs (Berkhout et al., 2006). Of relevance to learning is why and how organisations change their behaviour. Organisational learning can be seen as a cycle that begins with a stimulus leading to the generation of variation through experimentation and search. It proceeds with a process of internal selection, articulation and codification, followed by the replication and enactment of new routines across the organisation. Finally, the cycle returns to the beginning of a new cycle of innovation by virtue of a new stimulus (Berkhout et al., 2006). Like many other NRM systems, MPA management contains both formal and informal institutions. At the institutional level, integrated institutional structures may be better able to support the inclusion of stakeholders in decision-making processes and to ensure that their needs can be addressed by as wide an audience as possible (Tompkins and Adger, 2004). The core learning attributes that are vital to building an organisation that can truly 'learn' are based on five converging component technologies (Senge, 2006:1).

1. **Systems thinking** involves thinking about, and understanding a system as a whole and not necessarily any individual component. It is a language for describing and understanding the behaviour of a system through the forces and interrelationships that shape the system.
2. **Personal mastery** is where individuals continually assess and extend their personal vision to create the results they most truly desire. It is the cornerstone of a learning organisation. An organisation's commitment and capacity to learn can be no greater than that of its members.
3. **Mental models** are deeply ingrained pictures, images, assumptions and generalisations on how we perceive and conceptualise the world to improve our personal actions and decisions.
4. **Building shared vision** involves individuals building a shared vision of the future through genuine commitment and enrolment, rather than compliance.
5. **Team learning** involves the shared dialogue that allows team members to 'think together' as a unit. This is vital because teams are the fundamental learning unit in organisations, not the individual members.

### 15.4.3 Adaptive capacity

Adaptive capacity, particularly from a systems perspective, has been described as the ability to learn from mistakes, to generate experience of dealing with change and the capability for innovation (Tschakert and Dietrich, 2010). More specifically, adaptive capacity refers to the ability of a socio-ecological-economic system to cope with the current and anticipated changes and having sufficient room for responding to future uncertainties. Learning within institutions helps promote adaptive capacity and resilience by proactively building the capacity of individuals, communities, industries or nations (Folke, 2006) through shared learning. Present and future vulnerabilities have strong social elements because both are a function of adaptive capacity, which is in turn dependent on social capital, institutions, and resources and their distribution.

### *15.4.4 Adaptive management*

Adaptive management processes, informed by iterative learning about the ecosystem and earlier management successes and failures, increase present-day resilience, which can in turn increase the ability to respond to threats (Tompkins and Adger, 2004). Although adaptive management is applied typically at a project-specific level in relation to monitoring the effect of management decisions (Steyer and Llewellyn, 2000), it seeks to amass information to alter practice (Michel, 2009). Historically, low success rates in applying adaptive management have been attributed to cost, time scale and institutional barriers (Steyer and Llewellyn, 2000), including learning barriers between management authorities and other stakeholders. In particular, elected officials are notorious for their short time-horizons and their concern with personal gain (votes and 'rents') over effective policy, and this works directly against an adaptive management approach (Thompson, 2006).

### *15.4.5 Co-management*

The adaptive facet of co-management is captured in the incremental and iterative 'learning-by-doing' process where system understanding, action and evaluation are updated and refined every time new information is available (Tschakert and Dietrich, 2010), inspiring the co-production of knowledge. Co-management necessarily requires the integration of two distinct modes of knowledge (Iwasaki-Goodman, 2005). One is traditional ecological knowledge (TEK) gained by resource users from experience, and the other is scientific ecological knowledge (SEK) based on the accumulation of scientific data. Governments usually regulate the traditional resources that form part of the public good, whereas co-management most often involves vertical linkages and shifts in rights and responsibilities from government to local resource users (Berkes, 2002). Principle elements for an effective co-management system include sharing responsibility, balancing power, cooperation, participation, discussion, education and sharing of information, communication, consensus, flexibility and the use of TEK and SEK (Iwasaki-Goodman, 2005). Two issues that commonly arise in co-management of MPAs are the difficulty of establishing learning spaces for stakeholders to share experiences and learn, and how to deal with power inequalities. Co-management is an appropriate framework for effective MPA management because, in principle, it focuses on promoting power sharing between state and local people (Jones et al., 2011a) and, as a result, can improve the effectiveness of MPA management through the sharing of learning between managers and other key stakeholders while avoiding complex power sharing issues.

### *15.4.6 Adaptive governance*

Adaptive governance, which relies on the collaboration of a diverse set of stakeholders operating at different social and ecological scales in multi-level institutions and

**Table 15.2** Incentives and steering in top-down, bottom-up and market-based MPA governance systems (adapted from Jones et al., 2011a. United Nations Environment Programme).

| Incentives | Process | Steering |
|---|---|---|
| Economic | Using economic and property rights approaches to promote the fulfilment of MPA objectives. | Market steer |
| Interpretative | Promoting awareness of the conservation features of the MPA, the related objectives for conserving them, the policies for achieving these objectives and support for related measures. | Supporting all three approaches |
| Knowledge | Respecting and promoting the use of different sources of knowledge (local/traditional and expert/scientific) to better inform MPA decisions. | Supporting all three approaches |
| Legal | Establishment and enforcement of relevant laws, regulations, etc as a source of 'state steer' to promote compliance with decisions and thereby the achievement of MPA obligations. | State steer |
| Participative | Providing for users, communities and other interest groups to participate in and influence MPA decision-making that may potentially affect them in order to promote their 'ownership' of the MPA and thereby their potential to cooperate in the implementation of decisions. | People steer |

organisations (Olsson et al., 2004), is a major extension of conventional resource management and consists of at least four essential components:

1. understanding ecosystem dynamics;
2. developing management practices that combine different ecological knowledge systems to interpret and respond to ecosystem feedback and continuously learn;
3. building adaptive capacity to deal with uncertainty and surprise, including external drivers; and
4. supporting flexible institutions and social networks in multi-level governance systems.

MPAs can gain from combining the benefits of state control and binding legislation (top-down) and community-based approaches in management that empower local people and involve them in decision-making (bottom-up) process (Jones et al., 2011b). There are many types of strategies that managers can apply to steer other key stakeholders towards complying with MPA regulations (Jones et al., 2011b). Different types of incentives are used under different modes of governance (Table 15.2) but knowledge incentives support all three governance systems of top-down, bottom-up and a combination of them.

## 15.5 Integration, participation and network

The integration of stakeholders into the decision-making process expands their space of engagement, which provides incentive to continue to work together. This approach is now widely accepted as the preferred form of knowledge acquisition and strategy building for environmental management (McFadden et al., 2009). The integration of the different stakeholder groups, with learning by the different agents involved (Tompkins and Adger, 2004) in vulnerable resources management, such as in MPA management, may contribute to a general sense of enhanced capacity to manage the problem, both its causes and consequences. Although there are success stories of participatory multi-stakeholder/institutional bodies in managing natural resources, such as fisheries (Berkes, 2002; Iwasaki-Goodman, 2005), the level of participation has been weak, and the role of the institutions in empowering local communities has been limited (Kalikoski et al., 2010). It is increasingly recognised that for science and policy to generate effective working relationships, they must emerge from a co-productive iterative process with a requirement for constant interaction through networks (McFadden et al., 2009).

## 15.6 Stakeholder dynamics in MPA management

Co-learning may support fundamental changes to MPA management practices through creating the environment where stakeholders progress as they learn together for their mutual benefit. The groups would build social capital as they learn and develop as a group through:

- acquisition of a high level of personal self-confidence by individual members;
- getting to 'know' each other as individuals, developing shared values and trust;
- coming to regard each other as credible sources of support and advice; and
- commitment to fellow members, or being prepared to help each other out.

The process of co-learning requires attitudinal changes in all stakeholders, and skills in working together to make it effective. Skills in working together include communication, conflict resolution and group decision-making, and integrating scientific information and traditional ecological knowledge. Negotiating shared values and developing trust between stakeholder groups is an important and critical task. Likewise, building trust among stakeholders, as they share communication, is also fundamental to successful outcomes. Sound relationships and the reinforcement of values among MPA stakeholders are necessary for collective action to achieve socially optimal use of resources (Synapse Research and Consulting and CapitalAg Consulting, 2001). Developing effective stakeholder groups requires paying attention to building individual self-confidence and group communication processes, visions and internal and external networks.

## 15.7 Co-learning framework

Co-learning for MPA management improvement includes learning through stewardship, partnerships and community engagement in biodiversity conservation and natural resource management decision-making. It includes both formal and informal processes for creating sustainable options for the future and makes explicit links between co-learning and planning and actions for sustainable outcomes. The proposed co-learning framework is a combination of building blocks of formal and informal learning, fine-tuned through stakeholders' experiencing, sharing, updating and an updated information sharing processes (Figure 15.3). Reflection from experience and the literature reveals that NRM systems,

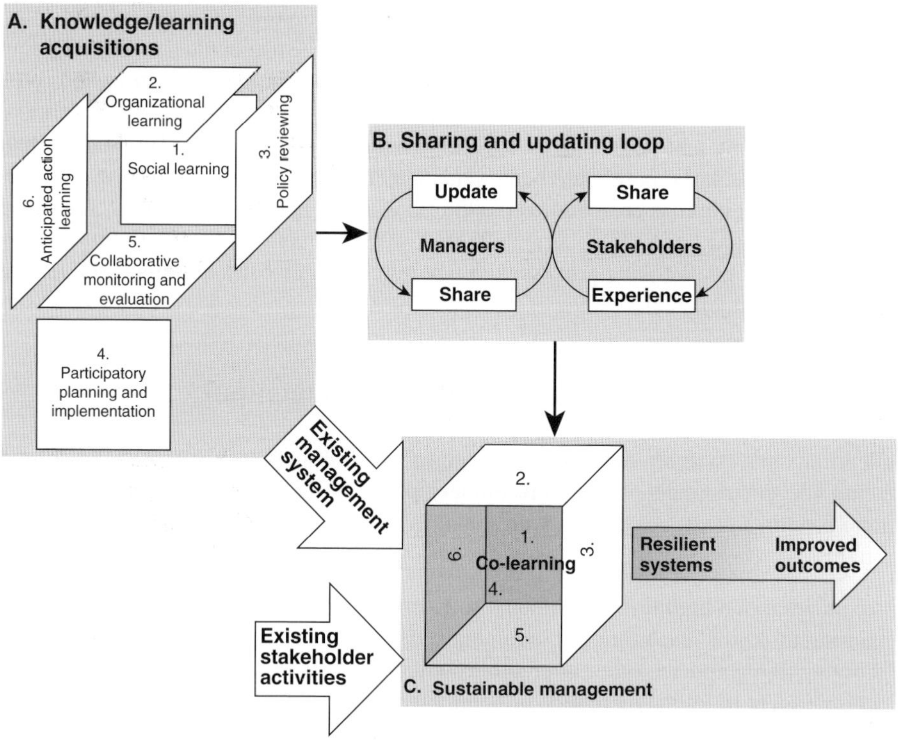


**Figure 15.3** Co-learning framework for MPA management improvement.

such as MPAs, already have different forms of information and learning in scattered form (A in Figure 15.3). An effective system requires an integrated continuous process of sharing-updating/experiencing-sharing (B in Figure 15.3), through which scattered learning would exist as a formal integrated compartment (C in Figure 15.3). This structured form of learning is denoted as the co-learning frame, which will help improve management outcomes through improving compliance with stakeholders' satisfaction.

## 15.8 Conclusion

A major driver of marine management performance is the sum of human population effects (i.e., fishing, deforestation, pollution), which require understanding the social and economic basis of these activities, their ecological impacts and the subsequent ramifications for ecosystem services. Understanding the knowledge base for decisions and the process of co-learning can assist marine system managers (government and non-government) with the flexibility needed to learn over time and inform marine management beyond protected area boundaries. The flexibility and inclusiveness of co-learning, inherent in the adaptive management approach, is appealing from a political and practical standpoint, because it allows stakeholders (i.e. regulatory and non-regulatory; government and non-government) to understand the implications of different objectives and options. Through improved understanding of how to foster co-learning will come improvement in the sustainable use of coastal resources. Like ecological niches (Hutchinson, 1957), for managers and other key stakeholders of an MPA, there may be different multi-dimensional learning spaces in terms of scientific, traditional, experiential, organisational, social and expectations of 'action learning' regarding MPA management. Understanding the knowledge–learning interface and flow within and between stakeholders is essential for adaptive management and for ensuring the productivity of the protected area and sustainability of its ecosystem services. This will contribute to increasing the capacity to better respond to future coastal environment challenges. It will create a new platform for considering actors' past mistakes and new information for opportunities of learning and informing future decisions.

## Acknowledgements

We acknowledge the constructive feedback from two anonymous reviewers. We also thank the adaptive learning theme of the CSIRO Coastal Cluster Project for providing funding support for Mr Siddique's participation in the 2nd International Symposium on Integrated Coastal Zone Management held in Arendal, Norway.

## References

Bardsley, D.K. and Sweeney, S.M. (2010) Guiding climate change adaptation within vulnerable natural resource management systems. *Environmental Management* **45**, 1127–1141.

Belfiore, S., Cicin-Sain, B. and Ehler. C. (2004) *Incorporating Marine Protected Areas into Integrated Coastal and Ocean Management: Principles and Guidelines*. IUCN, Gland, Switzerland and Cambridge, UK. viii + 38 pp.

Berkes, F. (2002) Cross-scale institutional linkages for commons management: perspectives from the bottom up. In: E. Ostrom, T. Dietz, N. Dolsak, P.C. Stern, S. Stonich, and E.U. Weber (Eds). *The Drama of the Commons* (pp. 293–321). National Academy Press, Washington, DC.

Berkhout, F., Hertin, J. and Gann, D.M. (2006) Learning to adapt: Organisational adaptation to climate change impacts. *Climatic Change* **78**, 135–156.

Brand, R. and Gaffikin, F. (2007) Collaborative Planning in an Uncollaborative World. *Planning Theory* **6**, 282–313.

DERM (2010) Building nature's resilience, a draft biodiversity strategy for Queensland. Department of Environment and Resources Management, Queensland Government, p 66.

Dietz, T., Ostrom, E. and Stern, P.C. (2003) The struggle to govern the commons. *Science* **302**,1907–1912.

Fard, H.D., Rostamy, A.A.A. and Taghiloo, H. (2009) How types of organisational cultures contribute in shaping learning organisation. *Singapore Management Review* **31**, 49–61.

Folke, C. (2006) Resilience: The emergence of a perspective for social-ecological systems analyses. *Global Environmental Change-Human and Policy Dimensions* **16**, 253–267.

Growcock, A.J., Sutherland, E.F. and Strathis, P.T. (2009) Challenges and experiences in implementing a management effectiveness evaluation program in a protected area system. *Australasian Journal of Environmental Management* **16**, 218–226.

Hockings, M. (2003) Systems for assessing the effectiveness of management in protected areas. *BioScience* **53**, 823–832.

Hockings, M., Stolton, S. and Dudley, N. (2000) *Evaluating Effectiveness: A Framework for Assessing the Management of Protected Areas*. IUCN, Gland, Switzerland and Cambridge, UK. x + 121pp.

Holmes, N., Davis, D. and Dutton, I. (1992) Information needs in coastal management: the role of science. In O.N. Crimp (Ed.). *Moreton Bay in the Balance* (pp. 115–127). The Australian Littoral Society, Moorooka, Qld.

Hutchinson, G.E. (1957) Concluding remarks. *Cold Spring Harbour Symposium on Quantitative Biology* **22**, 415–427.

IUCN-WCPA (2008). *Establishing Marine Protected Area Networks – Making It Happen*. IUCN World Commission on Protected Areas (IUCN-WCPA), National Oceanic and Atmospheric Administration and The Nature Conservancy, Washington, DC. 118 pp.

Iwasaki-Goodman, M. (2005) Resource management for the next generation: co- management of fishing resources in the Western Canadian Arctic region. In: Kishigami, N. and Savelle, J. (Eds). *Indigenous Use and Management of Marine Resources* (pp. 101–121). Osaka: The National Museum of Ethnology.

Jones, P., Qiu, W. and De Santo, E.M. (2011a) Governing Marine Protected Areas—Getting the Balance Right. Technical Report, United Nations Environment Programme.

Jones, P., Qiu, W. and De Santo, E.M. (2011b) Finding the balance: Strengthening MPA governance by mixing top-down, bottom-up, and other approaches. *International News and Analysis on Marine Protected Areas* **12**,1–8.

Kalikoski, D.C., Neto, P.Q. and Almudi, T. (2010) Building adaptive capacity to climate variability: The case of artisanal fisheries in the estuary of the Patos Lagoon, Brazil. *Marine Policy* **34**, 742–751.

Keen, M., Bruck, T. and Dyball, R. (2005) Social learning: a new approach to environmental management. In: Keen, M., Brown, V. and Dyball, R. (Eds.), *Social Learning in Environmental Management: Towards a Sustainable Future* (pp. 3–21). Earthscan, London.

Lave, J. and Wenger, E. (1991) *Situated Learning: Legitimate Peripheral Participation*. Cambridge University Press, Cambridge.

Lorenzoni, I., Jordan, A., O'Riordan, T., Turner, R.K. and Hulme, M. (2000) A co-evolutionary approach to climate change impact assessment: Part II. A scenario-based case study in East Anglia (UK). *Global Environmental Change-Human and Policy Dimensions* **10**, 145–155.

Mascia, M.B. (2003) The human dimension of coral reef marine protected areas: recent social science research and its policy implications. *Conservation Biology* **17**, 630–632.

McFadden, L., Penning-Rowsell, E. and Tapsell, S. (2009) Strategic coastal flood-risk management in practice: Actors' perspectives on the integration of flood risk management in London and the Thames Estuary. *Ocean & Coastal Management* **52**, 636–645.

MEAM (2010) Adaptive management: What does it look like in practice? *Marine Ecosystem and Management*, international news and analysis on marine ecosystem-based management **4**(3): 1–8, December 2010–January 2011.

Michel, D. (2009) Foxes, hedgehogs, and greenhouse governance: Knowledge, uncertainty, and international policy-making in a warming World. *Applied Energy* **86**, 258–264.

Muro, M., and Jeffrey, P. (2008) A critical review of the theory and application of social learning in participatory natural resource management processes. *Journal of Environmental Planning and Management* **51**, 325–344.

Myers, S.A., Blackmore, M.J., Smith, T.F. and Carter, R.W. (2012) Climate change and stewardship: strategies to build community resilience in the Capricorn Coast. *Australasian Journal of Environmental Management* **iFirst**: 1–18. doi:10.1080/14486563.2011.646755.

Nagoya Oceans Statement (2010) Co-Chairs' Statement: Emanating from the Oceans Day at Nagoya, held on 23October 2010 during the Tenth Meeting of the Conference of the Parties to the Convention on Biological Diversity (CBD COP 10), Nagoya, Japan.

NOAA (2004) Stakeholder participation: A synthesis of current literature. Coastal Service Center, National Oceanic and Atmospheric Administration, National Marine Protected Areas Center, Silver Spring, Maryland, USA.

Olsson, P., Folke, C. and Hahn, T. (2004) Social-ecological transformation for ecosystem management: the development of adaptive co-management of a wetland landscape in Southern Sweden. *Ecology and Society* **9**(4): 2. http://www.ecologyandsociety.org/vol9/iss4/art2/ (accessed 10 September 2012).

Orbach, M. and Karrer, L. (2010) *Marine Managed Areas: What, Why, and Where*. Science and Knowledge Division, Conservation International, Arlington, Virginia, USA.

Pahl-Wostl, C. (2002) Towards sustainability in the water sector—The importance of human actors and processes of social learning. *Aquatic Sciences*, **64**, 394–411.

Pahl-Wostl, C. and Hare, M. (2004) Process of social learning in integrated resources management. *Journal of Community and Applied Social Psychology* **14**, 193–206.

Pimbert, M.P. and Pretty, J.N. (1995) Parks, People and Professionals: Putting 'Participation' into Protected Area Management, Discussion Paper No. 57, United Nations Research Institute For Social Development, Geneva.

Pomeroy, R.S. (1995) Community-based and co-management institutions for sustainable coastal fisheries management in Southeast Asia. *Ocean & Coastal Management* **27**, 143–162.

Reed, M.S., Evely, A.C., Cundill, G., Fazey, I., Glass, J., Laing, A., Newig, J., Parrish, B., Prell, C., Raymond, C. and String, L.C. (2010). What is social learning? *Ecology and Society* **15** (4): r1. http://www.ecologyandsociety.org/vol15/iss14/resp11/ (accessed 10 September 2012).

Sandstrom, A. and Rova, C. (2010) Adaptive co-management networks: a comparative analysis of two fishery conservation areas in Sweden. *Ecology and Society* **15**, 14. http://www.ecologyandsociety.org/vol15/iss13/art14/ (accessed 10 September 2012).

Senge, P.M. (2006) *The Fifth Discipline: The Art & Practice of The Learning Organisation*. Random House Business Books, London.

Singh, S. (2007) Role of emotional intelligence in organisational learning: an empirical study. *Singapore Management Review* **29**, 55–74.

Smith, T.F., Carter, R.W., Thomsen, D.C., Mayes, G., Nursey-Bray, M., Whisson, G., Jones, Dovers, R.S. and O'Toole, K. (2009) Enhancing science impact in the coastal zone through adaptive learning. *Journal of Coastal Research* **56**, 1306–1310.

Steyer, G.D. and Llewellyn, D.W. (2000) Coastal wetlands planning, protection, and restoration act: A programmatic application of adaptive management. *Ecological Engineering* **15**, 385–395.

Storbjork, S. (2010) 'It takes more to get a ship to change course': Barriers for organizational learning and local climate adaptation in Sweden. *Journal of Environmental Policy and Planning* **12**, 235–254.

Synapse Research and Consulting and CapitalAg Consulting (2001) Improving natural resource management behaviour at the farm and regional levels. http://www.synapseconsulting.com.au/docs/improving_natural_resource_management.pdf (accessed 15 June 2011).

Thompson, A. (2006) Management under anarchy: The international politics of climate change. *Climatic Change* **78**, 7–29.

Tompkins, E.L. and Adger, W.N. (2004) Does adaptive management of natural resources enhance resilience to climate change? *Ecology and Society* **9**.

Tschakert, P. and Dietrich, K.A. (2010) Anticipatory Learning for Climate Change Adaptation and Resilience. *Ecology and Society* **15**(2): 11. http://www.ecologyandsociety.org/vol15/iss2/art11/ (accessed 10 September 2012).

Wenger, E. (1998) *Communities of Practice: Learning, Meaning, and Identity*. Cambridge University Press, Cambridge.

Chapter 16

# Reflexively Mapping the Science–Policy Interface for Coastal Zones

*Anne Blanchard and Scott Bremer*

**Abstract**

Coastal management has long promoted the mobilisation of knowledge in support of decision-making across a 'science–policy interface'. As coastal management has evolved, the science–policy interface has been increasingly represented as a complex political setting. The interface has become the focus of a vast and confusing literature, from which have been distilled a demanding, and at times conflicting, suite of principles for an 'ideal' interface. Against this background, coastal actors are confused on how best to give effect to the science–policy interface in practice. Therefore, this chapter asserts that coastal management initiatives would benefit from a reflexive deliberation among actors on how best to shape the science–policy interface to address their specific issues. To this end, this chapter proposes a deliberation support tool in the shape of the 'science–policy map', to help coastal actors make sense of this complexity and chart their course of action.

## 16.1 Introduction

This chapter is about how coastal communities mobilise knowledge in support of their collective decision-making, and more specifically about how communities deliberately shape the 'science–policy interface'. The broader scholarship and practice of coastal management has long promoted institutions within which knowledge is mobilised for decision-making, with a particular focus on the 'science–policy interface' setting. Conventionally, coastal management has looked to science alone to support policy (Cicin-Sain and Knecht, 1998; Vallega, 1997). However, in recognition of the uncertainty, plurality and high political stakes complicating most meaningful coastal issues, the science–policy interface literature has evolved from these rather simplistic beginnings and now presents a confusing array of different approaches. Today coastal actors are confronted with a vast literature on the many

*Global Challenges in Integrated Coastal Zone Management*, First Edition. Edited by Erlend Moksness, Einar Dahl and Josianne Støttrup.

ways of mobilising knowledge across the science–policy interface; from participatory fora (Human and Davies, 2010), to systems modelling (Turner, 2000), consultancy, or knowledge partnerships (Stojanovic and Barker, 2008) for instance. At the same time, drawing from this diverse literature, coastal academics and those funding coastal management initiatives demand a science–policy interface that is at once more participatory, integrated, science-rich, ecosystem-based, adaptive and critically reflexive (Forst, 2009; Knol, 2010; McFadden, 2007; Turner, 2000) though giving effect to these principles in practice presents enormous challenges.

It is no wonder that coastal actors operating within the science–policy interface struggle to make sense of the complex political reality of supporting decisions with knowledge, and feel lost when addressing questions of how best to re-structure the interface: Which knowledge should we admit? How do we integrate different types of knowledge? How do we judge the quality of knowledge for decision-making? How do we deal with the value-bias and politics inherent in the policy interface? Too often these questions go unanswered.

It is the assertion of this chapter that coastal communities would benefit from deliberately entering into reflexive debate on how best to mobilise knowledge in support of their collective decisions; on how to shape the science–policy interface. While coastal management has long advocated deliberation for coastal decision-making, this has not typically extended to the shaping of the science–policy interface. Here we argue that coastal actors, from scientists to policy-makers to wider stakeholders, require the means to critically reflect on the state of the interface and to agree on its evolutionary trajectory. To improve the quality of this critical reflection and deliberation, this chapter presents a reflexive deliberation support tool in the form of the 'science–policy map'.

The chapter is structured around two main parts. Part 1 explores the evolution of the science–policy interface for coastal zones, and emphasises the need for reflexive deliberation. Part 2 presents the 'science–policy map' and its uses.

## 16.2 A science–policy interface for coastal zones

Coastal management, and this chapter, starts from the assumption that it is desirable to support decisions with knowledge. One way to improve decisions is to ensure that the best quality knowledge concerning the issue and potential solutions is available to fuel deliberation. 'Better' knowledge does not guarantee better choices, but it does provide a relevant basis for guiding decisions and holding decision-makers accountable (Reid et al., 2006). Importantly, this does not necessarily imply 'more' knowledge in terms of quantity, but rather that a 'high quality' of knowledge is made available in a form able to support decision-making. To this end, coastal management has from the beginning placed a central importance on the successful integration of knowledge, particularly from the natural and social sciences, with the political decision-making process across the science–policy interface.

### *16.2.1 The evolution of the science–policy interface for coastal zones; from one to many*

In the early days of coastal management, the emphasis was on 'science-based management'. From its origins, coastal zone management sought improved management through advances in science and technology (Cicin-Sain and Knecht, 1998), which was deemed to provide detailed insights into the components of the coast and allowed predictions of its future. According to a linear and mechanistic model, scientific products were fed to coastal managers who transformed them into policies and rules. This model externalised values and politics, and assumed that any coastal problems could be overcome with more science.

Throughout the 1970s, under the influence of ecology and systems science, the coastal zone became increasingly viewed as complex and interconnected. With this came a realisation that collecting science for the coast must be similarly complex and interconnected, necessitating a multi-disciplinary approach (Stojanovic et al., 2004). Through the 1980s, 'coastal zone management' had become 'integrated coastal zone management' (ICZM). This transition was shaped by four key influences, with implications for the science–policy interface: (a) an increasing recognition of 'global changes'; (b) a more advanced discussion on the complexity and uncertainty of the coast, leading to a focus on interdisciplinarity; (c) an increasing emphasis on placing science in the service of 'sustainability' rather than 'progress'; and (d) a recognition of the plurality inherent within society, with the consequence that no one perspective can claim absolute 'truth' (Vallega, 1997). By 1994, the Joint Group of Experts on the Scientific Aspects of Marine Environmental Protection explored the role of the science–policy interface in coastal management. It recommended that 'more science' may not be enough in the face of complex and uncertain issues and argued that "scientists and managers must work together as a team" to implement adaptive coastal management (GESAMP, 1996).

Following this evolution of coastal management, two observations can be made. First, the science–policy interface has become recognised as a highly complex and politicised arena, where knowledge is debated alongside values. Second, coastal actors are today faced with a much less clear picture of the science–policy interface; characterised by a vast literature of the multiple overlapping and conflicting theories, principles, tools and processes. However, across this literature, some common principles of 'best practice' have emerged.

### *16.2.2 Principles of the 'ideal' science–policy interface for coastal management*

From across the coastal management literature have emerged a number of common principles for shaping the 'ideal' science–policy interface. Briefly, this interface should be:

- Ecosystem-based: Ecosystem-based management in the mid 1990s led coastal actors to address the coast as a complex and dynamic social-ecological system (Forst, 2009). So framed, the science–policy interface is charged with understanding the connectivity within a coastal system, and the health of key ecosystem variables.
- Adaptable: Recognising the complexity associated with an ecosystem-based approach implies recognition of the significant degrees of instability and uncertainty inherent in managing such a system. This denies a static model of the coast, and requires adaptive research, through the constant revision of theories as coastal systems change. Management interventions thus constitute experiments; furnishing the science–policy interface with an on-going stream of information (Knol, 2010).
- Science-rich: For a number of commentators (McFadden, 2007; Turner, 2000), coastal management's focus on complex ecosystems, and the experimental nature of adaptive

management, necessitates a strong scientific presence to inform decision-making. While we may admit other forms of knowledge, it is important to ensure that scientific evidence retains a place at the table and is nurtured as a credible, salient and legitimate source of knowledge.

- Participatory: In parallel with science-based coastal management, another group of scholars and practitioners (Knol, 2010; Runhaar and van Nieuwaal, 2010) have advocated a more participatory science–policy interface. In recognition of complexity and uncertainty, science is relativised alongside other types of knowledge relevant to decision-making, including local and traditional knowledge. Piecing together these diverse perspectives aims to produce a uniquely comprehensive picture.
- Integrated: It has become widely recognised that coastal management needs to be informed by knowledge that reflects the complexity and interconnectedness of the coast. In this regard, reductionist disciplined science alone is inadequate. Coastal management demands an integrated approach to knowledge that demonstrates the complex and co-evolutionary interactions between social and ecological systems, and how management interventions may alter this relationship. This has led many to assert that well-informed coastal management requires integration both across disciplinary boundaries, and across the science–policy interface (Cicin-Sain and Knecht, 1998; McFadden, 2007).
- Critically reflexive[1]: Within coastal management, critical reflection is translated by increasing calls to reframe the interface as a 'governance' setting, in recognition of the political nature of knowledge and the 'power' exerted in the interface. Authors such as Boesch (1999) expose the previously privileged position of scientists, and advocate opening the interface to the full spectrum of stakeholders and their diverse forms of rationality to give effect to models of participatory democracy.

The above principles are often discussed as complementary; however, they do not always co-exist easily within a science–policy interface setting, resulting in some uncomfortable trade-offs. Typically, the clash between principles of 'participation' and a 'science-rich' interface illustrates this, with many coastal management initiatives either committing to the scientific model wholly, or completely abandoning it in favour of local and traditional knowledge (Hommes et al., 2009). Therefore, giving effect to these principles in practice requires a sophisticated balancing act which further contributes to the complexity of the interface.

We argue in the second half of this chapter that traversing the confusing literature and practice on the science–policy interface, and seeking to give effect to the above principles, necessitates reflexive deliberation between coastal actors. By collectively exploring the various ways in which to shape the science–policy interface with the aid of a deliberation support tool, it is anticipated that coastal actors can steer themselves toward an interface better serving their needs.

[1]Here reflexivity and critical thinking is discussed in terms of the Frankfurt School. According to the 'critical theorists' of the Frankfurt School, a scientific theory should not only seek the understanding of the world, but should also help transform society by creating social and political conditions more conducive to human flourishing than the present ones (Adorno and Horkheimer, 1944).

### 16.2.3 *The need for reflexive deliberation support*

As noted, the 'ideal form' of the science–policy interface for dealing with complex issues is exhaustively described for coastal management. However, this discussion is scattered across a vast literature, such that it is relatively disconnected, with little clarity on how to give effect to it. Finding their way demands that coastal actors reflexively deliberate on how they think the science–policy interface should look for their specific issues. This requires support and guidance.

First, support for reflexive deliberation would contribute to improve the quality of the deliberation, in terms of accessing the diversity of perspectives, and ensuring reflexive participation. Indeed, coastal actors would be invited to structure their early collective reflections, by taking distance on their current position at the science–policy interface, and scrutinising the grounding assumptions behind it: Where are we now? Why are we here? Is it where we want to be? This reflexive questioning also has the advantage of facilitating open and respectful behaviour (Jollivet-Blanchard and Blanchard, 2004), as well as the acceptance of the legitimacy, and even preferability, of other forms of the science–policy interface.

Second, support for reflexive deliberation would encourage the iterative scrutiny of the means to achieve the preferred interface, following cycles of action and reflection. By visualising the evolution of the science–policy interface, and the way in which subsequent iterations learned from prior decisions and actions, coastal actors would be able to implement incremental, long-term and adaptive changes.

Third, by documenting and stabilising the lessons learned, support for reflexive deliberation aims at helping coastal actors to derive recommendations for future moves at the science–policy interface; thus increasing their capacity to navigate within this complex interface, by visualising the potential barriers that may occur; and reinforcing the flexibility, openness and adaptability of their institutions.

As implementing reflexive deliberation is a long-term, complicated endeavour, support is needed in order to insure the quality of the process and help coastal actors to concretely give effect to the science–policy interface needed for their specific issues, in their given context. To this aim, we propose a reflexive deliberation support tool called the 'science–policy map', described in the following section.

## 16.3 A reflexive deliberation support tool for coastal zones: the 'science–policy map'

The reflexive deliberation support tool proposed in this chapter is the science–policy map. This map, presented in Figure16.1, interposes two knowledge-related scales: (a) the degree of knowledge integration, from disciplinarity to transdisciplinarity, and (b) the objectives of knowledge mobilisation within the coastal zone, from an 'epistemological' to a 'social' pole. This map is designed as a simplified heuristic that enables coastal actors to position the state of their science–policy interface relative to these two measures. In this way, it presents a focus for reflection and point of departure for discussion.

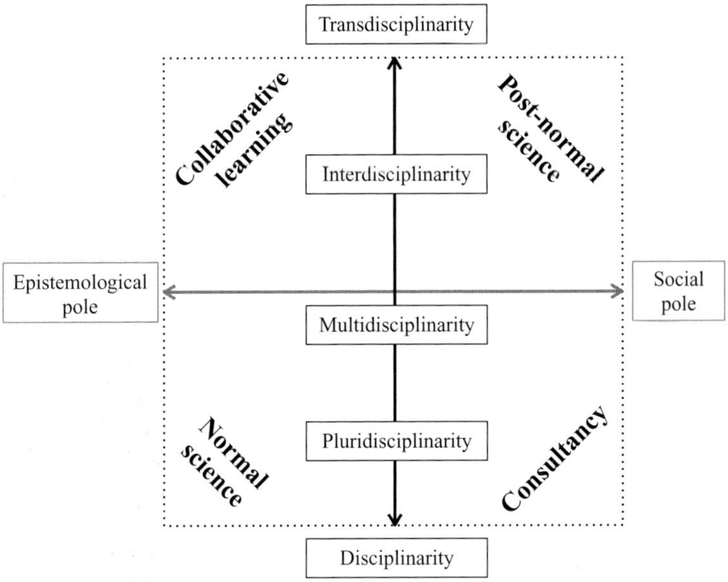


**Figure 16.1** The 'science-policy map', a reflexive deliberation support tool for mapping the science-policy interface for coastal zones.

### *16.3.1 A vertical axis representing the level of knowledge integration*

The vertical axis of the science–policy map represents the degree of knowledge integration, characterised across five different levels that actors involved in coastal management must differentiate between. As it is beyond the scope of this chapter to give an in-depth presentation of each of these five levels, we accompany them with a list of suggested further readings.

- Disciplinarity[2] is a form of knowledge creation that takes place within the boundaries of a single scientific discipline. There is no interaction with other disciplines, as the issue or aspect under scrutiny can be apprehended through only one disciplinary perspective; it is sufficient for the scientists of that discipline to use their own tools, data and approaches to understand their object.
- Pluridisciplinarity[3], like disciplinarity, is an approach that strictly belongs to the scientific community. It encourages several disciplines to coexist within the same entity, without necessarily requiring exchanges between them; for instance, the various departments of a university.

[2]For further insights into disciplinary science, see, e.g., Kuhn, 1962; Hunt, 1994; Fuller, 2007.
[3]For further insights into pluridisciplinarity, see, e.g., Blanchard and Vanderlinden, 2010.

- Multidisciplinary[4] approaches remain as well in the strict scientific domain. It is the meeting of distinct disciplines around a common study theme, though each is permitted to retain its specific rules, methods and tools. In this sense, the IPCC research on climate change, organised around three working groups studying its various aspects, is multidisciplinary.
- Interdisciplinarity[5], another scientific way of producing knowledge, is a long-term co-operation process whereby scientists, in order to mobilise knowledge around a complex and multi-facetted topic, conjugate complementary disciplinary approaches, in their ontological, epistemological, methodological and social dimensions, through the sharing of tools, methods, approaches and values. The SPICOSA project, as an integrated modelling approach, may be an example (Tett et al., 2011).
- Transdisciplinarity[6] is a form of knowledge production that transcends the boundaries of science, and extends to the integration of non-scientific insights, in order to reach a more complete comprehension of the complex world. Some authors also talk about transdisciplinarity in terms of the emergence of a new 'meta-discipline', resulting from the assembling of several knowledge sources, with its own codes and tools.

This vertical axis forms a continuum, where many positions are possible. As we evolve along the axis, we perceive gradual changes in the form of knowledge production. Indeed, disciplinarity is restricted to the scientific community, and in this sense is rather non-transparent, with a highly formalised language that makes it impermeable to interactions with other forms of knowledge. At the opposite end of the axis, science becomes one form of knowledge negotiated within a pluralistic dialogic space, with a focus on society's concerns. Knowledge production is transparent, participatory and open to public scrutiny. As seen in Section 16.2.2, this form of knowledge production admits the various scientific, social and political actors in a common endeavour of knowledge co-construction.

### *16.3.2 A horizontal axis representing the objectives of knowledge mobilisation*

The horizontal axis represents the objectives of knowledge mobilisation, to be evaluated along the continuum between the 'epistemological' and 'social' poles (Creutzer, 2002).

If the objectives of knowledge mobilisation are situated around the epistemological pole, then scientific inquiries are motivated by curiosity, and implemented for the sake of learning and 'puzzle-solving' (Kuhn, 1962). Scientists work toward increasing scientific knowledge, for instance through achieving a better understanding of complexity, or plugging the holes that are created by uncertainty. At the extremity of this pole is a quest for 'perfect' knowledge, guided by the assumptions that uncertainties can be eliminated or controlled, and knowledge can be complete (Rommetveit et al., 2010).

---

[4]For further insights into multidisciplinarity, see, e.g., Agrawala, 1998; and the report of the WMO & UNEP, 2004.

[5]For further insights into interdisciplinarity, see, e.g., Klein, 1990; Jakobsen, et al., 2004; Blanchard and Vanderlinden, 2010.

[6]For further insights into transdisciplinarity, see, e.g., Jantsch. 1972; Nicolescu, 2002.

At the other end of the continuum, the objectives of knowledge mobilisation are social, meaning that science is issue-based, or, in other words, concerned with socio-economic and political concerns. It pursues an attempt to help society dealing with its complex and persistent issues.

In practice we often see that the mobilisation of knowledge is situated somewhere along the spectrum between these two poles. Strictly epistemological knowledge, in the form of a 'hard' science research project for instance, may be appropriated by coastal actors to inform their decisions, though it retains its very scientific character. On the other hand, scientific funding is often contingent on the 'public good' derived from it, and in this way the scientific community often use an issue as an opportunity to justify their own curiosity-based endeavours.

### *16.3.3 Crossing the two scales; four quadrants: four generic types of science–policy interface*

By crossing these two knowledge-related scales, the map reveals in four quadrants, four particular ways in which knowledge is mobilised at the science–policy interface. Each quadrant represents four broad families of literature and practice on the different ways in which the science–policy interface can be framed.

The lower left quadrant is represented by the area formed by the crossing of low knowledge integration and epistemological objectives to knowledge mobilisation. It can be thought of in terms of Kuhn's (1962) normal science: a highly insular and disciplined study of specific scientific questions, and includes both pure and applied science.

Conversely, the lower right quadrant is represented by the crossing of low knowledge integration and social objectives, and refers to the broad family under the label of consultancy. According to Funtowicz and Ravetz (1993), knowledge for consultancy is mobilised by expert witnesses or consultants; for instance engineers. These experts employ science in the service of a specific problem, as defined by their employer.

The upper quadrants are defined by a high level of knowledge integration, extending beyond science to include other knowledge systems. The left quadrant, characterised by epistemological objectives, represents the broad literature on social learning, whereby scientists and stakeholders interact to promote learning around a subject. It encompasses theories of collaborative learning for instance (Daniels and Walker, 2001).

The upper right quadrant, promoting social objectives, is the domain of the family of approaches promoting new norms of knowledge mobilisation for complex issues. Knowledge is used for and built with society and decision-makers. These approaches espouse inclusiveness, integration, a dialogic epistemology, and attention to knowledge quality as the guiding principle of the science–policy interface. Coastal actors are encouraged to evaluate quality as part of an 'extended peer community'. Here it is labelled post-normal science (Funtowicz and Ravetz, 1993), but could equally refer to 'Mode 2' science or 'Sustainability science'.

As noted, the categorisation of the science–policy interface according to four quadrants is not exhaustive, but rather indicative. However, as the aim of the science–policy map is foremost to launch and structure reflexive and critical discussions as a basis for implementing concrete actions, the wide categories have the role of supporting broad discussions within a structured framework. In this way, the oversimplification within this map is at once its strength and its weakness.

### 16.3.4 How to use the 'science–policy map'

The science–policy map is a visual support for reflexive deliberation, on which coastal actors can locate their existing interface, according to the characteristics described above, and chart a course to a more desirable interface setting. It is designed to be printed on a large paper format, or re-drawn on an easily-changeable surface, such as a blackboard, to facilitate open and flexible debates among the various actors.

To this point the 'science–policy map' has found limited practical application. It was first introduced as a prospective tool in a pilot study in Palmerston North, New Zealand in June 2009, undertaken with a group comprising both coastal management practitioners from local government, and scholars in the field from the university. The pilot saw the presentation of the science–policy map, and an enriching discussion on its application, and concrete means of giving effect to it. It is on the basis of this pilot study that the discussion below is based.

Once the various levels of both axes and the four quadrants are defined and understood by the coastal actors, the first question that the science–policy map raises is: where are we now? This question aims at launching the debates among the various actors, by encouraging them to collectively understand the mechanisms of their current science–policy interface: on which kind of knowledge it is based, which set of objectives does it pursue and which actors are involved. This is a rather descriptive phase. Several crosses can be drawn on the map representing the variety of perspectives on the situation of the current science–policy interface among the coastal actors. Each actor or group of actors can explain why they have situated their cross as such, and debate it with each other. When the actors come to a collectively satisfying definition of their current interface, they can position a cross on the map, or a 'cloud' of crosses, if important indeterminacies or conflicting views remain. However, placing this cross or cloud of crosses is not as easy as it might sound, as it necessitates the understanding by all actors of the various components of the map, and as the 'ice-breaking' period before having authentic and lively discussions can be long; with side-effects such as weariness and disengagement.

The agreed-upon location of the existing science–policy interface on the map prepares the ground for actors to scrutinise the assumptions behind this position, and analysing the challenges faced by their current interface in dealing with their particular coastal issues.

Once the challenges are collectively discussed and prioritised, the coastal actors can reflect together on a preferable shape for their science–policy interface, better suited to manage their particular coastal issues in their own context. We enter into a more normative phase. At this stage, it is important to ensure that each actor or group of actors share their perspective on how the coastal issues should be addressed, and the role of the science–policy interface in supporting this. Actors justify their position in light of their own experiences and knowledge of the issue, from scientific to local and traditional knowledge. The various positions expressed will reflect the diverse interests co-existing around the issue, from social, to environmental, economic and aesthetic. Here again, several crosses can be situated on the map, thus reflecting the various desires and interests of the actors concerned by the issue.

At this stage, the coastal actors will have two crosses (or clouds of crosses) on their map: one showing the existing location of the science–policy interface, and a second showing the collectively desired interface; with an arrow between the two, indicating the route to follow. By tracing this route, the actors can visualise some of the challenges, as well as facilitators, in their way.

However, the arrow from the initial to the preferred location is a falsely simple representation of what this evolution actually means. Indeed, the passage, for instance, from a rather disciplinary way of producing knowledge, to a more participatory, adaptive and critical one, can take many policy cycles. It implies changes in coastal and research institutions, socio-economic attitudes, perceptions, values and languages. Coastal actors involved in the reflexive mapping of their science–policy interface should be aware that they are involved in a long-term process, and that the arrow drawn on the science–policy map represents long-term, incremental and evolutionary changes.

Importantly, mapping a route to a preferred science–policy location should not imply an endless quest for more or 'perfect' knowledge. Indeed, many would argue that the current knowledge base for coastal zones is sufficient now as a platform for debate and actions at the science–policy interface. Rather, the purpose of this mapping exercise is for coastal actors to reflect on how best to package and deliver quality knowledge in a form best able to support decision-making for their issues.

## 16.4 Conclusion

This chapter asserts that coastal management initiatives would benefit from a reflexive deliberation among actors on how best to shape the science–policy interface to address the specific issues faced by a specific coastal context. The chapter has proposed a novel deliberation support tool in the form of the 'science–policy map', with the objective of helping coastal actors to collectively make sense of what is happening at the science–policy interface, navigate their way through the vast literature, give effect to their preferred interface, and improve the quality of their deliberation. This tool is by no means perfect, and has to find further concrete application within a deliberative setting. However, it does nonetheless lay down the challenge for similar such tools and set the foundations for a more sophisticated discussion on this subject.

## References

Adorno, T.W. and Horkheimer, M. (1944) *Dialektik der Aufklarung*. New York: Social Studies Association Inc.

Agrawala, S. (1998) Structural and process history of the intergovernmental panel on climate change. *Climatic Change* **39**, 621–642.

Blanchard, A. and Vanderlinden, J.-P. (2010) Dissipating the fuzziness around interdisciplinarity: The case of climate change research. *Surveys and Perspectives Integrating Environment and Society (SAPIENS)* **3**, 1–6.

Boesch, D.F. (1999) The role of science in ocean governance. *Ecological Economics* **31**, 189–198.

Cicin-Sain, B. and Knecht, R. W. (1998) *Integrated Coastal and Ocean Management: Concepts and Practices*. Washington, DC: Island Press.

Creutzer, M. (2002) Aspects de l'Interdisciplinarité: Essai de Reconstruction d'un Débat. In Gélineau, L. and Mailloux, C. (Eds.) *L'interdisciplinarité et la Recherche Sociale Appliquée,*

*Réflexions sur des Expériences en Cours*. Montréal: Université de Montréal, Université de Laval, pp. 7–39.

Daniels, S. and Walker, G. (2001) *Working Through Environmental Conflict: The Collaborative Learning Approach*. Westport: Praeger Publishers.

Forst, M.F. (2009) The convergence of Integrated Coastal Zone Management and the ecosystems approach. *Ocean & Coastal Management* **52**(6), 294–306.

Fuller, S. (2007) *The Knowledge Book: Key Concepts in Philosophy, Science and Culture*. Stockfield: Acumen Publishing Limited.

Funtowicz, S. and Ravetz, J. (1993) Science for the post-normal age. *Futures* **25**, 739–755.

GESAMP (1996) The Contributions of Science to Integrated Coastal Management (Vol. 61), GESAMP (IMO/FAO/UNESCO-IOC/WMO/WHO/IAEA/UN/UNEP Joint Group of Experts on the Scientific Aspects of Marine Environmental Protection) Reports and Studies.

Hommes, S., Hulscher, S.J.M.H., Mulder, J.P.M., Otter, H.S. and Bressers, H.T.A. (2009) Role of perceptions and knowledge in the impact assessment for the extension of Mainport Rotterdam. *Marine Policy* **33**, 146– 155.

Human, B.A. and Davies, A. (2010) Stakeholder consultation during the planning phase of scientific programs. *Marine Policy* **34**, 645–654.

Hunt, L. (1994) The virtues of disciplinarity. *Eighteenth Century Studies* **28**, 1–7.

Jakobsen, C., Hels, T. and McLaughlin, W. (2004) Barriers and facilitators to integration among scientists in transdisciplinary landscape analyses: a cross-country comparison. *Forest Policy and Economics* **6**, 15–31.

Jantsch, E. (1972) Towards interdisciplinarity and transdisciplinarity in education and innovation. In: *Problems of Teaching and Research in Universities* 1st edn. Paris: Organisation for Economic Cooperation and Development (OECD), pp. 97–121.

Jollivet-Blanchard, C. and Blanchard, E. (2004) *L'Expérience de la Coopération en Education: Pourquoi? Comment*? Paris: L'Harmattan.

Klein, J.T. (1990) *Interdisciplinarity: History, Theory, and Practice*. Detroit: Wayne State University Press.

Knol, M. (2010) Scientific advice in integrated ocean management: The process towards the Barents Sea plan. *Marine Policy* **34**, 252–260.

Kuhn, T. (1962) *The Structure of Scientific Revolutions*. Chicago: University of Chicago Press.

McFadden, L. (2007) Governing Coastal Spaces: The Case of Disappearing Science in Integrated Coastal Zone Management. *Coastal Management* **35**, 429–443.

Nicolescu, B. (2002) *Manifesto of Transdisciplinarity*. New York: State University of New York (SUNY) Press.

Reid, W.V., Berkes, F., Wilbanks, T.J. and Capistrano, D. (2006) Introduction. In Reid, W.V., Berkes, F., Wilbanks, T.J. and Capistrano, D. (Eds.) *Bridging Scales and Knowledge Systems: Concepts and Applications in Ecosystem Assessment.*, Washington, DC: Island Publishing.

Rommetveit, K., Funtowicz, S. and Strand, R. (2010) Knowledge, democracy and action in response to climate change. In: Bhaskar, R., Frank, C., Hoyer, K.G., Naess, P. and Parker, J. (Eds.) *Interdisciplinarity and Climate Change*. Abingdon: Routledge, pp. 149–163.

Runhaar, H.A.C. and van Nieuwaal, K. (2010) Understanding the use of science in decision-making on cockle fisheries and gas mining in the Dutch Wadden Sea: Putting the science–policy interface in a wider perspective. *Environmental Science and Policy* **13**, 239–248.

Stojanovic, T., Ballinger, R.C. and Lalwani, C. S. (2004) Successful integrated coastal management: measuring it with research and contributing to wise practice. *Ocean and Coastal Management* **47**, 273–298.

Stojanovic, T. and Barker, N. (2008) Improving governance through local Coastal Partnerships in the UK. *The Geographical Journal* **174**, 344–360.

Tett, P., Sandberg, A. and Mette, A. (2011) *Sustaining Coasal Zone Systems*. Edinburgh: Dunedin Academic Press.

Turner, R.K. (2000). Integrating natural and socio-economic science in coastal management. *Journal of Marine Systems* **25**, 447–460.

Vallega, A. (1997) *Fundamentals of Integrated Coastal Management*. Kluwer Academic Publishers, Dordrecht, Netherlands; Boston, MA.

WMO and UNEP (2004) *Introduction on the IPCC*. World Meteorological Organisation, Geneva.

Chapter 17

# Bridging the Science–Policy Divide in the Coastal Zone: Is There a Role for Learning Processes?

*Melissa Nursey-Bray and Nick Harvey*

**Abstract**

The coastal zone has multiple institutions, actors and issues. It is also under pressure from climate change and social change. Science is often translated into policy so that multiple pressures are managed in an ad hoc and uncoordinated manner. We argue that in order to ensure science gets disseminated and used in implementation of policy in practice, we need to understand how learning occurs – individually and collectively across cultures and stakeholders. This chapter, using examples from Australia to exemplify key points, is a conceptual reflection on the utility of learning processes when adopted formally within coastal management, and whether it can help build bridges between science and policy. We conclude with a suggested learning framework based on the elements of time, partnerships, incentives, cross-cultural knowledge exchange, identity, local–global linkages and scale.

## 17.1 Introduction

The coastal zone is typically an area with multiple jurisdictions, multi-party investments, multiple habitats and scales, and many competing interests. Uncertainty partly caused by climate change impacts and partly by the pressure of increasing population adds complexity. A global response to this has been the advancement of integrated coastal management (ICM) which has different definitions (Harvey, 2004) but basically seeks to integrate all those competing pressures in a holistic and realistic manner. It is also often interchanged with the term ICZM which is used in this paper. Notwithstanding the adoption of ICZM across the world, there remains the stark reality that the environmental integrity of the world's coastal resources is still deteriorating (World Ocean Review, 2010). Diedrich et al. (2010) note that scientists still conduct research without considering the needs of decision-makers, while the latter make decisions based on political premises. McFadden (2007) notes that scientific

*Global Challenges in Integrated Coastal Zone Management*, First Edition. Edited by Erlend Moksness, Einar Dahl and Josianne Støttrup.

inputs are often lost within coastal management. Moreover, many scientists find that their work, even when explicitly aimed at informing policy, can be left on the shelf and ignored. In large part, this situation arises due to what is termed the 'implementation gap' (Cash et al., 2006); that is, the 'challenge associated with translating the basic principles of ICZM into management action'. There remains a gap and/or separation between the science about the coast and effective management, and the policy frameworks for the coast. Cash et al. (2006) advance the notion that this gap can be understood also as a knowledge – governance gap, caused by structural, functional, cultural and political disjuncture between knowledge and governance which constrain enablers from implementing knowledge into management regimes. We argue that this gap is in fact in and of itself largely caused by the lack of connection between science and policy. Nonetheless, 'when facts are uncertain, values in dispute, stakes are high and decisions urgent, scientists can provide useful input only by interacting with the rest of society' (Garmedia and Stagl, 2010: 1712). How they may do this is the subject of this chapter. We suggest that consideration of what role and what type of learning processes can contribute to enabling the transition of science into policy (which thus in turn informs coastal management in practice) is an important but as yet undervalued element in this discussion. Our paper is primarily a conceptual exploration into this area, and we use Australia as a case study to exemplify our points. The chapter does not report results of empirical research, but is the result of an institutional and documentary analysis which we use as the starting point for theoretical exploration into the science–policy divide. As such we hope our ideas will provide some food for thought in the coastal management space.

Our methodology was driven by a formal review of the history of coastal management in the literature consistent with social science practice (Walter, 2007). We used scopus to find resources using the terms 'learning', 'social learning', 'adaptive learning', 'coasts', and 'coastal management.' We also undertook a documentary analysis of the history of learning and coastal management in Australia over the past 20 years. This review included analysis of government documents, documents from natural resource management bodies, media reports and articles drawn from the literature. This was anchored by an institutional analysis of coastal management arrangements in Australia with a view to determining their flexibility for and current incorporation of learning into coastal practice. We found overall, that knowledge about the coast played a key role in the development of policy but not in its implementation, that is on ground management. As such, we see a clear role for learning methodologies to be embedded within coastal management and conclude this chapter with suggestions on the areas that might be included.

## 17.2 Key questions

In clarifying the role of learning, a number of questions arose that are pertinent to the integration of science and policy for coastal management. This includes consideration of how learning styles differ from each other, how learning works in practice, where and how is it particularly relevant, how one documents how people learn, how knowledge is understood and valued and why it matters in coastal management. How does adaptive learning, fit within other concepts such as adaptive capacity, adaptive management, communities of practice and social learning? These are grand questions, and we do not assert a naive claim to definitively answer these questions here. Rather we suggest that they are important questions, for unless the actual notion of learning is understood *per se*, it will not be possible to incorporate relevant learning methodologies into ICZM and achieve results. Moreover, while it is acknowledged that other areas, such as the terrestrial also experience massive biological and other pressures, the reality is that since 2008, over 50% of the world's population live within the coastal zone, a figure that will only rise over time. Learning matters in coastal management because 'attention to the processes and structures is necessary to involve a heterogeneous set of actors in analysing and making decisions about complex, multifaceted and value laden problem situations' (Woodhill, 2010: 62) is necessary.

In Australia the need to incorporate learning into management is necessitated by its scale, diversity and impacts. The coastal zone in Australia is typically perceived as a place of sun, surf and adventure. The Australian coast is managed using a complex three-tier government system (Federal, State and local), with different coastal legislation for each state (Harvey and Caton, 2003). The coast also supports a variety of recreational, economic and societal activities. In this context, the threats to Australia's coastal zone are both numerous and significant and include population rise and movement, increasing pressure on resources and environmental change such as sea-level rise, increased severity and frequency of storms and coastal flooding. Australia is also experiencing a population migration to non-metropolitan and regional coastal areas. Research into this phenomenon has been conducted by Burnley and Murphy (2004), who describe it as a 'sea change' in population movement which has subsequently been the subject of a number of studies (Beeton et al., 2006; Gurran and Blakely, 2007). In turn this creates pressure on coastal biodiversity (Nursey-Bray and Shaw, 2010), and agricultural land as it gets rezoned to residential title (Beeton et al., 2006). The social implications for this 'seachange' movement have been discussed by Harvey and Clarke (2007) and the impact on aesthetic value of coastal properties because of increasing pressures on the coast has been described by Mosadeghi et al. (2009).

Given the complexity of coastal systems, the scientific uncertainty associated with predicting the magnitude and timing of environmental change and high-decision stakes that are associated with many coastal zones (Smith et al., 2009), the planning and implementation of sustainable management strategies is both difficult and important. For example, planning for future coastal land-use settings and socioeconomic scenarios under climate-related change will prove difficult because of the uncertainty associated with predicting the rate and direction of change (Otter and Capobianco, 2000). Fletcher et al. (2008) examine pedagogy, or how learning takes place, in coastal management. They argue that because 'capacity building is essential to deliver effective ICM, pedagogy is a vital area of research for ICM' (p. 378). They use Kolb's experiential learning model (Kolb, 1985) and learning style methodology to test variation in learning styles for capacity building in ICM, principally through tertiary training courses in the UK. They conclude that ICM capacity building activities would benefit from more pedagogically informed learning models in order to maximise their capacity building activities by accommodating different learning styles. To achieve this aim, the incorporation of different learning strategies within the coastal zone is of paramount importance.

## 17.3 Learning – what is it?

To fully appreciate how learning may be embedded within coastal management, and offer something to resolving the divide between science and policy in practice, one must understand how learning *per se* has been constructed. The theories of learning have had a long history particularly in the fields of psychology, educational psychology and cognitive science with studies debating the idea of learning, how humans learn, and why it is important. Some authors have analysed components of learning such as Kolb (1985) and others have proposed different types of learning such as adaptive learning and social learning. One key concept is that of social learning, defined by Garmedia and Stagl (2010: 171) as: 'Deliberative approaches that enhance collective learning processes among a diverse group of social actors, with different types of knowledge and perspectives, are thus central in the creation of new responses to threats for socio-ecological systems'.

One of the first proponents of social learning Bandura (1974) focussed on the idea of learning as the observation of the behaviour of others. Kolb (1985) builds on these ideas arguing the link between the creation of knowledge and the transformation of experience. We grasp reality in concrete and abstract ways, and we then transform what we grasp by reflexive observation and active experimentation: 'Experiential learning is thus a recurrent process of adaptation to change, based on a rigorous process of transformation' (Blackmore, 2010: 47). Argyris and Schön (1978: 196) discuss the concept of double loop learning. Hall (1993) and Sabatier (1988) conceive policy making as a process of social learning, one that shifts dominant ideas and belief systems that drive policy making. Social learning as such is a key governance process driven by wider agendas than how individuals learn or experience things. It also encapsulates the idea that processes are more important than states, and is related to the concept of bounded rationality (Snyder and Wenger, 2010).

Dosi and Pavitt (1990) note that learning is greater than information acquisition *per se* – but also constitutes the development of changes in mental models of the world. Dewey (1938), also influential in the area of social learning, argued that scientists should facilitate citizen ability to make sensible political judgements. In this sense, social learning implies the capacity to reframe major problems in ways that incorporate values, cultures and politics. It is an interactive process that also acknowledges the limits of knowledge creation, and encourages self reflexivity. The capacity to be reflexive is discussed by many authors, as is the capacity to go beyond the individual to capture the activity of wider social and actor networks. Many argue that social learning is a core part of the road towards attaining sustainability. Garmedia and Stagl (2010: 1712) note that 'social learning processes in sustainability appraisals are poorly conceptualised'. Pahl-Wostl et al. (2007a, 2007b) argue the case for social learning in natural resource management, noting in the case of water management that it is crucial to move beyond command and control approaches to shift paradigms about water governance *per se.* Social learning in this context is not just learning by managers but occurs in situations whereby managers also function as facilitators of community learning, thus leading to optimum community management of resources. The role of participative processes is considered crucial here, and as such social learning plays a role: 'management cannot be based on optimizing in a predictable environment but requires many instances of learning in a fast changing world' (Pahl-Wostl, 2008: 485). She argues that social learning must be embedded in management systems so as to ensure flexibility and robustness within social–ecological systems, to build capacity for communities to work across and within cultural boundaries and to ensure that relational qualities as well as technical ones are brought into the management space. Social learning can contribute an integrative function by enhancing network governance, alleviating uncertainty and building the structural conditions for governance.

Further analysis (Pahl-Wostl et al., 2007a) highlights the idea of institutional 'lock in' when mutual dependencies develop in unhealthy ways between actors and institutions which in co-evolving becomes inefficient: 'To escape lock in, actors need to learn to recognise how their own frame of reference influences and constrains their thinking, and that other legitimate frames of reference exist' (Pahl-Wostl et al., 2007b: 30). Ultimately, 'In terms of our critical learning system approach then, we need to facilitate the transformation of communities in learning systems which are sufficiently self referential that they will be able to learn about their own learning' (Bawden, 2010 43).

In an analysis of Landcare in Australia, Woodhill (2010: 62) agrees, noting that social learning requires 'attention to the processes and structures necessary to involve a heterogeneous set of actors in analysing and making decisions about complex, multifaceted and value laden problem situations, such as NRM [natural resource management]'. It requires conscious design, and does not happen by accident. He goes further to suggest the following design principles for social learning that could be applied in management contexts: coordination, institutional diversity, local–global dialectics, multi-layered democratic participation, autonomous and integrated knowledge systems, meta-reflexivity, self organisation and cultivation of social capital. Learning processes can also be conceptualised as being global systems with three dimensions – an active learning capacity, cross boundary representation and cross level linkages (Snyder and Wenger, 2010).

Wenger (2000) notes that to belong to a social learning system, there are three components: (i) *engagement*: communicating with each other, doing things together, and engaging with each other and the world to shape our experience of who we are and how to respond to our actions, (ii) *imagination:* constructing an image of ourselves, our communities, and how we perceive the world, to re-orientate ourselves and reflect on the situation and explore possibilities and (iii) a*lignment:* alignment of local activities with other processes so they can be effective beyond our own engagement. This review of the learning literature highlights many elements that are of use in considering how to progress learning in coastal management, and it is to this aspect that we now turn.

## 17.4 Learning and coastal management

Given the impact of high decision stakes, uncertainty and increasing environmental pressure, adaptive capacity and social capital will be reinforced by effective, deep learning at all levels in coastal management. Many of the learning styles or notions described earlier have potential for application in a coastal context. Social learning when operationalised can support individuals to go beyond themselves and mobilise collectively to be forces for change; it can promote institutional flexibility and malleability and help transform knowledge as facts into knowledge in practice. Olsen et al. (1998) discuss learning in coastal management where there are five essential features of coastal management: (i) issue identification and assessment, (ii) program preparation, (ii) formal adoption and funding, (iv) implementation and (v) evaluation. He then advances a learning cycle advanced as a continuing cycle of action and reflection, with the following elements – experiencing, processing, generalising and applying. Torell (2000) apply these cycles to five case studies of coastal management in Tanzania to identify first to fourth world governance frames, where one starts out with a problematic coast and ends up with a sustainable one.

The capacity of social learning techniques to also support adaptive management has meant that it is often constituted in current discourse as adaptive learning, and is used in conjunction with the notions of adaptive capacity and management to imply an ability to learn from challenges and then amend or adapt to that challenge. This is crucial to our argument – the emerging pressure of climate change along the coast, will necessitate adaptation at a wide scale, so utilising how to encourage individual and collective learning in adaptive ways will enhance coastal management processes. Indeed, Smith et al. (2009) argue that complexity, uncertainty and the high decision stakes within the coastal zone in Australia make adaptive learning an essential tool to managing the coast effectively, and they identify a number of factors that make up a learning organisation: (i) proactive seeing of current information and use of multiple sources/forms of data to guide decision making, (ii) openness to change in practice and the rewarding of applications of learning, (iii) formalised processes for monitoring and evaluation, (iv) having formalised processes for reflection on management and practice modification and (v) being proactive in seeking knowledge sharing partnerships. Other commonalities in the literature also emerged as useful for framing how learning can be constructed within Australian coasts.

Reflexivity is a key element in most studies, highlighting the importance of being able to reflect back on the journey taken and learn from it. Knowledge sharing is an important facet. Flexibility, that is the ability for individuals, institutions and collectives to respond to change and apply lessons learned. occurs in most types of learning frameworks.

The importance of evaluation is clear. The notion that learning needs to be experiential is reiterated over and again. In the coastal space this is crucial because if multiple stakeholders can be brought together through mutual experiential learning, then management systems can become more robust and coastal networks more resilient.

Our review also highlighted that Indigenous peoples in Australia are not only part of contemporary coastal management, but have been implementing coastal policy based on Indigenous learning styles and techniques. In Australia, Indigenous Australians have the rights and often the opportunity to become directly involved in coastal management. For example, in the Northern Territory, over 90% of the coast is owned outright by Indigenous peoples. Moreover, Indigenous knowledge about the coast is extensive, and the result of up to 60 000 years of learning. Indigenous learning processes are oral, they are experiential and they are place based. Moreover, Indigenous learning processes are the vehicle by which Indigenous knowledge is transmitted into society and how the processes of environmental decision making – or in this case 'caring for country' – are facilitated. Indigenous ways of learning, however, differ substantially from Western modes, being more circular, more oral and more embracing of multiple forms of knowledge. As Sillitoe et al. (2002: 9) note:

> Indigenous knowledge is community based, embedded in and conditioned by local tradition. It is culturally informed understanding inculcated into individuals from birth onwards, informed continually by outside intelligence. Its distribution is fragmentary. Although more widely shared locally on the whole in specialised scientific knowledge, no one person, authority or social group knows it all. It exists nowhere as a totality, there is no grand repository, and hence no coherent overall theoretical model . . . It is the heritage of practical everyday life, with its Functional demands, and is fluid and constantly changing, being dynamic and subject to ongoing local, regional and global negotiation between people and their Environments.

## 17.5 Looking forward: a learning framework

What does our review mean for learning in coastal practice? What link can be forged between learning and governance, so that in part at least, the gap between coastal science and policy when implemented in practice can be bridged?

As a first step, we suggest the utilisation of learning in all its forms and based on integration of cultural knowledge systems and the utilisation of learning processes as a formal part of coastal governance. We suggest that our thinking about learning in coastal management could draw on some of the principles within Indigenous learning processes; that is, a focus on place, include oral means of learning, and again, accent the experiential nature of learning – on site and in place.

However, making space for the inherent creativity and flexibility offered by different learning processes is a key challenge for institutions, already constrained by established structure, rules and norms. We propose a framework for learning based around seven key elements, none of which require major change for institutions, but rather provide an opportunity to build on and strengthen existing elements, often intangible, by being more conscious and articulated about the specific role of learning. The seven elements are: (i) time, (ii) incentives, (iii) partnerships, (iv) scale, (v) identity, (vi) communication and (vii) integration.

Firstly, **time** is clearly crucial. Institutional decision makers need to allow staff time to develop learning programmes, and to participate in them, to enable translation of the lessons, mistakes and benefits of learning and time to build communications and joint collaborations between the coastal learning organisations, policy outcomes and implementation in practice. In this context it is important to note that there are different time-frames between long-established coastal communities, short-lived special interest groups, established governmental organisations and relatively short-term elected governments at local, state and federal levels.

Secondly, there needs to be a **system of incentives** that will encourage the participation of individuals and then collectives to work together in coastal management enterprises. Their survival relies on, among other things, motivated drivers who must have time to coordinate their activities. Community champions often play a pivotal role in this context, and need to be given added rewards or incentives to encourage ongoing participation, especially given that so many individuals are actually giving their time voluntarily. In this context, learning is more often than not an implicit activity, but the provision of incentives may enable its tangible manifestation as part of the management enterprise.

**Fostering partnerships** also means decision makers need to engender the cooperation of and interaction between stakeholders, and between their own agencies. Key people within social networks can play vital functions as brokers between different cultural and community boundaries in this sense. One of these roles would be in negotiating the zones of conflict that inevitably occur within coastal management, but also in navigating pathways between the different knowledge systems that occur within a coastal zone. Indeed until **common agreements** are made about what forms of knowledge are to be prioritised or identified within the management endeavour, the effectiveness of implementing adaptive learning strategies that may enhance how science is adopted into policy will be impeded.

**Scale** is an important determinant of the success (or not) of learning in practice and place – whether it is formal or informally constituted. Scope (of the activity) not scale (as in numbers participating) needs to be at the heart of any work. As such, the size of the community is important. While integrated coastal zone management occurs at local scales, it is also part of a global management experiment, and as such the tensions between **local and global learning** could be investigated and harnessed in productive ways. As Snyder and Wenger (2010, 121) add: 'each locality constitutes a local learning experiment that benefits from and contributes to the overall learning system'. In this sense, organisations and bureaucracies engaging with local learning organisations and movements need to be careful not to scale up too fast, be too top down in their approach and ensure that trust, at all levels – local to Federal – is built up over time.

Building bridges between different groups and forging a common purpose and **identity** is another performative role that undertaking mutually explicit learning activities can play. There is a deep connection between identity and practice, particularly important in the case of the Australian coastal zone, and as such the formation and support of a learning community is also an exercise in identity formation and practice. Identity can be defined in many ways – as negotiated experience, as community membership, as a learning trajectory, as a nexus of multi-membership, and as a relation between the local and global (Wenger 2000: 133).

Finally, **communication** is a key element in negotiating; (i) how people learn and (ii) ensuring common understandings of the issues at hand. Given the diversity of interests, peoples and environments in the coastal space, the forms by which people communicate will need strengthening and due consideration. Some starting principles might include choosing language that is appropriate to the audience (a good first step), concentrating on what is feasible for different groups (important) and maximising opportunities by aligning coastal management as an issue with other contemporary issues that resonate with local interests and local agendas (Nursey-Bray and Ferrier, 2009).

Communication materials and strategies need to work within the culturally accepted discourse at local levels. Ultimately, communication strategies must also be based on solid guidelines.

Finally, the **integration of different knowledge systems** and culturally informed ways of learning is an important part of any learning framework. For example, integrating Indigenous ways of knowing and doing can play a powerful role in facilitating the importation of different types of knowledge into Western management regimes, which normally wouldn't construct that knowledge as science, but which is an important vehicle for decision making nonetheless. For example, every year the Yolgnu people of Arnhem Land, in the Northern Territory, have the Garma Festival. The word Garma essentially means the pooling of fresh and sea water bodies, and the Yolgnu use this concept as the vehicle by which to hold a cultural summit each year, designed to facilitate two way communication and learning (or pooling) of knowledge around a particular theme. Themes have included environmental (including coastal) management, education and culture. Essential to the success of Garma, however, is the fact the knowledge exchange is posited around Indigenous styles of learning. When attending Garma, learning is oral, place based and experiential, and results in the formation of networks and communities that last long after the conclusion of the summit.

## 17.6 Summary: what does this mean for coasts?

A learning framework can contribute to the discourse about the science–policy divide by presenting a suite of ideas that may be used in practice. It is not suggested as a 'be all and end all solution', but hopefully provides some ideas that add value to current practice. Learning is more often than not an implicit driver, hard to monitor and harder to 'prove', but nonetheless is essential to achieve on ground outputs. User outputs need to be culturally palatable – especially scientific ones – but if they are not presented or disseminated in a way that is sympathetic to different cultures (or subcultures as in different stakeholder groups), then they will not be absorbed. Our work provides an important insight – that learning, currently a hidden and implicit part of coastal management, in fact plays a highly performative role and thus needs to be more explicitly incorporated into coastal management practice.

The explicit incorporation of learning into management practice via such a framework can also help enable adaptive governance of the coast. We characterise adaptive governance as being reflexive, building partnerships, having well-established communication mechanisms, and based on adaptive management. In so being, the integrative function of coastal management is facilitated.

A learning framework also performs other important functions. It makes explicit a process that is so often implicit and therefore overlooked. Supporting learning as a management process *per se,* can also support individuals to go beyond themselves and be forces for change. Institutional flexibility is enabled. Moreover, the transformation of the construction of knowledge *as science* to the construction of knowledge *as practice* and residing within and across cultures will widen the scope of management considerably. Importantly, the application of learning processes can also assist in preventing the institutional 'lock in' forecast by Pahl-Wostl (2002). Luhmann's ideas about the way in which institutional systems develop their own programs and codes and consequently give meaning only to those data that pass their filters in order that the system can maintain itself, are of note in this context also (Luhmann, 1989).

Finally, the application of learning processes *per se,* rather than the adoption of any one style, ensures that managers do not fall into unproductive discursive frameworks: management 'red herrings'. For example, 'adaptive learning' is the latest iteration in the nomenclature that characterises learning processes, but there is a danger in adopting the terminology as the 'end' point rather than the 'means' for the underlying processes. We argue that it is important to use a framework that picks up on the best of all learning processes as a valuable first step. In these ways, the adoption of learning processes, as a formal means of informing management, will create some of the linkages and bridges needed at the science and policy interface.

## Acknowledgements

We would like to acknowledge the Institute of Marine Science, Norway and particularly Erlend Moksness, for the opportunity to present a draft of this chapter at the conference in Arendal and subsequent opportunity to be part of this important publication. We would also like to thank the three reviewers for their comments, which have strengthened this paper.

## References

Argyris, C. and Schön, D. (1978) *Organizational Learning: A Theory of Action Perspective* Addison Wesley, Reading, Mass.

Bandura, A. (1974) The process and practice of participant modeling treatment. In: Cullen, J.H. (Ed.), *Experimental Behaviour: A Basis for the Study of Mental Disturbance*. Irish University Press, Dublin, Ireland.

Bawden, R. (2010) Messy issues, worldviews and systemic competencies. In Blackmore, C. (Ed.) *Social Learning Systems and Communities of Practice*, Chapter 6. Springer, Dordrecht.

Blackmore, C. (2010) *Social Learning Systems and Communities of Practice*. Springer, London.

Beeton, R.J.S., Buckley, K.I., Jones, G.J., Morgan, D., Reichelt, R.E. and Trewin, D. (2006) Australia State of the Environment 2006. Independent report to the Australian Government Minister for the Environment and Heritage, Department of the Environment and Heritage, Canberra.

Burnley, I. H. and Murphy, P. (2004) *Sea Change: Movement from Metropolitan to Arcadian Australia*. UNSW Press, Sydney.

Cash, D.W., Borck, J.C. and Patt, A.G. (2006) Countering the loading dock approach to linking science and decision making: Comparative analysis of El Niño/Southern Oscillation (ENSO) forecasting systems. *Science, Technology, and Human Values* **31**(4), 465–494.

Dewey, J. (1938) *Experience and Education*. Simon and Schuster, New York.

Diedrich, A., Tintore, J. and Navine, F. (2010) Balancing science and society through establishing indicators for integrated coastal zone management in the Balearic Islands, *Marine Policy* **34**, 772–781.

Dosi, G. and Pavitt, K. (1990) *The Economics of Technical Change and International Trade*. Harvester Wheatsheaf, London.

Fletcher, S., Potts, R. and Ballinger, R. (2008) The pedagogy of integrated coastal management. *The Geographical Journal* **174**(4), 374–386.

Garmendia, E. and Stagl, S. (2010). Public participation for sustainability and social learning: Concepts and lessons from three case studies in Europe. *Ecological Economics* **69**(8), 1712–1722.

Gurran, N. and Blakely, E. (2007) Suffer a sea change? Contrasting perspectives towards urban policy and migration in coastal Australia. *Australian Geography* **38**(1), 113–132.

Hall, P. (1993) Policy paradigms, social learning and the state: the case of economic policy-making in Britain, *Comparative Politics* **25**, 275–296.

Harvey, N. (2004) Integrated Coastal Management. In: Goudie, A. (Ed.) *Routledge Encyclopedia of Geomorphology*, Routledge, London, pp. 567–570.

Harvey, N. and Caton, B. (2003) *Coastal Management in Australia*. Oxford University Press.

Harvey, N. and Clarke, B. (2007) Policy implications for Australian coastal communities affected by sea-level rise. *Just Policy*, **7**.

Kolb, D.A. (1985) *Learning Style Inventory: Technical Manual*. McBer, Boston.

Luhmann, N. (1989) *Ecological Communication*. Polity Press, UK.

McFadden, L. (2007) Governing coastal spaces: the case of disappearing science in integrated coastal zone management. *Coastal Management* **35**, 429–423.

Mosadeghi, R., Tomlinson, R., Mirfenderesk, J., Warneken, H. (2009) Coastal Management Issues in Queensland, and Application of Multi-Criteria Analysis Techniques, *Journal of Coastal Research* **56**, 1252–1256.

Nursey-Bray, M. and Shaw, J. (2010) Australia, 'Climate Change and the Sea Change Phenomenon', *The International Journal of Environmental, Cultural, Social and Economic Sustainability*, **6**, 15.

Nursey-Bray, M. and Ferrier, T. (2009) Risk assessment and local government, Tasmania: Applying an inter-disciplinary approach to climate change adaptation. In Mannke, F. (Ed.) *Inter-disciplinary Aspects of Climate Change*. Peter Lang Scientific Publishers.

Olsen, S.B., Tobey, J. and Hale, L. (1998) A learning-based approach to coastal management. *Ambio* **27**(8), 611–619.

Otter, H. and Capobianco, M. (2000) Uncertainty in integrated coastal zone management. *Journal of Coastal Conservation* **6**, 23–32.

Pahl-Wostl, C. (2002) Polycentric integrated assessment. *Integrated Assessment* **3**(2–3), 220–232.

Pahl-Wostl, C., Sendzimir, J., Jeffrey, P., Aerts, J., Berkamp, G. and Cross, K. (2007a). Managing change toward adaptive water management through social learning. *Ecology and Society* **12**, 30. http://www.ecologyandsociety.org/vol12/iss2/art30/ (accessed 10 September 2010).

Pahl-Wostl, C., Craps, M., Dewulf, A., Mostert, E., Tabara, D. and T. Taillieu. (2007b) Social learning and water resources management. *Ecology and Society* **12**, 5. http://www.ecologyandsociety.org/vol12/iss2/art5/

Pahl-Wostl, C., Tàbara, D., Bouwen, R, Craps, M., Dewulf, A., Mostert, E., Ridder,D. and Taillieu, T. (2008) The role of social learning and culture for water resources management. *Ecological Economics* **64**, 484–495.

Sabatier, Paul. (1988) An Advocacy Coalition Framework of Policy Change and the Role of Policy-Oriented Learning Therein, *Policy Sciences* **21**,129–168.

Sillitoe, P., Bicker, A. and Pottier, J. (2002) *Participating in development: approaches to indigenous knowledge*. ASA Monographs 39. London: Routledge.

Smith, T., Carter, B., Thomsen, D., Mayes, G., Nursey-Bray, M., Whisson, G., Jones, R., Dovers, S. and O'Toole, K (2009) Enhancing science Impact in the coastal zone through adaptive learning. *Journal of Coastal Research* **56**, 1–4.

Snyder, W. and Wenger, E. (2010) Our world as a learning system: a communities-of practice approach. In: Blackmore, E. (Eds), *Social Learning Systems and Communities of Practice*. Springer, London, pp.124–144.

Torell, E. (2000) Adaptation and Learning in Coastal Management,: the experience of five east African initiatives. *Coastal Management* **28**, 353–365.

Walter, M. (2007) *Social Research Methods, an Australian Perspective*, Oxford University Press.

Wenger, E. (2000) Communities of practice and learning systems. *Organization* **7**, 225–246.

Woodhill, J. (2010) Sustainability, social learning and the democratic imperative: Lessons from the Australian Landcare movement. In: Blackmore, C. (Ed). *Social Learning Systems and Communities of Practice*. Part II, 57–72. London: Springer.

World Ocean Review (2010) at *www.maribus.com* (accessed 10 September 2012).

Chapter 18

# Perspectives of Social and Ecological Systems

*Paul Tett, Audun Sandberg, Anne Mette, Denis Bailly, Marta Estrada, Thomas Sawyer Hopkins, Maurizio Ribera d'Alcalà and Loraine McFadden*

**Abstract**

ICZM depends on an understanding of the coastal zone system to be managed. During the SPICOSA project, the authors' task was to provide teams engaged in applications of the 'Systems Approach Framework' with a theoretical basis for understanding these systems in terms of ecological, economic, and social structures and processes. This chapter briefly considers disciplinary approaches to conceptualizing coastal zones, and draws on these and the ideas of a number of twentieth-century thinkers, including Habermas, Luhmann, and Popper, to propose an interdisciplinary model of coastal 'social–ecological' systems. The model takes the form of a three-dimensional, animated object which can be seen at www.coastal-saf.eu. The use of the conceptual model is illustrated briefly in an application to a system in western Scotland that is losing its integrity under globalization.

## 18.1 Introduction

This chapter concerns ways of conceptualizing coastal zone systems and their problems. It results from the authors' experience in preparing guides for application of the SPICOSA project's 'Systems Approach Framework' (SAF) (Hopkins et al., 2011). The SAF is a toolbox of methods (Tett et al., 2011a) for bringing together stakeholders, scientists, and representatives of governance in a multi-stage procedure (Figure 18.1) to identify a dysfunction in a coastal zone and evaluate scenarios for its amelioration.

One of the challenges in preparing the guides was to understand and reconcile discipline-based accounts of coastal zones. Doing this required a move from the multidisciplinary perspectives of ecology, economics, oceanography, political science, and sociology, to a more integrated, which is to say interdisciplinary, perspective. Thus the guides, in particular that for the stage called 'System Design', went through numerous drafts in order to integrate

*Global Challenges in Integrated Coastal Zone Management*, First Edition. Edited by Erlend Moksness, Einar Dahl and Josianne Støttrup.

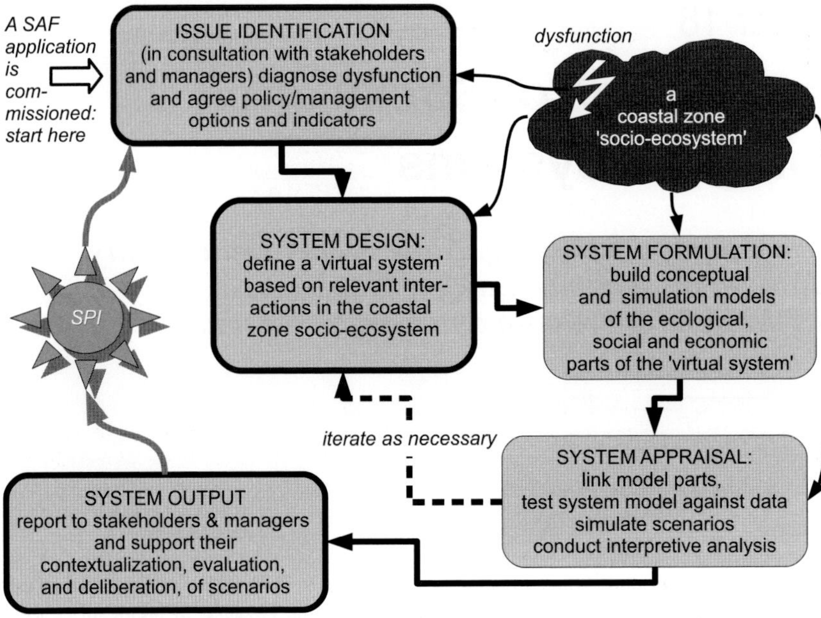


**Figure 18.1** The SPICOSA 'Systems Approach Framework' or SAF; SPI 'Science-Policy Interface' (modified from Tett and Sandberg, 2011a). Reproduced with permission of Dunedin Academic Press, Edinburgh.

the authors' differing and subject-based perceptions of coastal zone systems. Observing the use of the guides helped improve the synthesis that is presented here.

This chapter is not intended as a review of the many papers that have been written about the co-dependency of social and ecological systems. Instead, it brings together some ideas that appear to apply across all our disciplines, as well as examples of those that relate more particularly to individual disciplines. What had been envisaged for presentation at the Arendal conference was merely a somewhat obvious explanation of why each academic discipline saw a coastal zone system in its own particular way. However, the synthesis that is presented here may have explanatory power in its own right, for example as it applies the ideas of Luhmann (1989) concerning the problems of 'Ecological Communication'.

## 18.2 Coastal Zone systems

We can roughly sort theories of coastal zone, and similar, systems into three groups, depending on their main perspective:

1. **Ecologists** tend to see 'nature', or natural ecosystems, or 'the environment', as most important, and humans—once part of these systems—as now causing them problems because of excessive removals (such as fish) and noxious inputs (such as waste discharges) and because the human 'slice of the pie' is expanding at the expense of nature's share (Figure 18.2). The 'Driver-Pressure-State-Impact-Response' (DPSIR) paradigm (Luiten, 1999) provides an environment-centred framework that links human pressures

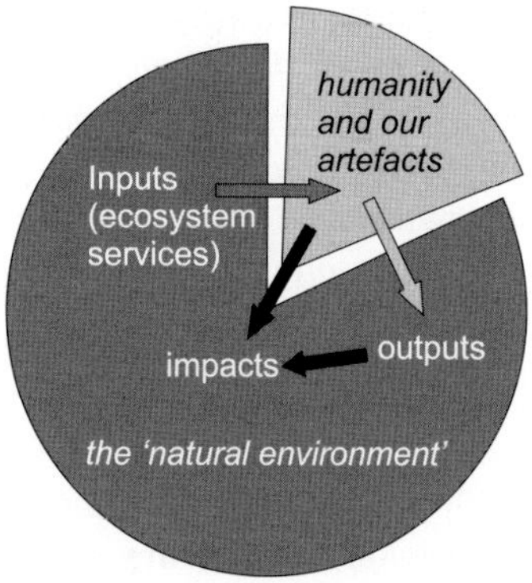


**Figure 18.2** Humans and 'the environment'. Several ideas are suggested here. One concerns the way in which human societies have increasingly seen themselves as distinct from the 'natural world'. Another is the expansion of the 'human world' (or the 'human footprint') at the expense of the 'natural world'. The third is the set of interactions between the two components that are sometimes conceptualized in terms of the DPSIR framework and sometimes as 'ecosystem services'.

on ecosystem state to their impact on the human sector. Much EU environmental legislation has aimed to protect species and habitats, or to restore and maintain good ecosystem status, against the anthropogenic pressures. A problem with this perspective is that the action necessary to 'protect the environment' tends to depend on social factors as much as on ecological arguments (Fischer, 2004).

2. Environmental **economists** consider the human-use values of ecosystem services and natural capital (Costanza et al., 1997), often comparing scenarios (e.g. Millennium Ecosystem Assessment: MEA, 2005). The expression of these values in dollars or euros can help users to appreciate the ecosystem services on which industries such as fisheries or agriculture depend. It can provide a common currency (both metaphorically and literally) with which to evaluate scenarios, as exemplified for climate change (Stern, 2006; Weitzman, 2007). A further step is to create property in, and a market for, an ecosystem service. This should, in principle, result in an exchange value that optimally relates human demand and ecosystem supply. In practice, the value tends to be distorted by non-economic considerations and the market often fails to anticipate resource depletion. The alternative, however, is the 'tragedy of the commons' (Hardin, 1968).
3. **Social and political scientists** tend to focus on the institutional (including governmental) control of access to ecosystem services and protection of natural capitals, and the content and codes of communication amongst actors about these matters. Olsen (2003) wrote about the 'four orders of outcomes' in coastal zone governance. The first order '*requires building the constituencies and the institutional capacity to undertake* [and implement] *integrated coastal planning*'. If successful this will lead to changes in behaviour by institutions and individuals (second order), and eventually to the fourth order outcome of sustainable coastal development. Ostrom (2007) proposed analysing a '*social–ecological system*' as composed of '*a resource system . . . , resource units generated by that system . . . , users of that system, and . . . the governance system*', existing in a '*social, economic and political setting*', with the aim of matching '*governance*

*arrangements'* to specific problems. Halliday and Glaser (2011) conceptualized a *'social ecological system'* in terms of *'four functional subsystems . . . : natural . . . , worldview . . . , control/management . . . , and technology . . . Emergent systems properties are conceptualized as arising through exchanges of matter and meaning between subsystems, and between the system as a whole and its environment*'. Management, or the '*"institutional subsystem" . . . represents "humans outside nature"*'.

## 18.3 Ethics

The aim of SPICOSA was to help improve the (ecological) **sustainability** of ecosystem services in the coastal zone, the (economic) **efficiency** with which these services are used, and the (social) **equity** with which they are distributed amongst members of society. In all these, of course, the non-human creatures and natural processes that provide these services have a purely instrumental value. Rolston (1994) contrasts this (the '*value of nature*') with the intrinsic worth of organisms to themselves (the '*value in nature*'), adding that there are also systemic values in nature, which are emergent properties of ecosystems that cannot be localized in any particular component.

## 18.4 Systems

The SPICOSA SAF is based on General Systems Theory (von Bertalanffy, 1968) which asserts that '*there exist models, principles and laws that apply to generalized systems or their subclasses irrespective of their particular kind, the nature of the component elements, and the relations or "forces" between them*' (von Bertalanffy, 1972). At the commencement of the project, there was a debate: when we spoke of 'coastal zone systems', were we dealing with 'real' or 'hard' systems that truly existed in the physical coastal zone and its associated economy and society, or was it better to see the project as using a 'soft systems' approach (Checkland and Scholes, 1990) in which the 'system' is merely a mental construct used for understanding and problem solving? We took the view that: (i) real coastal zone systems exist, with biophysical and social components and the properties described by General Systems Theory; and, (ii) the conceptual and simulation models built to study dysfunctions in this real coast zone, were 'virtual' or 'soft' systems, partial heuristic devices.

It was found helpful in aiding understanding between biophysical scientists, who by and large took the hard systems view and aspired to a single best model, and social scientists, who mostly took a soft systems view in which there could be multiple alternative and valid models, to introduce the '3-worlds cosmology' of Popper (1972). The 'worlds' (Popper, 1978), are:

1. the physical (natural) world, that '*consists of physical bodies: . . . of plants and of animals; but also of radiation, and of other forms of physical energy*'.
2. '*the mental or psychological world, the world of our feelings . . . thoughts . . . decisions . . .* [and] *perceptions . . . ; in other words, the world of mental or psychological states or processes, or of subjective experiences*'.
3. '*the world of the products of the human mind, such as languages; tales and stories and religious myths; scientific conjectures or theories, . . .* '

**Figure 18.3** Words on a beach. (Anne Mette and Simona Dalla Riva). As 'world 1' phenomena, these are local changes in the elevation of the sand surface. As 'world 3' phenomena, they convey information that is meaningful to any 'world 2' mind equipped with the English language. Extra meaning is available to those who took part in the SPICOSA meeting in Faro in 2008 and were thus part of a finite sub-system having *'a distinctive identity that is constantly reproduced in its communication and depends on what is considered meaningful and what is not'* (en.wikipedia.org/wiki/ Niklas_Luhmann).

Objects or phenomena can belong to both 'world 3' and 'world 1', as exemplified in Figure 18.3. In this photograph, there are, in 'world 1', local differences in beach topography; in 'world 3' there are words and meanings. This raises the question of how the former get converted to the latter, and how events in 'worlds 2 and 3' lead to actions that perturb 'world 1'.

Popper's concern is ontological, i.e. with 'how things are'. Taking an epistemological approach – i.e., dealing with 'how we know things', Habermas (1984:333; 1987:120, re-ordered by us) – proposes a similar cosmology in relation to 'validity claims':

1. the **objective world**, *'the totality of entities about which true statements are possible'*;
2. the **subjective world**, *'the totality of experiences to which a speaker has privileged access and which he can express before a public'*, and
3. the **social world**, *'the total of legitimately regulated interpersonal relations'*, claims about which *'can be contested under the aspect of rightness'* at the level of *'a given action in relation to a given norm, and, at the next level, the rightness of the norm itself'*.

The distinction between worlds 1 and 2 has been debated at least since Descartes wrote in his *Discourse on the Method* (1637) that *'je pense, donc je suis'*. The addition of 'world 3' seems useful in understanding social–ecological systems, for in this world there can legitimately be different opinions about 'world 1'. Habermas shows how positions held by individuals or social groups might be harmonized, through *'communicative action'* and *'discursive redemption'* of validity claims against physical and social data.

Within the social world, Habermas (1987) distinguishes between **lifeworld** and **'system'**. The former provides the domain in which *'communicative actors move in the medium of a natural language, draw upon culturally transmitted interpretations, and relate simultaneously to something in the one objective world, something in their common social world, and something in each's own subjective world*' (Habermas, 1984:392). Large and complex societies, however, need an additional framework of impersonal, institutionalized, functions. In Habermas' view, the institutional components of the '*system*' of modern states operate using two main *'steering media'*, money and power, which are '*delinguistified*', that is stripped of the need for 'communicative action'. Luhmann (1989) takes this analysis further, seeing society as a set of sub-systems of communication, each of which consists of a *program* (or set of rules) that filter information for internally relevant meaning according to binary *codes*. Each social sub-system remains distinct only if it performs a unique and necessary function.

## 18.5 A unified view of social–ecological systems

The label *social–ecological system* has become widely used for:

> *linked systems of people and nature. The term emphasizes that humans must be seen as a part of, not apart from, nature—that the delineation between social and ecological systems is artificial and arbitrary.*

(attributed to Berkes and Folke, 1998, by the 'Resilience Dictionary' of the Stockholm Resilience Centre, www.stockholmresilience.org). Given Odum's definition (1959) of an *ecosystem* as

> *Any area of nature that includes living organisms and nonliving substances interacting to produce an exchange of materials between the living and nonliving parts.*

*a social–ecological system* can be defined as:

> **a spatially-bounded region containing an ecosystem and a social system interacting with each other.**

Figure 18.4 shows our conceptual model of such a system. This has been conceived as a dynamic three dimensional object, and the figure shows still images from different stages in the animation at www.coastal-saf.eu/design-step/refs.shtml. The two viewpoints show how perspectives of a Coastal Zone system can be partial. Although most if not all social–ecological systems have open boundaries, trans-boundary fluxes have been omitted in the interests of clarity, and so this conceptual object is more a model of 'spaceship earth' than of an open coastal zone.

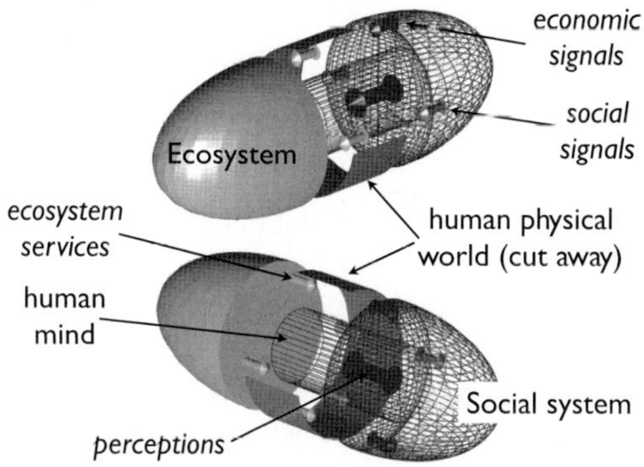


**Figure 18.4** Two views of the conceptual model of a social-ecological system. 'Ecosystem' and 'human physical world' exist in Popper's 'world 1'; 'Human mind' in 'world 2'; and 'Social System' in 'world 3'.

The model has a four-part structure, the parts linked by arrows representing signals flowing between them. Two parts lie in Popper's 'world 1', made up of the systems described by natural science, and including the physical aspects of human life. However, adopting the usual anthropocentric viewpoint, we have divided the 'world 1' component into an 'ecosystem' and a 'human physical world'. In principle, the latter is the space that humanity occupies and in which 'natural' ecosystems are modified for human purposes. The former includes the natural biogeochemical cycles and trophic networks that now function under anthropogenic pressures. Interactions between the two parts are labelled as 'ecosystem services' and 'impacts'. As parts of 'world 1', these are tangible things.

Ecosystems also provide intangible services to humans; these flow directly to minds or society. In the diagram, 'mind' ('world 2') is centrally placed because it is here that values originate and here that, according to a hedonistic ethic, lays the well-being that a utilitarian economic calculus aims to maximize. Although only one subjective world is shown, there should be many, intercommunicating with each other through the social network of 'world 3'. The space between 'mind' and 'social system' in Figure 18.4 might be interpreted as Habermas' 'lifeworld', the network comprising his 'system' of institutions.

Within 'world 3', the signals can be categorized as economic or social. For example, price signals can provide information about the willingness of humans to eat farmed fish as well as about the supply of fish feed and the costs of mobilizing human effort and capital into the industry. The corresponding social signals deal with whether eating salmon is perceived as healthy, whether there is concern for the welfare of the caged fish or the farmers, and what the farmers must do to maintain good environmental conditions and satisfy the industry's regulators. The relevant information is propagated and transformed by various institutions, including non-governmental organizations and the 'communications media' of newspapers, television, etc. 'Science' is a sub-set of institutions with special rules for selecting and processing information, and a claim to a high level of reliability. In Luhmann's terms its primary function is to produce new knowledge for society.

Signals arising from the sub-systems of Government employ the steering medium of power. However, governance is in principle a process of social decision making, and insofar as the law represents social norms or has been formulated according to higher-level norms, most people will see the coercion as legitimate. The point is not that coercion is bad, but that it is delinguistified (Habermas, 1987): courts decide whether in a given case a norm has been obeyed or breached; they do not discursively redeem validity claims about what is the law (Luhmann, 1989).

In 'world 1' the economy comprises human labour and physical capital, and the tangible services that the human physical economy draws from the supporting natural ecosystem, for example in the form of harvested fish. In 'world 3' stands the valuations that we place on the human products and the ecosystem services, and the norms concerning their allocation. In 'world 2' is the human well-being that is to be maximized by economic processes.

Diverse human societies have found diverse ways to make such valuations and to allocate the goods amongst their members. Examples include hunting and gathering, subsistence farming, gift exchanges, or central planning. In much of the modern world, however, the non-physical part of 'economy' is dominated by the institutions of money, property and the market. In principle, the latter could act in the interests of sustainability, providing a program and codes for the receipt, processing and transmission of information about resource availability and human demands. Sometimes, however, as in the present epoch, the money economy becomes a dominant and destabilizing run-away sub-system, overwhelming environmental signals about the sustainability of natural resources, and dangerously damping the ability of other social sub-systems to 'resonate' (Luhmann, 1989) to vital messages from natural ecosystems.

Each component in the model of Figure 18.4 has a structure (which in systems theory is the network of relationships amongst subcomponents) plus a set of fluxes (whatever it is that actuates the relationships). In the case of an ecosystem, the structure is that of the energy-passing trophic relationships and the biogeochemical cycles that process the elements (especially carbon, nitrogen and phosphorus) necessary to life (Lindeman, 1942). Biological species are in essence gene-pools, that is dynamically self-perpetuating stores of information stored in nucleic acids. So we can summarize the 'currency' of the natural component as mass, energy and information.

In the case of society, the network is that of institutions (Habermas, 1987; Luhmann, 1989). Between the institutional network and the minds of individuals is the 'lifeworld' (Habermas, 1984) in which the medium is language, processed by individuals, on their own and through 'communicative action', to enable understanding of the (social–ecological) world. Figure 18.4 supposes no particular model of 'mind' except that, in order to relate human action to economic theory, mental state needs to include the 'well-being' that individuals seek to satisfy. No doubt there are other mental components of importance, such as identity and associated sets of values, which (to use Luhmann's terms) provide the codes by which information is converted to meaning. Thus the 'currency' of the world 2 and world 3 components is essentially information that is filtered for meaning within each sub-system, but it seems useful to distinguish economic information (e.g. prices) and information originating in the governance sub-system (e.g. laws) from the information that is the notional 'currency' of institutions such as the mass media and science, and that which is carried by language within the lifeworld.

In attempting to describe a spatially limited system, a modeller must decide where to put the boundaries. Sometimes they can be sited at natural discontinuities, such as the margin of the lakes studied by Lindeman (1942). However, in ICZM the obvious discontinuity, between land and sea, is an essential part of the system to be analysed. Luhmann (1989) deals with

*autopoiëtic systems that reproduce all the elementary components out of which they arise by means of a network of these elements themselves and in this way distinguish themselves from an environment . . .*

The distinction between such systems – which may be ecosystems, minds, or systems of social institutions – and what lies outside, is very general: systems operate 'programs' and 'codes' that filter meaning or structure from a chaotic, highly entropic, environment. They are not necessarily spatially distinct from that environment (e.g. mind is not located elsewhere from body, nor social systems elsewhere than the physical environment). However, while the money-steered global economy is diffuse, power-steered political systems are, by and large, spatially defined units that reflect the historical mechanics of power projection, and the physical boundaries of a coastal zone system might be found the edge of the continental shelf and the watershed around the rivers draining into it.

Thus, as examined in Box 18.1, in few cases do human-drawn borders match natural boundaries. This is not an insuperable difficulty, because the system modeller's 'toolbox' includes the idea of asymmetrical 'boundary conditions'. What lies beyond the boundary, wherever placed, is allowed to influence what happens inside, but what lies outside is made independent of the modelled system. Nevertheless, the consequence of a mismatch between the spatial scale of the ecological component and that of the social–economic component may be that a model has difficulty in capturing the social–ecological feedback loops that are important for system resilience.

### Box 18.1 Strathclyde: a Scottish example of a social-ecological system in the coastal zone

Strathclyde has existed twice as a polity. From the fifth to the eleventh century it was a Brythonic kingdom in the valley of the River Clyde in western Scotland. Between 1975 and 1996 it was a local government region headquartered in Glasgow and including seven Scottish counties around the Firth of Clyde, a large fjord. Its artificial boundaries define a social–ecological system.

During the nineteenth century a heavily urbanized and industrialized area developed around the Clyde estuary. Local coal fuelled iron and steel works that in turn supplied shipbuilding. The Firth of Clyde supported a fishery for cod and herring that helped feed the city dwellers. Many buildings were roofed with slates quarried on the western coast and brought in by water. A local bank, the Clydesdale, issued its own Scottish banknotes. Rich urbanites kept houses for their families in the purer air of the coasts of the Firth, and many poorer residents took their annual holidays there.

The governance system was dominated by the council of the city of Glasgow and the economic system by the local financial, fishing, manufacturing and mining sectors. The human-influenced physical sector was most evidently that of the cities and the mining towns, but in fact most of the land surface was impacted by agriculture, forestry and hunting, and even the deep waters of the Firth had suffered some biotic changes as a result of pollution. A Marine Station, established in 1897 on one of the islands in the Firth, provided scientific information on conditions in the local sea.

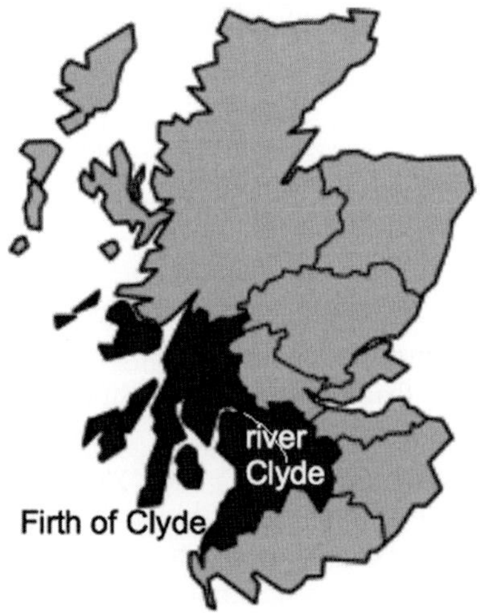


This sketch describes a unit, which while open at the boundaries to wider natural and economic influences, would have been susceptible to integrated and democratic management of the environment in relation to society. In the late nineteenth century the Glasgow city council responded to grossly polluted and disease-inducing conditions by improvements to piped water and sewerage. Between 1955 and 1996, the Clyde River Purification Board, directed by local politicians and stakeholder representatives, monitored and worked to reduce pollution of the river Clyde, the Firth, and the smaller rivers that drained the western coast. By the time the Strathclyde region was formed, however, heavy industry was in decline, and it was economic and social problems that were seen as critical. And when the region was dissolved, governance was dispersed both upwards, to a reformed Scottish government and parliament, and downwards, to reformed county councils.

The new Scottish Environment Protection Agency (SEPA) assimilated the river boards and re-organized away from local connectivity in the interests of efficiency and uniform standards. The local authorities were mostly too small to afford great expertise in, for example, marine systems, and in many cases their activities were tightly controlled by Scottish and UK governments. No longer did the city-dwellers use the Firth for their holidays, or eat fish caught there: they flew on cheap airlines to warmer and sunnier destinations, and in supermarkets bought fish from all over the world. Increasingly, with the exception of special-interest groups such as divers and hill-walkers, most of civil society grew more detached from their natural environment, and at the same time the local economy, culture and governance became increasingly globalized. As an example, the Clydesdale Bank was bought by the English Midland Bank in 1919 and sold to the National Australia Bank in 1987.

When in 2007, the local SPICOSA team sought stakeholder partners for a SAF application in a part of the Firth, they found them in a Strathclyde legacy organization, the Firth of Clyde Forum. However, most of the Forum's member organizations were concerned at the time with the need to implement the marine spatial planning aspects of the Scottish Government's forthcoming Marine Bill, itself a response to the European Marine Strategy Framework Directive (MSFD) of 2008.

This account is of necessity a sketch, intended only to illustrate the application of the model of Figure 18.4, and not a serious attempt to diagnose and document the past and present states of this social–ecological system. Haig (1986), Macleod (1997), Tett et al. (2012), and Thurstan and Roberts (2010) give more details. What the sketch suggests is that the information flow from the local ecosystem into the local social–economic system has become more attenuated as the latter has become more fully integrated into the global system. European Directives, such as the MSFD, help rebalance governments' powers to maintain sustainability against the large-scale powers driving economic growth; however, they also lengthen the 'Coastal Zone Feedback Loop' (Hopkins et al. 2011). This poses challenges to organizations such as SEPA, which must now look up the institutional ladder as well as down to local environments and societies. Finally, it becomes harder to match scales in describing or modelling the ecological and social components. The SAF application in part of the Firth of Clyde was able to simulate an ecological system with feedback loops within boundaries defined at least in part by the terrain and seascape. However, the socio-economic system, defined on the same scale, was linear– because of highly open boundaries – and hence apparently unable to add to total system resilience.

Finally, a social–ecological system is a complex system, typically including both feedback loops that lead to non-linear responses to external pressures, and hierarchies of lower-level systems. Each sub-system contributes emergent properties to higher levels and is provided with boundary conditions by these higher levels. Analysis in such terms allows sub-systems to be studied in conceptual isolation. Complexity may include self-reference: the conceptualization of the 'real' social–ecological system in Figure 18.4, or the simulation models developed during SPICOSA, are not only part of 'world 3' but are also potentially capable of engendering change in the social–ecological system of which they are a part. Indeed, that is the point of the SPICOSA SAF.

## 18.6 Discussion

This model of social–ecological systems is, of course, soft: it is only one way to conceptualize the complex pattern of relationships amongst tangible and intangible things in a coastal zone region. It has the advantage of blurring the old disciplinary boundaries. Instead of, for example, ecology being concerned with 'nature' and the social sciences with 'the human world', the key distinction is between:

(a) the physical 'world 1', which is finite in space and resources, in which the main components are mass and energy, and the main 'signals' (into the other system components) are (tangible) 'ecosystem services' and 'information'; here are 'nature', the human physical world and the physical economy;
(b) the non-physical 'worlds 2 and 3', which are mutable, possibly infinite, composed of rule-sets, and in which the main 'signals' are 'information' filtered for 'meaning'; here are 'mind', 'lifeworld' and social institutions.

It is possible to go a step further and see the entire system as one in which information is the common currency (see Appendix) and in which the signals from one part of the system to another flow through negative feedback loops providing resilience in the face of externally imposed pressures. Thus our model reinforces the argument that *adaptability* and *resilience* (Holling, 1973) are properties of the integrated system (Holling and Sanderson, 1996). As the human footprint continues to increase (Hails et al., 2008) at the expense of the 'natural world', the latter has less room for adapting to changed conditions. Thus, it would seem that adaptation must take place largely in the social system. But this requires:

(a) rapid and correct economic signals (i.e. with externalities properly costed);
(b) rapid and correct responses by governance using regulatory (i.e. power) signals;
(c) rapid and correct provision of information by Science and the Communications Media;
(d) proper interpretation of these signals in terms of required actions by individuals and institutions.

Part of the motivation for the SPICOSA project was to address these issues, so that early warning signs could be acted on before change became difficult to reverse. The technical methodology included the use of 'quick-assembly' computer models to simulate and evaluate the consequences of different scenarios of change in, and adaption by, the 'real' system (Tett et al., 2011b). The social methodology included stakeholder engagement, which we have argued (Mette, 2011) improves reliability and legitimacy of communications and decisions (the latter seen by the model as rule-bound transformations of the former).

Finally, the idea that '*value in nature*' (Rolston, 1994) includes '*systemic value*' would seem to apply also to social–ecological systems, making explicit two arguments: that nature's value is not a luxury for humans but essential for human social survival; and that human social adaption is essential for the maintenance of the biosphere in anything like its 'natural' form.

## 18.7 Conclusions

Amongst the objectives of SPICOSA was that of overcoming '*two critical challenges facing multidisciplinary science, that of creating a working* ***science-policy interface*** *and that of qualifying and quantifying* ***complex systems***'. Although the tasks of the 'System Design' and related work-packages in SPICOSA seemed initially technical (i.e. the preparation of manuals), the deliberative exchange of disciplinary information and views that took place during the work has driven progress in conceptualizing the complex social–ecological systems of the coastal zone. If the problem of managing coastal zones for sustainability can be generalized as that of resolving conflicts between 'world 1' biophysics, the need for human well-being in 'world 2', and the current set of social norms and institutionalized rule-sets in 'world 3', then the analysis in this chapter shows the need to change some of the 'world 3' rules. SPICOSA investigated these matters on the scales of lagoons (Melaku Canu et al., 2011), fjords (Moksness et al., 2011) and river basins (Mongruel et al., 2011).

However, there remains a need to scale up such methodologies to the larger scales increasingly imposed by globalization on coastal zone social–ecological systems. We hope that the model sheds some light on what is needed to understand and re-design the functions of science and governance in relation to ecological signals, as well as in relation to understanding the role of other social institutions such as the communications media and education. In general, we think that there is much to be gained by understanding social–ecological systems better as networks of communication, as argued by Luhmann (1989).

Finally, we conclude that during the SPICOSA project we were able to progress from multidisciplinarity to interdisciplinarity by seeking perspectives, such as that developed here, which embraced ecology, economics, sociology and communication science. We believe that a further step is necessary: towards a transdisciplinary perspective of social–ecological systems, to be achieved by adding to a science-derived model the 'discursively redeemable validity claims' of members of the other institutions of society, including those of the lifeworld's public sphere and of the media-steered sub-systems.

## 18.8 Appendix

Drawing on the discussion of Luhmann's work by Leydesdorff (2000), we can specify the behaviour of any system thus:

$$[output,\ TF,\ state] = TF(input,\ state)$$

Where *input* is information received across the system boundary, *state* is the current state of the system, *TF* is a transformational function, and *output* is transformed information. The transformation is supposed to result in the extraction of locally meaningful information, which may or may not be transmitted onwards as altered information or as action in the physical world. The equation suggests that the result of transformation can include a change in the local state of the system and in some cases in the transformational rules. (Thus, in the most general terms, the system is conceived of as a Turing (1937) machine.) If $output/input >= 1$, then the system 'resonates' (Luhmann, 1989) to the signal.

Because our model of the social–ecological system allows it to contain a hierarchy of sub-systems, the equation can apply to the whole system or to any part – for example a small set of interacting biological species, or a institution such as an environment protection agency. Both the parts and the whole can have emergent properties. The property of *resilience* results from negative feedback loops within a system, so that the resulting outputs and state changes are minimized compared with input perturbations. The property of *adaptation* involves significant changes in state and *TR* that help to maintain the viability of the system (cf. Walker et al., 2004, based on a state-space approach).

## References

Berkes, F. and Folke, C. (eds.) (1998) *Linking Social and Ecological Systems. Management Practices and Social Mechanisms for Building Resilience*. Cambridge University Press, Cambridge.

Checkland, P.B. and Scholes, J. (1990) *Soft Systems Methodology In Action*. John Wiley & Sons, Ltd, Chichester.

Costanza, R., d'Arges, R., de Groot, R.S., Farber, S., Grasso, M., Hannon, B., Limburg, K., Naeem, S., O'Neill, R.V., Paruelo, J., Raskin, R.G., Sutton, P. and van den Belt, M. (1997) The value of the world's ecosystem services and natural capital. *Nature* **387**, 253–260.

Fischer, F. (2004) Citizens and experts in risk assessment: technical knowledge in practical deliberation. *Technikfolgenabschätzung* **2**, 90–98.

Habermas, J. (1984) *The Theory of Communicative Action. Volume 1: Reason and the Rationalization of Society (translated McCarthy, T.).* Boston, MA, Beacon Press.

Habermas, J. (1987) *The Theory of Communicative Action. Volume 2: Lifeworld and System: a Critique of Fundamentalist Reason* (translated McCarthy, T.). Beacon Press, Boston, MA.

Haig, A.J.N. (1986) Use of the clydc estuary and firth for the disposal of effluents. *Proceedings of the Royal Society of Edinburgh* **90** B, 393–405.

Hails, C., Humphrey, S., Loh, J. and Goldfinger, S. (eds.) (2008) *Living Planet Report 2008.* WWF-World Wide Fund for Nature, Gland, Switzerland.

Halliday, A. and Glaser, M. (2011) A management perspective on social ecological systems: A generic system model and its application to a case study from Peru. *Human Ecology Review* **18**, 1–18.

Hardin, G. (1968) The tragedy of the commons. *Science* **162**, 1243–1248.

Holling, C.S. (1973) Resilience and stability of ecological systems. *Annual Review of Ecology and Systematics* **4**, 1–23.

Holling, C.S. and Sanderson, S. (1996) Dynamics of (Dis)harmony in Ecological and Social Systems. In: Hanna, S., Folke, C. and Mäler, C.-G. (eds) *Rights to Nature, Ecological, Economic, Cultural and Political Principles of institutions for the Environment.* Island Press, Washington, DC, pp. 57–85.

Hopkins, T.S., Bailly, D. and Støttrup, J.G. (2011) A systems approach framework for coastal zones. *Ecology and Society* **16**(4):25.

Leydesdorff, L. (2000) Luhmann, Habermas and the theory of communication. *Systems Research and Behavioral Science* **17**, 273–288.

Luhmann, N. (1989) *Ecological Communication* (translated by Bednarz, J.). Cambridge UK/Chicago, Polity Press/University Of Chicago Press. xvii + 187 pp.

Luiten, H. (1999) A legislative view on science and predictive models. *Environmental Pollution* **100**, 5–11.

Macleod, C. (1997) Implementing pollution control policy in Scotland: present trends, future prospects. *Scottish Affairs* (18):26–142.

Melaku Canu, D., Campostrini, P., Dalla Riva, S., Pastres, R., Pizzo, L., Rossetto, L. and Solidoro, C. (2011) Addressing sustainability of clam farming in the Venice Lagoon. *Ecology and Society* **16**(3).

Mette, A. (2011) Bridging thc gap between science and society. In: Tett, P., Sandberg, A. and Mette, A. (eds) *Sustaining Coastal Zone Systems.* Dunedin Academic Press, Edinburgh, pp. 103–149.

Millennium Ecosystem Assessment (2005) *Ecosystems and Human Well-being: Synthesis.* Island Press, Washington, DC.

Moksness, E., Gjøsæter, J., Lagaillarde, G., Mikkelsen, E., Moland Olsen, E., Sandersen, H.T. and Vølstad, J.H. (2011) Effects of fishing tourism in a coastal municipality: a case study from Risør, Norway. *Ecology and Society* **16**(3):11.

Mongruel, R., Prou, J., Ballé-Béganton, J., Lample, M., Vanhoutte-Brunier, A., Réthoret, H., Agúndez, J.P., Vernier, F., Bordenave, P. and Bacher, C. (2011) Modeling soft institutional change and the improvement of freshwater governance in the coastal zone. *Ecology and Society* **16**(4):15.

Olsen, S.B. (2003) Frameworks and indicators for assessing progress in integrated coastal management initiatives. *Ocean and Coastal Management* **46**, 347–361.

Ostrom, E. (2007) A diagnostic approach for going beyond panaceas. *Proceedings of the National Academy of Science of the United States of America* **104**, 15181–15187.

Popper, K. (1972) *Objective Knowledge: an Evolutionary Approach.* Oxford University Press, Oxford.

Popper, K. (1978) Three worlds. The Tanner Lecture on Human Values, delivered at the University of Michigan, 7 April 1978. University of Utah, Salt Lake City. Available at: www.tannerlectures.utah.edu/lectures/documents/popper80.pdf (accessed 12 October 2012).

Rolston III, H. (1994) Value in nature and the nature of value. In: Attfield, R.and Belsey, A. (eds) *Philosophy and the Natural Environment*. Cambridge University Press, Cambridge, pp. 13–30.

Stern, N. (2006) *Stern Review Report on the Economics of Climate Change* (prepublication edition). U.K. Government: H.M. Treasury, London. Archived at: webarchive.nationalarchives.gov.uk/ + / http://www.hm-treasury.gov.uk/sternreview_index.htm (accessed 10 September 2012).

Tett, P., Sandberg, A. and Mette, A. (eds) (2011a) *Sustaining Coastal Zone Systems*. Dunedin Academic Press, Edinburgh, pp. 1–28.

Tett, P., Ribera d'Alcalà, M. and Estrada, M. (2011b) Modelling coastal systems. In: Tett, P., Sandberg, A. and Mette, A. (eds) *Sustaining Coastal Zone Systems*. Dunedin Academic Press, Edinburgh, pp. 79–102.

Tett, P.B., Valcic, T., Potts, C., Whyte, F., Culhane and T. Fernandes. (2012) Mussels and yachts in Loch Fyne, Scotland: a case study of the science-policy interface. *Society and Ecology* **17**(3):16.

Thurstan, R.H. and C.M. Roberts. (2010) Ecological meltdown in the firth of clyde, Scotland: two centuries of change in a coastal marine ecosystem. *PLoS ONE* **5**(7):e11767.

Turing, A.M. (1937) On computable numbers, with an application to the Entscheidungsproblem. *Proceedings of the London Mathematical Society s2* **42**, 230–265.

von Bertalanffy, L. (1968) *General Systems Theory: Foundations, Development, Applications*. George Braziller, New York.

von Bertalanffy, L. (1972) The history and status of General Systems Theory. *The Academy of Management Journal* **15**, 407–426.

Walker, B., Holling, C.S., Carpenter, S.R. and Kinzig, A. (2004) Resilience, adaptability and transformability in social–ecological systems. *Ecology and Society* **9**(2):5.

Weitzman, M.L. (2007) A review of the Stern Review on the economics of climate change. *Journal of Economic Literature* **45**, 703–724.

# Index

Access right, 145, 195
Adaptive capacity, 198–200, 219, 222, 223
Adaptive governance, 200, 226
Adaptive learning, 196, 219–21, 223, 224, 226
Adaptive management, 196, 198, 199, 202, 203, 220, 223, 226
Adaptive strategy, 8
Age–length relationship, 64
Amazonian coast, 49
Anthropogenic pressures, 231, 235
Appraisal step, 131
Area fee, 159
Australia, 101–04, 111, 160, 218–20, 222, 223, 225, 238

Baltic Sea, 46, 71–73, 77, 78, 80, 81
Biochemical oxygen demand (BOD), 174
Biodiversity, 5, 8, 13, 20, 22, 23, 27, 30, 32–34, 44–47, 98, 102, 146, 147, 153, 155, 161, 192, 201, 220
Biodiversity losses, 9
Biomass quantification, 73, 74, 76
Bio-shield, 85, 98
Bottom-fishing, 57
Brazil, 49–57

Capture per unit effort (CPUE), 49, 52
Carbonate sand, 23, 26
Carrying capacity, 33, 37, 49
Chemical oxygen demand (COD), 174, 176
Climate change, 3, 7, 17, 32, 84, 94, 98, 143, 212, 218, 223, 231
Climate Change Adaptation, 85
Coastal development, 28, 101, 102, 176, 177, 231
Coastal management, 101, 127, 129, 132, 137–139, 150, 172, 173, 177, 180, 183–85, 196, 206–11, 214, 215, 218–26
Coastal zone planning, 135, 164, 167
Coastal zone systems, 3, 5, 229, 230, 232
Cod, 11, 27, 36, 38, 40–42, 131–33, 136, 139, 161, 237
Collective decision-making, 179–81, 185, 187, 188, 206
Co-management, 181, 192, 193, 199
Conservation, 9, 11, 12, 51, 57, 62, 85, 97, 101, 136, 147, 151, 153–57, 159, 160, 161, 164, 167, 192, 195, 200, 201
Convention on Biological Diversity, 32, 144
Cost–benefit analysis, 2, 8
CZS degradation, 12

Decision making process, 8, 10, 89, 118, 130, 141, 154, 177, 183, 185, 188, 192, 193, 197, 200, 207
Deliberation, 9, 117, 121, 123, 141, 184, 186, 206, 207, 209–11, 214, 215
Design step, 130, 234

Economic value, 8, 9, 62, 97
Ecosystem services, 31, 32, 62, 129, 140, 141, 192, 196, 202, 203, 231, 232, 235, 236, 239
Ecosystem-based management (EBM), 127, 128, 129, 138, 141, 208
Eelgrass, 25, 26
Eelgrass meadows, 23, 25, 27
Environmental management, 31, 118, 155, 177, 200
Erosion, 173, 176
EU Water Framework Directive, 11, 22, 27, 38, 44, 172
European Marine Strategy Framework Directive (MSFD), 238
Eutrophication, 9, 13, 36, 40, 71, 175
Exclusive Economic Zone (EEZ), 50, 159, 175
ExtendSim, 131

Fish production, 62
Fishing tourism, 131
Formulation step, 130
Fostering partnership, 224

*Global Challenges in Integrated Coastal Zone Management*, First Edition. Edited by Erlend Moksness, Einar Dahl and Josianne Støttrup.

Gill-net fishery, 49, 58
Global bifurcation, 3
Global quotas, 161
Governance, 1, 10, 11, 14, 16, 89, 90, 91, 95, 117, 119, 123, 127, 128, 130, 135, 137, 139, 141–43, 145–48, 150–57, 177, 179, 180–83, 188, 194, 196, 199, 200, 209, 219, 221, 222, 224, 226, 229, 231, 235–40

Habitat mapping, 33
Harmful algal bloom, 119
HELCOM Biodiversity Assessment Tool, 45
Human activities (HAs), 11, 12, 16, 21, 22, 40, 130, 172, 176
Human capital, 185, 187

Ice marginal deposits, 23, 25
Image analysis, 88
Institutional framework, 127
Institutional quality, 185
Integrated coastal zone management (ICZM), 1, 2, 127, 128, 160, 167, 171, 172, 179, 192, 208, 225
Integrated management, 123, 177
Integrated Management Plans, 162
Interactional quality, 185
Interdisciplinary, 128, 132, 133, 137, 138, 229
Inter-institutional Management Authority (IMA), 151, 154
International Union for Conservation of Nature (IUCN), 58, 148, 150, 152, 192
Island community resilience assessment (ICRA), 93, 94
ISODATA algorithm, 88

Jurisdictional conflict, 119

Kelp forest, 23–25, 40, 42
Knowledge quality, 179, 180, 188, 213

Laminaria hyperborea, 24, 37, 40, 41
Landsat TM image, 85, 87, 88, 92–94

Macroalgae, 36, 38, 61–66, 71–75, 77, 78, 80, 81
Management measures, 49, 141
Management tool, 117, 119, 120, 123, 124, 130
Mangrove Ecosystem, 85, 86, 89, 151
Mangrove forest, 85, 86, 89–93, 94–98
Mapping, 21–28, 33, 85, 87, 92, 93, 97, 185, 211, 215
Marine megafauna, 101, 102
Marine Protected Area (MPA), 58, 130, 136, 145, 160, 192–195
Marine Spatial Planning (MSP), 142, 168, 238
Mesocosm, 61, 63–66
Migration season, 107
Millennium Assessment of Human Behavior (MAHB), 3
Mines, 176
Mitigative strategy, 8
Mobilisation of knowledge, 181, 206, 213
Mortality, 50, 61–66, 131
Multidisciplinary, 1, 4, 127, 194, 212, 229, 240
Mussel-farming, 117, 118, 120, 121, 124

Natural Heritage Areas (NHA), 148
Natural resource potential, 72, 73, 80, 81
Nature Index (NI), 31–3, 47
Norwegian aquaculture, 168
Nursery, 8, 13, 25, 62, 97

Otoliths, 64
Output step, 131
Oyster areas, 23, 26

Participatory management, 58, 119, 125, 145, 147, 150
Participatory Management Board (PMB), 154
Policy Issue, 1, 5, 6, 9, 11, 13, 118, 119, 121, 123, 125, 130–32, 134, 137, 138, 140, 142
Policy-making, 11, 129, 132, 136, 138, 141
Problem-driven methodology, 5
Production-sharing approach, 98
Property of adaptation, 241
Property of resilience, 241
Protecting biodiversity, 32

Qualitative approach, 138, 140

Reflexive deliberation, 206, 207, 209–11, 214
Resilience assessment, 89, 90, 93, 95, 97

Salmon farming, 162, 163
Scallop populations, 23, 26, 27
Sea space, 142, 160, 163, 165
Sea World Whale Watch, 104
Seagrass, 25, 61–66
Seasonal fisheries, 160
Serra Spanish mackerel, 50–52
Social learning, 182, 183, 187, 189, 196, 197, 202, 213, 219, 221–23
Socio-economic system, 12, 239
Soft sediments, 23, 25
Spatial predictive modelling, 21, 23, 26
Spawning areas, 23, 27, 134
Species distribution, 21
SPICOSA, 1, 2, 11, 124, 127, 128, 130–34, 137, 138, 140, 142, 183, 184, 187, 212, 229–33, 238–40

Stakeholder involvement, 117, 131, 132, 134, 141, 194, 195
Stakeholder participation, 10, 15, 132, 139, 145, 148, 154, 187, 188, 194
Stakeholders' collaboration group, 119, 121
Stakeholders' engagement, 118, 120, 122, 124
Stakeholders' meeting, 120–22
Sustainability science, 2–4, 14, 214
Sustainable development, 1, 2, 4, 5, 10, 14, 16, 21, 46, 82, 125, 177, 178, 180
Sustainable ecological state, 33
System of incentives, 224
Systems Approach Framework (SAF), 1, 2, 4, 5, 117, 229, 230

Total Allowable Catch, 161
Tourism, 11, 72, 131, 136, 137, 151, 162, 172, 173, 176
Tourism industry, 101, 111, 136
Transport sector, 161
Tropical cyclone, 97

Urban wastewater, 171, 174
Urbanised coastal waters, 101, 111
User-conflicts, 137, 140, 142

Wastewater treatment plants (WWTPs), 174
Whale watching, 102, 111

Zostera marina, 25, 61, 63